Biomaterials
in Food Packaging

Biomaterials
in Food Packaging

edited by
Mohd Yusuf
Shafat Ahmad Khan

Jenny Stanford
Publishing

Published by

Jenny Stanford Publishing Pte. Ltd.
101 Thomson Road
#06-01, United Square
Singapore 307591

Email: editorial@jennystanford.com
Web: www.jennystanford.com

British Library Cataloguing-in-Publication Data
A catalogue record for this book is available from the British Library.

Biomaterials in Food Packaging

ISBN 978-981-4877-98-5 (Hardcover)
ISBN 978-1-003-25678-6 (eBook)

Contents

Preface

This book presents fundamental to advanced knowledge and practical applications of the food packaging materials derived from natural resources. It not only encompasses a comprehensive overview of environment-compatible and biodegradable biomaterials but also highlights the recent trends in their applications in food packaging. The book successfully provides a cool answer to develop biodegradable materials as a promising alternative to their harmful synthetic counterparts. With 13 chapters prepared by prominent contributors from reputed institutes, it provides the reference material that includes current developments and the future outlook on the theme and discusses the advances made in bioactive, biodegradable, and edible films and nano-based smart materials for food packaging applications that can substitute their synthetic counterparts to enhance food's shelf life significantly.

Our sincere thanks go to the eminent authors for their valued and priceless contributions with informative coverage. The readers are welcome to provide us their valuable comments and suggestions, which will help to improve future volumes.

We are highly grateful to our research guide, Dr. F. Mohammad, ex-professor, Department of Chemistry, Faculty of Natural Sciences, Jamia Millia Islamia, India. We would also like to thank Jenny Stanford Publishing, especially Ms. Jenny Rompas and Mr. Stanford Chong, for inviting us to compile this volume.

March 2022

Mohd Yusuf, Editor
Associate Professor
Department of Natural and Applied Sciences
School of Technology, The Glocal University
Mirzapur Pole, Saharanpur, Uttar Pradesh 247121, India

Shafat Ahmad Khan, Coeditor
Assistant Professor
Department of Chemistry, Galgotias University
Greater Noida, Uttar Pradesh 203201, India

Chapter 1

Biomaterials for Food Packaging: An Overview

Fazilah Ariffin,[a] Hanisah Kamilah,[b] Kaiser Mahmood,[a] and Alias A. Karim[a]

[a] *Food Technology Division, School of Industrial Technology, Universiti Sains Malaysia, Penang, Malaysia*
[b] *Faculty of Agriculture, Science and Technology, Universiti Putra Malaysia, Bintulu Sarawak Campus, Bintulu, Sarawak, Malaysia*
fazilah@usm.my

The food industry is growing tremendously, where food packaging materials are being widely used. Currently, most of the food packaging materials are made of fossil-fuel sources, known as synthetic plastics. The passive degradation products of synthetic packaging materials are piling up pollution to the environment, typically to marine life. It is estimated that the synthetic plastics will occupy more than half of the sea by 2050, overriding the marine population. Indirectly, the food supply will be declined, which needs to be ameliorated. Hence, utilizing biomaterials is preferable for food packaging, since these are eco-friendly and biodegradable. In this chapter, a few major groups of biomaterials, which are potential alternatives to synthetic packaging materials, such as (i) directly extracted from biomass: polysaccharides, lipids, and proteins; (ii) produced by microbes:

Biomaterials in Food Packaging
Edited by Mohd Yusuf and Shafat Ahmad Khan

ISBN 978-981-4877-98-5 (Hardcover), 978-1-003-25678-6 (eBook)
www.jennystanford.com

polyhydroxyalkanoate (PHA), bacterial cellulose, and xanthan; and (iii) obtained by polymerization route: polylactic acid, will be discussed. The methods (solution casting, blending, extrusion, and others) and current trends (electrospinning and 3D printing) in developing bio-based active materials will also be summarized. The technical advancements in the field of food packaging will ensure sustainable food supply, conserve the ecosystem, and ultimately assist in accomplishing the goals of Industry 4.0.

1.1 Introduction

The synthetic plastics industry has been growing widely since World War II. The polymer industry has been developing since then to replace and fulfil the shortage of silk. In 2015, about 320 million tons of synthetic plastics were produced, globally, and the production is continuously increasing up to 5% annually. The sources of synthetic plastics are mainly petroleum and natural gases. Approximately 7% of the sources are being used to produce plastics, and 40% of the total production is oriented for packaging. Currently, plastic is synthesized to be resistant against water, air, sunrays, or any substance that a plastic may encounter in the environment. Synthetic plastics, such as polyethylene terephthalate (PET), polyvinylchloride (PVC), polyethylene (PE), polypropylene (PP), polystyrene (PS), and polyamide (PA), are in high demand, due to their high-resistance property. These synthetic plastics are preferable in packaging, as they provide convenience and containment to ensure the quality of packaged food products. The low price of the synthetic plastics with excellent mechanical properties, better tensile strength, good barrier to oxygen and carbon dioxide, heat sealability, and durability of the plastics are preferable for daily usage of food packaging [1, 2].

However, resistance and accumulation of synthetic plastics have caused tremendous pollution to the environment, including the marine habitat, as they require longer period to degrade. The swallowed plastic particles by fish may transfer to human by the food cycle and risk to cause various diseases. It is predicted that by 2050, the accumulation of synthetic plastics will override the marine population. Hence, the food supply will be reduced, and it is unfortunate that the human population is expected to increase thrice, reaching up to ~9 million by 2050 [3]. Currently, there is no efficient method to recycle used synthetic plastics. The presence of food waste and biological substances mixed with the used synthetic plastics make them difficult to be recycled and reused.

Thus, it is important to develop biomaterial-based packaging that has comparable properties as synthetic plastics, yet degradable and environmental-friendly [4]. Hence, this chapter will discuss the current findings on biomaterials as food packaging. It will also include technological advancements in biomaterials, as bio-based active materials to overcome packaging issues in the supply chain. This information will be crucial to prepare biomaterial-based food packaging, especially in pursuance of Industry 4.0.

1.2 Categories of Biomaterials

Biomaterials are vastly available in nature and few exist in the form of dry materials. Most of them are fully degraded in the environment and produce natural by-products, such as gases (nitrogen and carbon dioxide), water, biomass, and inorganic salts [5–8]. Based on the European standard EN 16575, a biomaterial is derived from biomass, which has biological origin and not a material that is embedded in geological formations (EN 16575, 2014) [9].

Nowadays, biomaterials have become the main focus in replacing synthetic plastics, especially for food packaging. This is due to the degradable characteristic and relatively short functional life (disposable) of the biomaterials. The lower carbon footprint or renewable resources are the key factors that lead to the development of biomaterials. Since 2013, biomaterials have replaced 5–10% of synthetic plastics in the market [8]. The biomaterials can be formed naturally or synthetically, and most of their composition are consisted of ester, amide, and ether functional groups that determine their properties [10].

Basically, biomaterials can be divided into two large groups: agro-polymers and biopolyesters. The agro-polymers are polymers that are derived from biomass, including polysaccharides and proteins. The structure of polysaccharides is formed by the presence of glycosidic bond, where the hemiacetal or saccharide is bonded to alcohol *via* loss of water. Cellulose, hemicellulose, and lignin are the foremost natural polymers [11, 12]. Modified natural polymers that consist of starch and cellulose derivatives are also the examples of agro-polymers. These polymers are available under the trade names as Mater-Bi®, Solanyl®, BIOpar®, Cellgreen® Cornpol®, and Cellgreen®. Amino acids are the main components that form proteins. The amino acids link to each other by the reaction of condensation and form peptide bonds that consist of amide functional groups. Table 1.1 shows the properties and applications of common agro-polymers in food packaging.

Whilst, the biopolyesters are polymers that are synthetically made by polymerization or extracted from microorganisms. Synthetically made biomaterials produced by chemical polymerization of

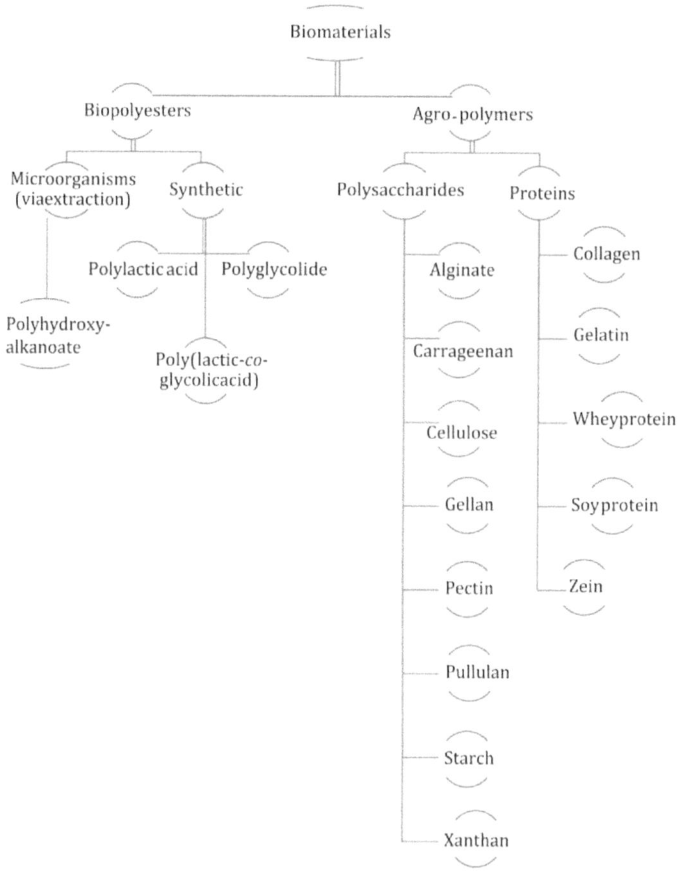

Figure 1.1 Classification of biomaterials [6, 7].

Table 1.1 Properties and applications of common agro-polymers in food packaging

Agro-polymer	Thermal property		Tensile strength (MPa)	Degradation period (days)	Structure	Application	Reference
	T_m (°C)	T_g (°C)					
Starch	60–80	77–90	3–20	15–21	Amylose (α-1,4 glycosidic linkage) 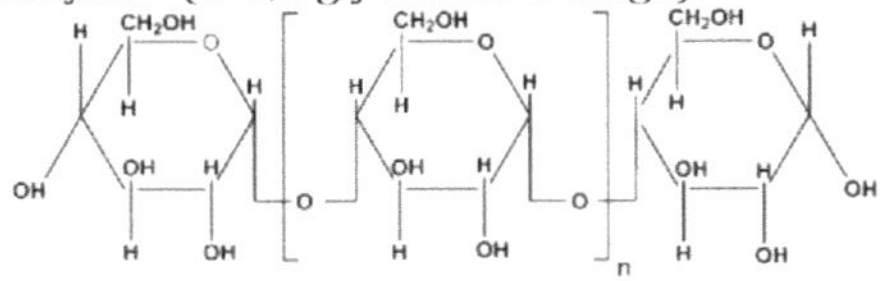 Amylopectin (α-1,6 glycosidic linkage) 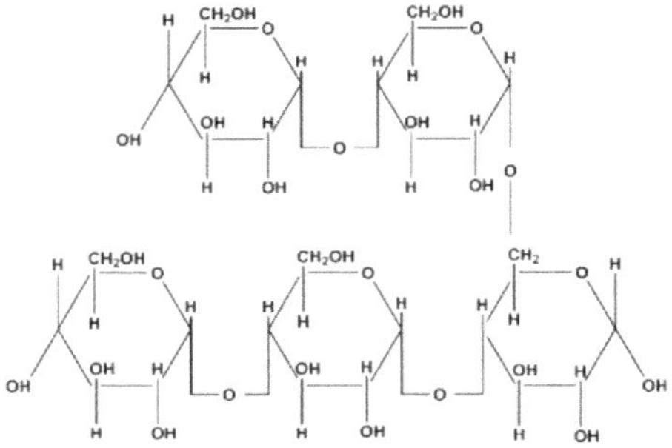	Paper making, food packaging, thickener, food ingredients, etc.	[13–16]

(*Continued*)

Table 1.1 (*Continued*)

Agro-polymer	Thermal property		Tensile strength (MPa)	Degradation period (days)	Structure	Application	Reference
	T_m (°C)	T_g (°C)					
Cellulose	260–270	180–230	200–4600	18–20	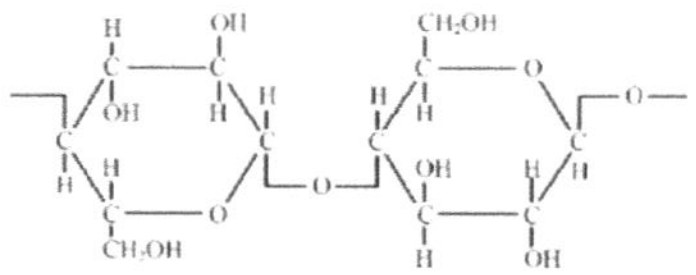	Food coating, food packaging, food additives	[17–19]
Gelatin	30–40	217–230	123–126	10–15	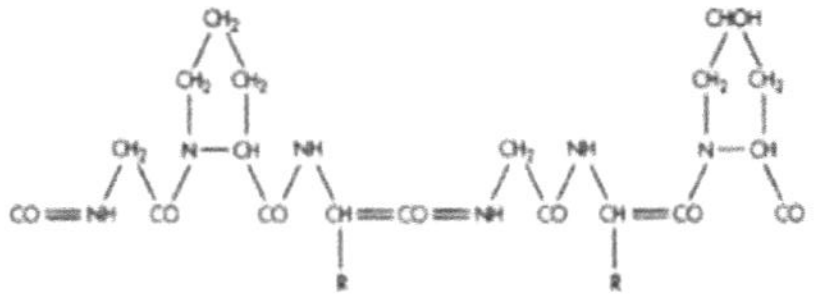	Food coating, paper making, food adhesives, etc.	[20–23]
Chitosan	290	140–150	50–80	~30	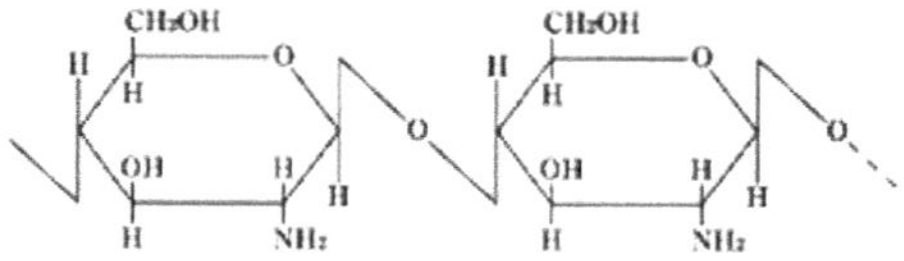	Food packaging, nanofilm, antioxidant film, antimicrobial film	[24, 25]

Table 1.2 Properties and applications of common biopolyesters in food packaging

Biopolyester	Thermal property		Tensile strength (MPa)	Degradation period (months)	Structure	H_2O permeability	O_2 permeability	Application	Reference
	T_m (°C)	T_g (°C)							
PLA	150–178	45–65	21–60	12–16	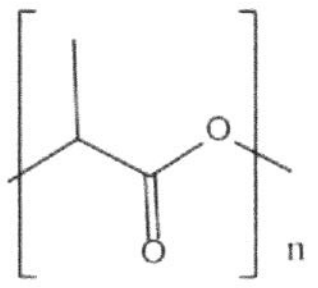	1.42–1.8 ×10^{13} (25 °C)	0.175–0.219 ×10^{13} (25 °C)	Food packaging, antimicrobial packaging, cold drink cups, cutleries, etc.	[26, 27]
PGA	225–230	35–40	60–99.7	6–12	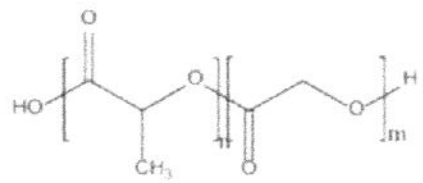	11 (g/m^2•day)	0.9–1.1 (cc/m^2•day•atm)	Food packaging, barrier material	[26, 28]
PHA	50–180	4–8	17–43	~2	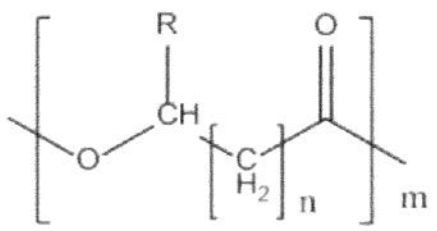	0.168×10^{13} (25 °C)	≈0.09 ×10^{13} (25 °C)	Food packaging, paper coating	[29, 30]

monomers are derived from biological processes, for example, poly(lactic acid) (PLA), poly(glycolic acid) (PGA), poly(lactic acid-*co*-glycolic acid) (PLGA), or poly(ε-caprolactone) (PCL) [6]. Table 1.2 shows the common applications of biopolyesters complemented with their properties. Biopolyesters are sold under the trade names of NatureWorks®, Hycail HM, Hycail LM, Lacea®, U'z, GS Pla, and Bionolla. PHA is an example of biopolyesters obtained *via* the extraction from microorganisms. It is sold under the trade names of Biomer®, Mirel™, Biogreen®, Biocycle®, and Biopol®. Figure 1.1 shows the overall biomaterials classification.

1.3 Advantages and Disadvantages of Biomaterials

The conventional plastics lack in degradability and caused environmental pollution. Biomaterials require around 65% less energy to produce 1 kg mass, as compared to synthetic plastics and reduce the emission of greenhouse gases (GHGs) about 30–80% [31, 32]. Biomaterials are also suitable for production of compost to deliver nutrients and organic matters to plants. Typically, biomaterials from starch degrades 10–20 times faster than that of the synthetic plastics [33]. Incineration of synthetic plastics generates toxic smoke and fumes, as compared to biomaterials, which decomposes into natural gases in the environment. The possibility of addition of various nanoparticles in the biomaterials results in improved material stability against heat, controlled permeability, tensile strength, and antimicrobial properties. Moreover, increasing demand for biomaterials may reduce the emission of GHGs. More land will be utilized to produce biomaterial feedstock, such as maize, oil palm, and seaweed. Thus, it will reduce the GHG emission with direct or indirect use of land [34, 35]. However, disadvantages of biomaterials still hinder their preferred use as an alternative for food packaging. The abundant usage of biomaterials may affect the agriculture sector by causing eutrophication and acidification of soils or water. The faster degradation of biomaterials may provide excessive nutrients to the plants that could lead to their unfavorable growth and loss of crops.

1.4 Degradation of Biomaterials

Biomaterials are preferable to be stable and durable; however, the degradation property is the most crucial criterion to be used in any application. The degradation of biomaterials normally begins from end groups, as most of these biomaterials have strong carbon backbone that is difficult to break. Thus, an area that has higher surface tends to degrade first, due to the easy access of chemicals, light, or microorganisms [6]. Commonly, biomaterials formed by chain branching as crosslinking decreases the number of end groups per unit weight.

The degradation of biomaterials forms gases, salts, and biomass. Complete degradation is occurred when there is no oligomers and monomers left [6, 7]. The major factors that influence the degradation rate of biomaterials are: bond type, solubility, copolymerization, pH, temperature, microorganisms, and humidity. The biodegradation process is divided into two mechanisms: physical decomposition and biological process. The physical decomposition occurs when the reactions of hydrolysis and photo degradation occurred. This may lead to partial or complete degradation. Whereas, the biological process is divided into two conditions: aerobic and anaerobic. The degradation that comes from aerobic condition (Eq. 1.1) produces carbon dioxide (CO_2) and water (H_2O). However, methane (CH_4) together with CO_2 and H_2O is produced when the degradation is exposed in an anaerobic circumstance (Eq. 1.2). Yet, within any degradation condition, carbon (C) residues and biomass are always obtained [7]. The general equations below show the summary of the explanation:

$$C_{polymer} + O_2 \rightarrow C_{residue} + C_{biomass} + CO_2 + H_2O \tag{1.1}$$

$$C_{polymer} + O_2 \rightarrow C_{residue} + C_{biomass} + CO_2 + CH_4 + H_2O \tag{1.2}$$

Numerous microorganisms could degrade natural polymers. The microorganisms will digest the polymer's fragments (oligomers or monomers) to produce energy, adenosine triphosphate (ATP), and biomaterials by-products (Eqs. 1.1 and 1.2). Oxidation and hydrolysis reactions take place for the conversion and assisted by the enzymes, which mainly are proteases, esterases, or glycosidases [7]. Few of biomaterials can be hydrolyzed in water, but they are not totally biodegradable, while the modified natural polymers can only be enzymatically degraded. Contrary to biopolyesters, PHA undergoes both the hydrolysis and enzymatic reactions [36].

It was reported that PGA degrades rapidly in aqueous solution. Deterioration of mechanical integrity was observed within two to four weeks. Addition of methyl group in the repeating unit of PLA may result it to be more hydrophobic, reducing the molecular affinity to water and leads to improve resistant against degradation. Thus, the compound inside the structure has major influence toward the potential degradation. Different biopolyesters have different degradation rates. PCL is known as the slowest to degrade, as compared to other biopolyesters. Thus, PCL is preferable to be used in long-term implant and controlled release applications. Yet, Woolnough and co-workers reported that PHA has the shortest period of degradation (40–50 days) as compared to other biopolyesters [37]. Having such unique properties, PHA are among the materials of interest that could be chosen to replace petrochemical synthetic polymers [29, 38, 39]. The degradation rates reduce in the following manner [40]:

$$\text{PHA} > \text{PGA} > \text{PLA} > \text{PCL}$$

Unfortunately, the petroleum-based synthetic plastics require hundreds of years to degrade. This is due to the long chain and unusual bonds, including different organic and inorganic materials such as carbon, hydrogen, chloride, oxygen, nitrogen, coal, and natural gases [41]. Yet, recently, a small number of microorganisms have been identified for their capability to degrade synthetic plastics (Table 1.3). The potential to degrade the synthetic plastics is suspected to come from the enzymes of microbes, itself. However, the study is still at the preliminary stage, thus, biomaterials are current alternatives to mitigate the issues. Table 1.3 shows the microorganisms that are capable to degrade biomaterials, including synthetic plastics. Biodegradable Product Institute, an organization that encourages recycling of biodegradable materials, has collaborated with American Society for Testing and Materials (ASTM) to provide ASTM standards (ASTM D6400, D6866). Thus, these standards are the principal database for the degradation practices at industrial scale [42].

Table 1.3 Microorganisms degrading polymers

Polymer	Microorganism	Reference
Synthetic polymer		
PE	*Brevibacillusborstelensis, Comamonasacidovorans* TB-35, *Pseudomonas chlororaphis, P. aeruginosa, P. fluorescens, Rhodococcuserythropolis, R. rubber, R. rhodochrous, Staphylococcus cohnii, S. epidermidis, S. xylosus, Streptomyces badius, S. setonii, S. viridosporus, Bacillus amyloliquefaciens, B. brevis, B. cereus, B. circulans, B. circulans, B. halodenitrificans, B. mycoides, B. pumilus, B. sphaericus, B. thuringiensis, Arthrobacter paraffineus, A. viscosus, Acinetobacter baumannii, Microbacteriumparaoxydans, Nocardia asteroides, Micrococcus luteus, M. lylae, Lysinibacillusxylanilyticus, Aspergillus niger, A. versicolor, A. flavus, Cladosporium cladosporioides, Fusarium redolens, Fusarium* spp. AF4, *Penicillium simplicissimum* YK, *P. simplicissimum, P. pinophilum, P. frequentans, Phanerochaetechrysosporium, Verticillium lecanii, Glioclodiumvirens, Mucor circinelloides, Acremonium Kiliense, Phanerochaetechrysosporium*	[43–47]

Polyvinyl chloride (PVC)	*Pseudomonas fluorescens* B-22, *P. putida* AJ, *P. chlororaphis*, *Ochrobactrum* TD, *Aspergillus niger*	[43, 47]
Polyurethane (PU)	*Comamonasacidovorans* TB-35, *Curvularia senegalensis*, *Fusarium solani*, *Aureobasidium pullulans*, *Cladosporium* sp., *Trichoderma* DIA-T spp., *Trichoderma* sp., *Pestalotiopsismicrospora*	[43, 44, 47]
Biopolyester		
Poly(3-hydroxybutyrate) (PHB)	*Pseudomonas lemoignei*, *Alcaligenes faecalis*, *Schlegelellathermodepolymerans*, *Aspergillus fumigatus*, *Penicillium* spp., *Penicillium funiculosum*	[43, 44, 47]
Poly(3-hydroxybutyrate-*co*-3-hydroxyvalerate) (PHBV)	*Clostridium botulinum*, *C. acetobutylicum*, *Streptomyces* sp. SNG9	[43, 44, 47]
PCL	*Bacillus brevis*, *Clostridium botulinum*, *C. acetobutylicum*, *Amycolatopsis*sp, *Fusarium solani*, *Aspergillus flavus*	[43, 44, 47]
PLA	*Penicillium roqueforti*, *Bacillus brevis*, *Rhizopus delemar*	[43]

Source: Reproduced from Pathak and Navneet [42].

1.5 Biomaterials vs. Synthetic Plastics for Food Packaging

Several efforts have been made to substitute synthetic plastics with biomaterials for food packaging. However, few obstacles and challenges need to be overcome and evaluated prior to the full or

Table 1.4 The comparison of synthetic polymers and biomaterials based on their properties (mechanical and barrier) and applications

Synthetic polymer								
Polymer	**Melting temperature, T_m (°C)**	**Tensile strength (MPa)**	**Elongation at break (%)**	**Young's modulus (GPa)w**	**H_2O permeability**	**O_2 permeability**	**Application**	**Reference**
Low-density polyethylene (LDPE) -Polyolefin	110	13.65	533.49	0.3	93×10^{13} (25 °C)	5.18×10^{13} (25 °C)	Trays, juice cartons, plastic wraps, etc.	[50, 51]
High-density polyethylene (HDPE) -Polyolefin	120–135	18	>70	1.095–1.25	137 (37.8 °C) (gm.cm/cm^2.s)	0.31×10^{13} (25 °C)	Oxygen barrier materials, 3D printing plastics, etc.	[52–54]
PP -Polyolefin	162–175	24.5	493	0.855	38.3×10^{13} (25 °C)	0.77×10^{13} (25 °C)	Microporous membranes, glame-retardant films	[55–57]
PVC -Vinyl	100	3.8	240	7.19	$1–2\times10^{13}$ (23 °C)	$0.02–0.09\times10^{13}$ (23 °C)	Electrical carriers	[58, 59]
PS -Styrenic	100	48.9	5.6	1.3	0.78–3.9 (g.mm/m^2.day)	$1.34–1.79\times10^{13}$	Filtration films and beads, etc.	[60, 61]

(*Continued*)

Table 1.4 (*Continued*)

Synthetic polymer								
Polymer	**Melting temperature, T_m (°C)**	**Tensile strength (MPa)**	**Elongation at break (%)**	**Young's modulus (GPa)w**	**H_2O permeability**	**O_2 permeability**	**Application**	**Reference**
PET -Polyester	270	55–75	50–150	2.8–3.1	$0.5–2×10^{13}$ (23 °C)	$0.01–0.06 ×10^{13}$ (23 °C)	Plastic bottles	[62]
Polybutylene Terephthalate (PBAT) -Polyester	223–250	54–143	25	2.8–10	2.27 (g 100 μm m^{-2} d^{-1} $mbar^{-1}$)	537 (cm^3 (STP) 100 mm m^{-2} d^{-1} bar^{-1})	Insulators, thermoplastics, etc.	[63, 64]
Polycarbonate (PC) -Polyester	190	55–75	80–150	2–2.4	$1400×10^{10}$ (25 °C)	$1.4×10^{10}$ (25 °C)	Electronic components, construction materials, etc.	[65]
Biomaterial								
Alginate	150	25.8	2.5–5	13	$25.745×10^{-6}$, (g.m/Pa.h.m^2)	$8.94×10^{-6}$(mL.mm/ m^2.day.pa)	Food coatings	[66]
Carrageenan	70	45.7–70.16	8.3–10.76	0.04	1.16 ($×10^{-9}$ g m/m^2 Pa s)	45 (cm^3/mm^2/24h)	Coatings, encapsulation aromatic compounds	[66]
Cellulose	260–270	118–120	14	0.07	$142.78×10^{-13}$ (25 °C)	$0.0015×10^{-13}$ (25 °C)	Cellophane packaging films	[67, 68]

(*Continued*)

Table 1.4 (*Continued*)

Synthetic polymer								
Polymer	**Melting temperature, T_m (°C)**	**Tensile strength (MPa)**	**Elongation at break (%)**	**Young's modulus (GPa)w**	**H_2O permeability**	**O_2 permeability**	**Application**	**Reference**
Chitin	>300	13–80	0.6–10	0.3–2	-	0.003	Packaging films, paper making, etc.	[69–71]
Chitosan	220–290	8.25–14.6	16.8–45.9	0.07	12×10^{-13} (25 °C)	0.00122×10^{-13} (25 °C)	Edible membranes and coatings, packaging films	[70, 71]
Galacto-mannan	100–110	1–7	35–70	0.2	6.78–10.69 $\times10^{-11}$ (g m^{-1} Pa^{-1} s^{-1})	1.64–2.94 × 10^{-15} g m (Pa s m^2)$^{-1}$	Edible coatings and films	[72–74]
Konjac Glucomannan	103–107	12	7.5–10	2.7–3	1.37 $\times10^{14}$ (kg Pa^{-1} s^{-1} m^{-1})	8–10 cm^3·μm/(m^2·d·kPa)	Edible films	[75, 76]
Starch	60–70	2.85–9.9	13–21.4	2.7	9.25×10^{-13} (25 °C)	0.0140×10^{-13} (25 °C)	Flexible packaging, rigid packaging as thermoformed trays	[77, 78]
Gellan gum	70–80	2.14–3.70	12.2–14.98	0.039–0.058	352–489 (g $m^{-2}d^{-1}$)	-	Edible coatings and flavor encapsulation materials	[79, 80]
Pullulan	95	32–41	3.2	1.8	7.053×10^{-6}(g.m/ Pa.h.m^2)	0.48×10^{-6}(mL.mm/ m^2.day.Pa)	Coating, seasoning bags	[81]

partial replacement of synthetic plastics. An ideal biopolyester as food packaging material should be resistant to environmental factors of food deterioration, such as temperature, oxygen or carbon dioxide, humidity, and light or external forces. The preferable food packaging is mostly relying on the type of food products intended to be packaged. Food products, such as beer or soft drinks, require an opaque and air impermeable packaging. However, fruits and vegetables require a permeable packaging for the transfer of oxygen and carbon dioxide, and moisture exchange [48]. Hence, the barrier, mechanical, and thermal properties are crucial in the selection of a food packaging. Generally, an ideal packaging material is with a lower rate of oxygen permeability and water vapor permeability. The characteristics of limited light transmission and resistance to mechanical or thermal risks are also preferable in food packaging [49]. Table 1.4 shows the comparison of biomaterials and synthetic plastics in term of properties and applications.

On the other hand, the costs of biomaterials and synthetic plastics have a substantial difference when the weight basis is considered. Biomaterials have higher density as compared to the synthetic plastics. PLA is preferable than PS, as the film thickness can be altered to less thickness of up to 0.66 mm [82]. It is expected that the price of synthetic plastics will rise, due to the fluctuation in petroleum prices. Thus, the biomaterials price will reduce with the circumstances of economic production and conversion into products. Table 1.5 shows the difference of price between the synthetic plastics and biomaterials.

Table 1.5 Price level of synthetic plastics and biomaterials in 2016 [9]

Biomaterial	Price level (€/kg)	Density (kg/m^3)
Plastic		
Cellulose acetate	5	1200–1300
Bio-polyamide	+10–20%	1040–1190
Bio-polyethylene	+20–40%	910–970
Bio-polyethylene terephthalate	-	1370–1390
Bio-polypropylene	+80–100%	900–920
Polypropylene (Certified bio)	+40–50%	900–920
Polybutylene adipate terephthalate	3.5	1250

(Continued)

Table 1.5 (*Continued*)

Biomaterial	Price level (€/kg)	Density (kg/m^3)
Plastic		
Bio-polybutylene succinate	4	1260
Polyhydroxyalkanoate	5	1200–1250
Polylactic acid	2	1250
Polytrimethylene terephthalate	4	1320
Starch blend	2–4	1250–1350
Synthetic plastic		
Low-density polyethylene	1.250–1.450	910–940
High-density polyethylene	1.200–1.500	930–970
High-impact polystyrene	1.350–1.525	1080
Polyethylene	0.850–1.050	1370–1390
Polypropylene	1.000–1.200	900–920
Polystyrene	1.250–1.430	1040
Polyvinyl chloride	0.800–0.930	1100–1450

1.6 Methods of Forming Biomaterial-Based Packaging

1.6.1 Film Preparation Methods

Biomaterial film is known as a dried and extensively interacting polymer network with three-dimensional gel structure. The film is formed by a spatially rearranged polymer matrix, which has been incorporated with plasticizer, solvent, and other additives. The formation processes of biomaterial films can be divided into two groups: wet and dry processes [83].

1.6.1.1 Wet process

It is also known as solvent casting technique that is based on the controlled drying of the film forming polymer solution (Figure 1.2). This method consists of three steps: (i) polymer solubilization, (ii) pouring or casting, and (iii) drying under controlled temperature and humidity [84]. In the solubilization of biomaterials, the choice of solvent is water, alcohol, or their mixture and some organic solvents. The solvent for food grade packaging development is water and alcohol or a combination of both. The polymer solution is casted through an applicator at a uniform thickness on a smooth surface of glass or Teflon. To obtain the cohesive matrix of the film, the formation of three-dimensional continuous network of polymer chains is very crucial. This depends on the biomaterial and solvent types, pH of the solution, drying temperature, rate of drying, and final moisture content. During the drying process, solvent evaporation enhances the polymer chain entanglement and results in a gelling state that shifts to a solid matrix after further drying [85]. Various additives are admixed to improve the processability and enhance mechanical or barrier properties of the casting films. Common additives include plasticizers, emulsifiers, antimicrobials, antioxidants, micro or nanoparticles, and cross-linkers. Generally, the solvent casting is conducted either in a continuous manner or isolated molds in batch process. Solvent casting is carried out with the solution of polymers, but not with the polymer melts, hence, the severe anisotropic effect is not observed in the films. Moreover, this technique does not exert any thermal stress on the polymer. Currently, various types of blended, layered and composite films are designed through casting, with different kinds of biomaterials and natural nanofibers or nanocrystals, or inorganic metals or organoclays are mixed to tailor the functional properties of the resulting films for packaging applications [86].

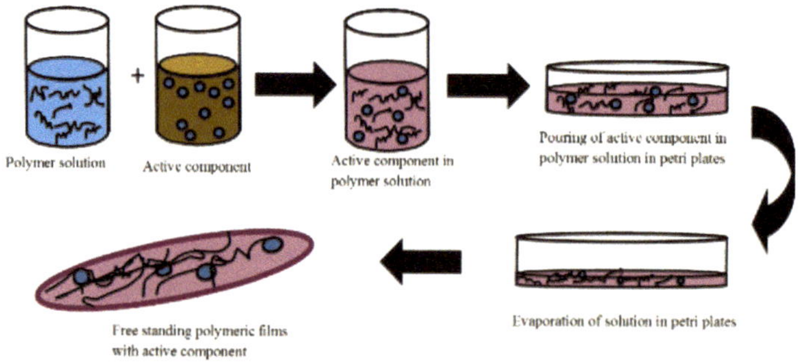

Figure 1.2 Casting method of biomaterial film.

1.6.1.2 Dry process

Dry process is based on the thermoplastic properties of biomaterials. Usually, the thermoplastic biomaterials are heated above the glass transition temperature (T_g) and in low moisture contents. Heat changes the amorphous polymers into soft and rubbery state, which assists the formation of films after cooling. The dry method requires more equipment as compared to wet process, yet the dry method provides the platforms of industrial implementation and reduces the solubility of a biomaterial film. This is resulting from highly cross-linked film network that is developed from the film forming process. Examples of dry process of biomaterial film formation methods are extrusion, thermoforming and coating [84, 87, 88].

1.6.1.2.1 *Extrusion*

The required time and space for preparation and drying of casting films make the process limited to be adopted in industries. Extrusion is an alternative technique that facilitates the commercial production process [88] (Figure 1.3). In this technique, the granular polymer material is processed after melting it. An extruder is used, which is a screw conveyer, and has heated barrel where feed material is melted and squeezed under pressure and temperature and forced through a die to get a sheet. There are several types of extrusions, such as hot extrusion, cold extrusion, warm extrusion, friction extrusion, and micro extrusion [84]. In the extrusion process, the molecular structure of biomaterials is modified by aligning due the flow. Additionally, the screw speed, feed rate, heat and mechanical energy, and die shape determine the extent of aggregation, denaturation, or crosslinking of the polymeric chains [89]. Similar to the process of casting, some additives or plasticizers are added in the extrusion process to improve the processability of polymers [88]. The extrusion and compression moldings have been applied for thermoplastic biomaterials, such as gelatin, soy protein, myofibrillar protein, whey protein, wheat gluten, zein, different starches, and banana flour [84, 87].

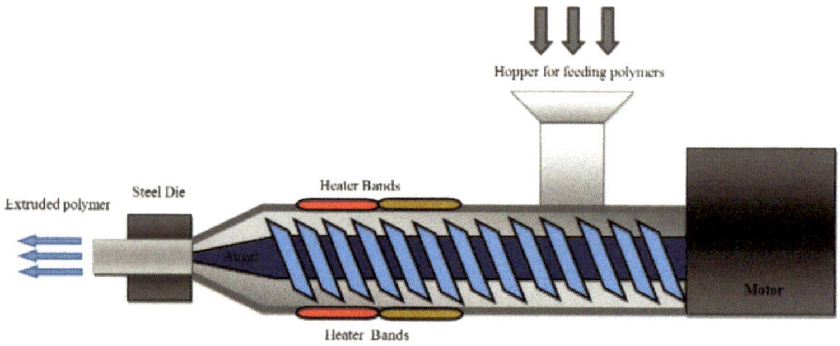

Figure 1.3 The extrusion process of a biomaterial.

1.6.1.2.2 *Thermoforming*

Thermoforming is one of the methods to fabricate packaging films, where a thermoplastic and viscoelastic polymer is heated under pressure and cooled to form a sheet. The films are stabilized by hydrogen bondings, covalent links, ionic interactions, or hydrophobic interactions [89]. The thermoforming or thermo-pressing unit consists of two thermostated plates that are used for pressing polymers into sheets. The parameters of thermo-pressing, such as working temperature, applied pressure, duration, moisture, and type and percentage of additives, regulate the final product properties. However, an adequately high temperature is mandatory to obtain a desired degree of aggregation or denaturation of homogeneous polymer films [90]. The denaturation sometimes favors the process by exposing the functional groups and establishing new interactions, leading to desired properties of the packaging materials [89]. In the case of sulfhydryl bonds in proteins, during the extrusion processing, the crosslinking reaction is favorable and it results in enhancement of physico-chemical properties of the packaging films [91]. Thermoforming could be used alone or in combination with the extrusion technique. This method makes possible to design single or multilayered sheets, trays, or pots, which could be of great interest in food packaging application. Biomaterials including proteins and starches from various sources have been processed through thermoforming method [84].

1.6.1.2.3 *Coating*

Coating is generally a thin layer of matrix forming biomaterial melts (carbohydrates, lipids, proteins, or their mixture) applied on food packaging films to improve the barrier or functional properties of the packaging [92]. Polymer melts or emulsions are directly applied on packaging for coating purpose. Coatings are sticky on the packaging materials that sometimes used to fabricate layer-by-layer type of packaging films with desired functionalities. However, adhesion of coatings to food packaging materials remains challenging in case of difference in the hydrophilicity of the both materials, hence addition of surfactants in coating dispersions helps in improving the adhesion and overall process of coatings [93]. Biomaterial-based coatings could be applied by spraying, spinning, dipping, and rubbing [92]. The coatings are applied externally to the packaging material or sandwiched between two layers of polymer films based on required functionalities. Overall, the matrix of coatings is formed by two basic mechanisms such as gelation/thermal coagulation and coacervation. In the thermal coagulation or gelation process, heat denatures the proteins or melts the starches followed by the precipitation and gelation, respectively. However, cooling of dispersion of proteins (gelatin) and polysaccharides (agar) induces gelation and results in matrix formation. While, in case of coacervation, a charged hydrocolloid is dispersed in water and later precipitated by adding a non-electrolyte anti-solvent, such as ethanol. The phase change is caused by the addition of an electrolyte either by crosslinking or salting out. When two solutions of oppositely charged hydrocolloids are mixed, the interaction results in precipitation of the resulting complex and is termed as coacervation [94].

1.6.2 Advancements in Packaging Material Development

1.6.2.1 Nanocomposites

Nanocomposites are the combination of polymers with non-polymeric substances (fillers) of smaller than 100 nm in at least one dimension to form a mixture with improved properties [95]. The resultant nanocomposite material displays improved mechanical, barrier, thermal, optical, electrical, and biodegradability properties, as compared to conventional composites [96]. Nanofillers could be classified as natural fibers from biomaterials or inorganic materials, such as metals and organoclays. Polymer-clay nanocomposites are usually obtained by using montmorillonite, kaolinite, saponite, hectorite [97], carbon nanotubes, graphene nanosheets, and others [98]. The greater surface area and higher aspect ratio (50–1000) of montmorillonite with minimum platelet thickness (1 nm) make it suitable for reinforcement of the nanocomposites [99]. Once the polymers are mixed with clays, three types of nanocomposites are obtained: intercalated, exfoliated, and tactoids. In intercalated nanocomposites, single-polymer chains are sandwiched between the clay layers, however, the system retains ordered stack type of arrangements. In the case of exfoliated nanocomposites, the layers of clays are delaminated and well dispersed between the polymer chains. The formation of intercalated and exfoliated composites depends on the type of the clays and the processing conditions [100]. In case of tactoids, the clay particles are dispersed in the polymer matrix without separating the layers. Among these three types, the exfoliated is suggested to present the most desirable properties of nanocomposites.

Nanocomposites with submicron size of natural fibers as fillers (microfibrils or nanocrystals) have been studied, where cellulosic fibers from cotton, rice husk, bagasse, flax, hemp, wheat straw, kenaf, bamboo, coir, date palm, oil palm, sisal, jute, curaua, and jowar have been tested [101]. The relative advantages of the natural fibers include low cost, biodegradability, and lightweight.

Cellulose microfibrils and nano-whiskers are two types of fillers that have been used in developing nanocomposites. Microfibrils have length in micrometers, while the diameter is in the range of nanometers [102]. Nano-whiskers are crystalline cellulose obtained after acid digestion of the amorphous components, and include cellulose nanocrystals and nanorods [103]. Nanocomposites are developed by three main pathways: (i) solution intercalation, (ii) melt intercalation, and (iii) in situ polymerization. In the solution intercalation method, firstly, the clay is thoroughly swelled in the solvent and then the polymer is solubilized in the solvent. Later both the components are combined, and the polymer chains intercalate the layers of clay by displacing the solvent [104]. In the melt intercalation, the polymer is heated above the melting temperature and the clay is added, and the temperature is kept higher under shear to accelerate the intercalation and exfoliation of the clay. In in situ polymerization, organoclays and monomers are combined, and the polymerization of the monomers is carried out, which precisely locks the exfoliated particles of the clays in the nanocomposites. Among these three methods, the melt intercalation is the most desirable approach, as it is compatible with the extrusion and other techniques of similar kinds. Moreover, it is benign as no solvent is used in the process [104]. On the other hand, the solution intercalation is just a type of solution casting that is considered suitable for preparation of nanocomposites of many biomaterials that degrade under high-processing temperature [105].

1.6.2.2 Electrospinning

Electrospinning is a nanotechnological method through which nanofibers are produced under the influence of electrostatic forces (Figure 1.4). It is a simple and versatile technology and it has enabled the development of encapsulated active components as food ingredients or for packaging. The nanofibers provide high-surface area for effective release and mass transfer, lower degradation and faster release, and control on the release by controlling the ratio of the polymer to active component [106, 107]. Various approaches in electrospinning have been adopted to obtain the desired encapsulation and release of active ingredient, including blend electrospinning, core-shell electrospinning, emulsion electrospinning, and surface loading of nanofibers. These approaches dictate the pattern of loading of active ingredient to get a desired release performance of the nanostructure for food packaging [108].

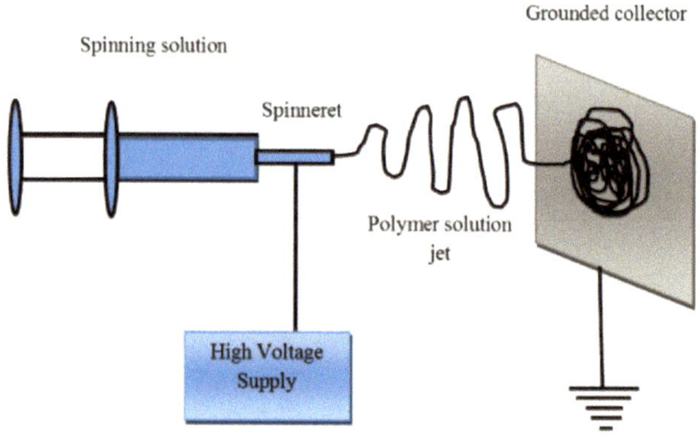

Figure 1.4 Basic electrospinning setup for fabricating nanofilms.

1.6.2.2.1 *Blend electrospinning*

In this method, a direct dissolution of the active ingredient in the polymeric solution is done and electrospinning is conducted (Figure 1.5). The uniformity of the electrospun nanofilm depends on the compatibility of both the polymer carrier and the active component, which later determines the release performance [109]. For instance, in case of higher affinity of active component to the polymer carrier than the solvent, it will migrate into the inner of the fiber than the surface. This technique is considered suitable for encapsulation of active components that are not sensitive to the solvents, however, it is not suitable for enzymes [110].

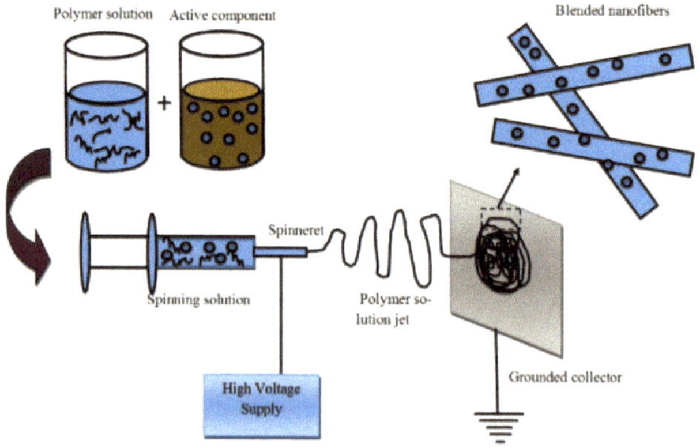

Figure 1.5 Blend electrospinning of active agent in polymeric nanofilms.

1.6.2.2.2 *Core-shell electrospinning*

The core-shell electrospinning provides nanofibers where the core is concentrated with the active ingredient and the shell that provides protection and modulates the release of active ingredient. The polymer and the active ingredient are kept separate by simultaneous spinning of two coaxial spinnerets to get a core-shell nanofiber structure [111] (Figure 1.6). This method provides more flexibility in designing the nanofibers because the core could be any active agent, which may be spinnable or non-spinnable [112, 113]. Additionally, it provides effective protection to the light sensitive active ingredient [114].

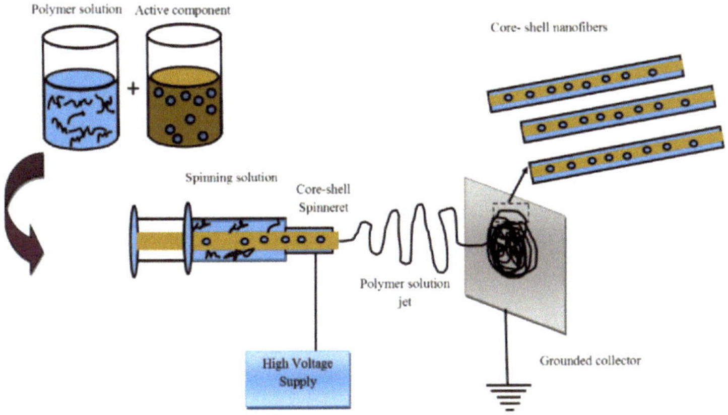

Figure 1.6 Core-shell electrospinning method.

1.6.2.2.3 *Emulsion electrospinning*

In this electrospinning method, the active ingredient, which is surrounded by the surfactants or emulsifiers, is loaded to the polymeric carrier (Figure 1.7). This helps to design a control or sustained release packaging system [115]. Contrary to the core-shell two feeding spinnerets, emulsion electrospinning provides a core-shell type of fibers with a single spinneret [108]. This technique serves as an alternative to core shell, as it can assist to incorporate a lipophilic active component in a hydrophilic polymeric carrier to avoid burst release [116]. It is suggested that the electrostatic forces influence the stretching of polymeric emulsions, induce de-emulsification, and creates core-shell fiber structures. The fabricated emulsion electrospun nanofibers could be tuned based on the choice of solvent, polymer, emulsifier, and surfactant, and by the electrospinning condition, such as applied electrostatic potential [115].

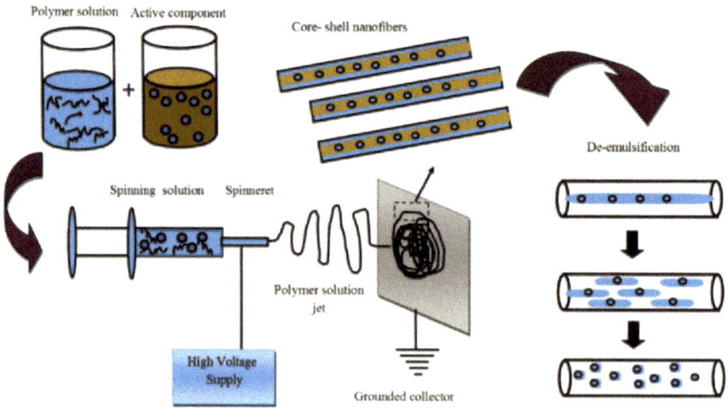

Figure 1.7 Emulsion electrospinning and de-emulsification of active agent in polymeric nanofibers.

1.6.2.2.4 *Surface loading of nanofibers*

This technique allows the post-spinning loading of active ingredient, where nanofibers act as supporting structures because of their higher specific surface area and porosity. Surface loading could be realized either by physical adsorption or chemical immobilization or by multilayer assembly. The physical adsorption is a straight-forward dipping of nanofibers within the solution or suspension of active ingredient. The adsorption is occurred by specific affinity or electrostatic forces on the surface of the nanofibers. However, this method is not suitable for controlled release of the active agent. In the chemical immobilization method, active ingredient is attached on the nanofiber surface by chemical bonding. The functional groups of primary significance are the hydroxyl, carboxyl, and amine groups that provide platform for chemical immobilization [117]. In the layer-by-layer assembly, the poly-electrolytes are adsorbed on the surface of the nanofibers to design a multilayer coating. Besides, hydrogen boding, covalent bonding, hydrophobic interaction, and molecular recognition are also used in developing multilayer coatings. Persano et al. published a comprehensive review on industrial up-scaling and commercialization of electrospinning, suggesting the electrospinning for the food industry [118].

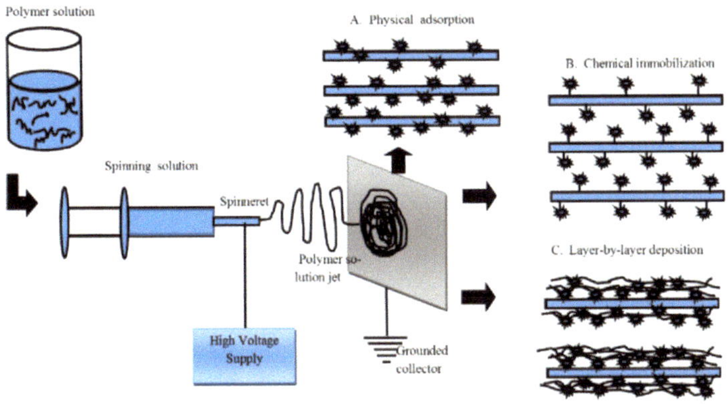

Figure 1.8 Surface loading of nanofibers: (a) physical adsorption, (b) chemical immobilization, and (c) layer-by-layer deposition.

1.6.2.3 3D printing

It is a process of fabricating 3D objects from the digital models by deposition of multilayer of biomaterials [119]. Use of natural biomaterials for 3D printing involves cellulose, starch, lignin, alginate, and chitosan, which are good in meeting demand and sustainable in terms of producing non-toxic by-products upon degradation [120]. Based on various materials, 3D printing technique involves extrusion, solidification of powder, direct-energy deposition, photopolymerization, and sheet lamination. The 3D printing technology for food applications would mainly involve extrusion-based printing, inkjet printing, binder jetting, and selective-sintering printing (with hot air or laser) [121]. However, different 3D techniques and applications need different types of biomaterials. The most common used thermoplastic material is PLA. Based on the desired applications, the biomaterials could be converted to products by blending, compounding, or copolymerization. In compounding, various additives like plasticizer, color pigment, nucleating agent, chain extender, and UV-stabilizer could be added.

Development of biopolymeric 3D-printed-smart materials is ongoing. Smart biomaterials have one or more character that changed with external stimuli such as temperature, pH, pressure, or light. Other materials change in size and shape with the heat or self-healing properties. This smart technology will help to fabricate whole lot of products for novel applications. A comprehensive review has recently published on biomaterials use in 3D printing [122]. 3D printing has been adopted more successfully to design food packaging of different textures and develop active or smart packaging materials. Technical advancement in fabricating biomaterial-based feedstock for 3D printing technology has presented success at lab scale, however, to relish the full potential of biomaterials in 3D printing, it is needed to thoroughly evaluate their processability, mechanical, bio-chemical, and degradation behaviors. Therefore, it could be speculated that future 3D printing materials will be biomaterial-based green feedstock.

1.6.2.4 Smart interactive active packaging

On the other side, research studies have been expanded to produce biomaterial-based smart-food packaging. Moving toward Industry 4.0, food packaging has become a crucial issue in the supply chain of the food industry. The consumers tend to purchase food packaging products in bulk and it is important to maintain the quality of the products during the storage. Human workforce is limited to fulfilling the complicated tasks of recording data and protecting the printed documents. At the same time, the increased pollution by synthetic plastics has become an obstacle to further utilize these materials. Hence, biomaterial-based smart-food packaging is expected to be the alternative. This packaging is in demand and is expected to increase due to Industry 4.0.

There are two different types of smart-food packaging. One is intelligent packaging, in which the packaging is done with the help of sensors, with or without the complementation of electronics. It provides relative humidity, temperature, and light data to identify the shelf-life of chilled food, including fruits, vegetables, or meat. Normally, intelligent packaging is presented as tag and QR code, and complemented with apps in smartphones. The intelligent packaging is not the focus of this literature. Another smart-food packaging is interactive packaging, in which it is interacting with food products. Normally, the active packaging is made up of biomaterials and complemented with functional additives, such as plant extracts, antioxidants, or antimicrobials. As the food comes in contact with biomaterials, the biomaterials indicate the change in color, odor, or graphic symbol. This interactive packaging is used to wrap vegetables and fruits [123].

Smart-food packaging can be alternatively produced from biomaterials such as gelatin, starch, chitosan, PHA, or PLA. This packaging is efficient in maintaining the freshness and the safety of the food. A smart-food packaging can be a moisture absorber, antimicrobial, antioxidant packaging, carbon dioxide emitter, and oxygen scavenging, which enhances product attributes (e.g. appearance, taste, flavor, aroma, and others) [123, 124]. This packaging is also cost effective and has potential to reduce food waste, especially when it is applied in large-scale production. Hence, the smart-food packaging is a potential alternative toward the development of Industry 4.0.

1.7 Safety Concerns of Biomaterial-Based Packaging

In recent packaging technology, the use of various polymers with functional additives or fillers of nanosize is very common. However, almost every innovative technology brings some unknowns to the consumers and society. Nanotechnology though has played a pivotal role in revolutionizing the packaging industry; however, thorough understanding of possible interaction between nanoparticles in packaging films with food or food environment is crucial. Poor packaging may result in issues regarding food deterioration and contamination, and may alter the appeal and organoleptic properties of the food product and render it inappropriate for consumer safety and acceptability [125, 126]. The most challenging part of the current advancement in packaging technology is the lack of comprehensive understanding of toxicity and possible fate of the nanoparticles and their impact on the human and environment [127].

On the other hand, the interaction of nanomaterials with the biological systems is not yet elucidated well. Nanoparticles get into the body *via* ingestion of food and interaction with biological macromolecules and stay in the body fluids, including blood. It is reported that the nanomaterials have the ability to penetrate biological systems and present incomplete clearance from the body based on their shape and size [127, 128]. Smaller nanoparticles with their high-surface area have found to be more toxic than the larger counterparts [129]. Besides the size, surface functionalization and charge on the nanoparticles worsen the situation. Cationic nanoparticles exceed in toxicity than anionic or neutral particles, resulting by their higher affinity toward the negative charge on the plasma membranes [130, 131].

In order to ensure there is no contamination of food that may harm the consumers, migration tests are performed on various packaging materials. The study of migration should include following aspects, in order to minimze the consumers' exposure to risk factors: The physico-chemical properties of biomaterials need to be evaluated and compared with synthetic plastics, as these may affect the migration rate of the compounds. Theoretical models should be applied in the aspect to estimate the migration of stabilizers and additives from food packaging to the food. Migration limits are defined in two possible ways: specific migration and overall migration limits. The specific migration limit of an individual substance is set based on the acceptable daily intake (ADI) established by food authorities. In the case of overall migration limit, a 60 mg of all the possible migrating substances in 1 kg of the food matrix is set by the European Food Safety Authority (EFSA) [132].

Many studies on the migration of nanoparticles in biopolymeric- and synthetic plastic-based nanocomposite packaging have been done for food or food simulants [133]. Usually, the nanomaterials in food matrix are in low concentration and to investigate them, very sensitive techniques are needed. Among the techniques currently being used in detection and analysis are: (i) *Imaging technique*: Scanning or transmission and atomic force microscopy,

(ii) *Spectroscopic technique*: X-ray diffraction and scattering, and dynamic light scattering, and (iii) *Quantitative technique*: Atomic absorption, and inductively coupled plasma with mass spectrometry.

Few other methods are used in combination based on the particles complexity [134–137].

However, the selection of correct set of experimental conditions and simulants or food matrices is very crucial in defining the migration issues more clearly. Studies on nanocomposites with most common fillers, i.e. nanoclays indicate the intercalated composites that provide less migration of clays than exfoliated structures. Migration of few compounds, such as lactic acid and dimer from PLA, was evaluated, however, no significant toxicological data was seen, and it was categorized as a Generally Recognized as Safe (GRAS) substance. Similarly, the migration of polyhydroxybutyrate (P3HB) was not in concern, as it is not harmful to the consumers. Other than that, a study has evaluated the migration of natural antimicrobials from thermoplastic starch. The antimicrobial concentration was very low, but still considerable. However, the toxicological profile of other monomers, such as bisphenol A (BPA), bisphenol S (BPS), vinyl chloride, acrylamide, nanoclay, pro-degradant, benzophenone, and dithiocarbamate, should be taken into consideration during the production or after migration. Only a standard concentration of additives can be used for specific application. Though the migration studies are providing better idea about possible transfer of additives or monomers, however, the suitability and compatibility of the simulant or food matrix, and the studied experimental conditions are very crucial to avoid under or overestimation of migration behavior of any substance of concern [132].

Other than that, with rapid increase in use of nanotechnologies in food packaging, it is direly needed to lay down legislation and regulation for the safety of the contained food products. Currently, partial regulations on safety assessment of nanomaterials are existing. Mostly the nanomaterials are regulated horizontally based on the food laws existing in different countries [138, 139]. Due to the lack of sufficient and inconclusive scientific evidences, mostly precautionary approaches are implemented in the regulations of nanotechnologies in food packaging [140]. The institute of Food Science and Technology (IFST) has mentioned the deficiency of regulations related to the impact of nanomaterials in food packaging. The Scientific Committee of the European Food Safety Authority (EFSA) has highlighted the hurdles in authenticating the possible risks of nanomaterials, including the difficulty in detecting and characterizing and lacking toxicology data [141]. Consumers must have adequate information about the presence of nanoparticles in the food packaging, processing methods of packaging, and the related safety concerns. Very few agencies have established regulations and standards to regulate the use of nanotechnology. Some cases of existing regulations on nanotechnology have been reviewed by Farhoodi (2015). Regulation (EC) No.1272/2008 deals with the classification, labeling, and packaging of various materials and their mixtures. The U.S. Food and Drug Administration (USFDA) has awarded "Generally Recognized As Safe" (GRAS) status to a few nanomaterials as food contact materials, including nanoclays, carbon blacks, zinc oxides, and alumina. While according to the European Regulations only silica, titanium nitride, and carbon blacks can be casted-off in food packaging. Other nanomaterials such as aluminum nanomaterials, zinc oxides, nanoclays, and silver materials are not authorized for food packaging [127].

The Codex Alimentarius Commission, a joint agency of the U.S. Food and Agriculture Organization (FAO) and the World Health Organization (WHO), is endorsing the food safety regulations and standards designed by governmental and non-governmental organizations, worldwide. The Commission should take into consideration the deliberate use of nanomaterials and nanotechnologies in food and food packaging to minimize the regulatory gaps. Hazardous materials presence should be informed to the consumers by new labeling systems, including new safety symbols where possible. Moreover, materials safety data sheets must be provided to the users and with changing the size to nano level, the substance could be classified differently. Unfortunately, there are very few international regulations on nanomaterials and nanoproducts. Additionally, most of the studies regarding the safety of nanomaterials are rarely focusing on food nanomaterials and similar aspects. Though the use of nanotechnology is rising massively in the food industry, yet the lack of clear definitions, food safety evaluations, and appropriate regulations and standards jeopardize its core potential.

As far biomaterials are concerned, a few prospects of regulations and standards are taken into consideration. In Europe, the application of biomaterials is under the regulation of Commission Regulation (EU) 10/2011. A biomaterial needs to fulfill the standards of compostability and biodegradability to be defined as biodegradable. There are three international standards of compostability: EN 13432:2000, ISO 17088:2012, and ASTM D6400-12. These standards are applying similar set of characteristics such as (i) material characterization, (ii) disintegrating ability of the material, (iii) biodegradation of the material into carbon dioxide, biomass, and water, within 6 months, and (iv) ecotoxicity tests of the material compost. As per the standards of biodegradability, the characteristics are directly related to compostability. The test of biodegradability is conducted in different environments, such as aquatic, marine, compost, soil, landfill, or under aerobic/anaerobic conditions.

It is crucial to measure the presence of carbon content in any biomaterial products. Thus, the standards for the quantification of carbon-14 (^{14}C) contents are ASTM D6866-12, ASTM D7026-04, and CEN/TS 16137:2011. Only ^{14}C can be found in biomaterials instead of fossil fuel. Lastly, the biomaterial producers are required to proceed with certifications and labeling. Certifications are based on the voluntary of the producers and verified by few organizers such as DIN CERTCO, Vinçotte (both EU), Biodegradable Products Institute (US), or Japan BioPlastics Association. Successful certified biomaterial products are given a verified 'seedling' logo that is provided by EN13432, DIN CERTCO, and Vinçotte. Usually, "OK compost" logo is issued by Vinçotte. The Japanese BiomassPla logo is only issued to the material that consists of more than 25% of bio-based resources.

1.8 Global Trends toward Biodegradable Polymers Packaging

The legislative banning of synthetic plastics in various countries resulted in initiatives to develop biodegradable plastics. Typical to the food packaging products, the factor of biodegradability is driving the demand for biodegradable biomaterials. However, the packaging foams and bags, in terms of volume, are on the top of the list. It is reported that a 67-fold rise in demand for biodegradable polymers is expected by 2050 [142]. Hence, in terms of demand, the annual growth for biodegradable plastics will be around 15%. The term sustainable development was introduced by the World Commission on Environment and Development in 1987 in a report entitled *"Our Common Future"*. The World Packaging Organization (WPO) raise a slogan *"Better quality of life, through better packaging, for more people,"* aiming to retain the nutritional quality and minimize the food spoilage by educating on packaging. Food packaging is sharing almost 50% of the market of the packaging industry. Thus, the USFDA is focusing on the ability of food packaging to preserve the quality of food and control the migration of unwanted chemicals to the food during the service life of the packaging [143].

The FDA is administering the packaging industry to follow the recyclability regulations regarding the threshold for food contact materials and packaging materials [144]. A few initiatives have been taken to promote sustainable packaging instead of focusing on the waste management [145]. Amongst in the first few attempts, the Sustainable Packaging Alliance of Australia defined the sustainable packaging [146, 147]. The European Parliament and Council Directive 94/62/EC' was made on 20th December 1994 to minimize the packaging waste (Council for the European Union, 1994). Later, the amended directives on packaging and its waste were: Directive 2004/12/EC, 2005/20/EC, and Directive 2013/2/EC, and the latest was Directive (EU) 2015/720 regarding the preferred use of lightweight plastic bags [145]. The expansions in sustainable packaging resulted in the startup of "*The Global Packaging Project*" in France, in collaboration with "*The Consumer Goods Forum*". In November 2008, more than 650 global retailers, producers, and stakeholders across 70 types of businesses were the part of this association. Across the globe, Latin America alone is producing 33% of the overall biodegradable materials for packaging. Similarly, the abundant feedstock in Brazil is paving way for the commercial production of bio-based and biodegradable materials for food packaging [146].

Commercial availability of cellulose derivatives was started, but interestingly in 1993, it was considered that these derivatives were non-biodegradable, however, fully biodegradable cellulose acetate was developed by Eastman Chemical Company, USA [148–150]. Polylactides, poly-hydroxyalkanoates, and polycaprolactones are biopolyesters whose good mechanical and barrier properties with the ease of processability render those suitable candidates in fabricating biodegradable packaging. The polylatcides alone shares almost half of the total demand for biodegradable polymers followed by starch (~41%). Cargill Dow and Mitsui are the two major companies producing PLA; Cargill is currently using genetically engineered yeast in conversion of raw sugars to the lactic acid [151]. PLA has its applications in developing food packaging films, short shelf-life bottles, thermoformed bowls and cups, and others. Polybutylene succinate is also one of the biodegradable polymers, which could be processed into various suitable shapes for food packaging applications. SUCCIPACK® is a European company developing several articles from the bio-based polybutylene succinate, including packaging films, trays, and pouches. Thermoformed PBS articles, such as cups, trays, and containers, were firstly patented by Li and Co-workers in 2013 [152].

Other than that, the emergent use of biomaterials as feedstock for food packaging may limit their supply for food production and result in an undesirable competition. In 2008, food prices were increased, as the edible feedstock supply was limited due to the usage of the feedstock in synthesis of biofuels. However, for biomaterials production, it is expected to utilize lower grade of edible feedstock. Thus, it is suggested that the production of biomaterials as food packaging may not affect the price of food. The consumption of biomaterial sources as feedstock is relatively small. The European Bioplastics data shows that approximately 0.02% of land was needed in 2018 to grow the feedstock of biomaterial-based plastics. Besides, it is expected that there will be alternative feedstock following the development of biomaterials [153]. Food waste such as spent cooking oil, organic fraction from municipal food waste, or interesting molecules from side streams of the agro industry (such as beet pulp) can be the alternative feedstock for biomaterials production [154, 155]. The side streams are favorable for biopolyesters production. With time, the applications of the biomaterials in food packaging are increasing because of the lack of the land and space to manage synthetic plastic waste. The current existing biodegradable polymers with myriad properties can combat this challenge; however, more scientific studies are mandatory to integrate the further developments in the properties of these bio-based plastics.

1.9 Conclusion

In overall, biomaterials for food packaging are gaining importance especially in the current time of Industrial Revolution 4.0. However, it is challenging to find the most appropriate alternatives to synthetic plastics that have similar properties. Hence, blending of biomaterials is actively tested and it showed that biomaterials have potential in the future of packaging. Yet the production methods are also a major challenge, the method that is efficient in production and economically feasible, especially in large-scale production, is preferred. Thus, this literature is able to provide various methods for the formation of biomaterials as future of food packaging.

Acknowledgement

The authors would like to acknowledge the Fundamental Research Grant Scheme; Malaysia Rising Star Award (Grant no: 203PTEKIND6711530) to support this research studies. Mahmood K. acknowledges the USM-Global Fellowship Scheme for supporting the study.

References

1. Geyer, R., Jambeck, J.R., Law, K.L. (2017). Production, Use, and Fate of All Plastics Ever Made, *Law Sci. Adv.*, **3**, pp. 1–5.
2. Barnes, D.K.A., Galgani, F., Thompson, R.C., Barlaz, M. (2009). Accumulation and Fragmentation of Plastic Debris in Global Environments, *Philosophical Trans. Royal Soc. B.*, **364**, pp. 1985–1998.
3. Avella, M., Vlieger, J.J.D., Errico, M.E., Fischer, S., Vacca, P., Volpe, M.G. (2005). Biodegradable Starch/Clay Nanocomposite Films for Food Packaging Applications, *Food Chem.*, **93**, pp. 467–474.
4. Sorrentino, A., Gorrasi, G., Vittoria, V. (2007). Potential Perspectives of Bio-Nanocomposites for Food Packaging Applications, *Trends Food Sci. Technol.*, **18**, pp. 84–95.
5. Leschine, S.B. (1995). Cellulose Degradation in Anaerobic Environments, *Annual Rev. Microbiol.*, **49**, pp. 399–426.
6. Avérous, L., Pollet, E. (2012). *Environmental Silicate Nano-Biocomposites*, Springer, London.
7. Bastioli, C. (2005). *Handbook of Biodegradable Polymers*, Rapra Technology, Shrewsbury, Shropshire, United Kingdom.
8. Chamy, R., Rosenkranz, F. (eds.) (2013). *Biodegradation – Life of Science*, InTech, Croacia, United States.
9. van den Oever, M., Molenveld, K., van der Zee, M., Bos, H. (2017). Bio-Based and Biodegradable Plastics – Facts and Figures: Focus on Food Packaging in Netherlands, Report No. 1722, Wageningen Food & Biobased Research, Dutch Ministry of Economic Affairs, Wageningen, Netherlands.
10. Lendlein, A., Sisson, A. (2011). *Handbook of Biodegradable Polymers: Isolation, Synthesis, Characterization and Applications*, Wiley-VCH, Weinheim.

11. Premraj, R., Doble, M. (2005). Biodegradation of Polymer, *Indian J. Biotechnol.*, **4**, pp. 186–193.

12. Perez, J., Dorada, J.M., Rubia, T.D.L. (2002). Biodegradation and Biological Treatments of Cellulose, Hemicelluloses and Lignin: An Overview, *Int. Microbiol.*, **5**, pp. 53–63.

13. Vu, H.P.N., Lumdubwong, N. (2016). Starch Behaviors and Mechanical Properties of Starch Blend Films with Different Plasticizers, *Carbohydr. Polym.*, **154**, pp. 112–120.

14. Zhang, B., Cui, D., Liu, M., Gong, H., Huang, Y., Han, F. (2012). Corn Porous Starch: Preparation, Characterization and Adsorption Property, *Int. J. Biolog. Macromolecules*, **50**, pp. 250–256.

15. Mairiza, L., Mariana, M., Ramadhany, M., Feviyussa, C.A. (2018). The effect of starch-garlic powder ratio on degradation rate of Gadung starch bioplastic, IOP Conference Series: Materials Science and Engineering, Indonesia, pp. 1–4.

16. Nafchi, A.M., Moradpour, M., Saeidi, M., Alias, A.K. (2013). Thermoplastic Starches: Properties, Challenges, and Prospects, *Starch/Starke*, **65**, pp. 61–72.

17. Lee, H., Sundaram, J., Mani. S. (2017). *Nanotechnology*, Chapter 1 "Production of Cellulose Nanofibrils and their Application to Food: A Review", Prasad, R., Kumar, V., Kumar, M., (eds.), Springer Nature, Singapore, pp. 1–33.

18. Huang, Q., Li, L., Fu, X. (2007). Ultrasound Effects on the Structure and Chemical Reactivity of Cornstarch Granules, *Starch/Stärke*, **59**, pp. 371–378.

19. Sakai, K., Yamauchi, T., Nakasu, F., Ohe, T. (1996). Biodegradation of Cellulose Acetate by Neisseria Sicca, *Biosci. Biotechnol. Biochem.*, **60**, pp. 1617–1662.

20. Mahmood, K., Muhammad, L., Ariffin, F., Kamilah, H., Sulaiman, S. (2016). Review of Fish Gelatin Extraction, Properties and Packaging Applications, *Food Sci. Quality Managem.*, **56**, pp. 47–59.

21. Fakirov, S., Cagiao, M.E., Calleja, F.J.B., Sapundjieva, D., Vassileva, E. (1998). Melting of Gelatin Crystals below Glass Transition Temperature: A Direct Crystal-Glass Transition as Revealed by Microhardness, *Int. J. Polym. Mater.*, **43**, pp. 195–206.

22. Hanani, Z.A.N., Roos, Y.H., Kerry, J.P. (2014). Use and Application of Gelatin as Potential Biodegradable Packaging Materials for Food Products, *Int. J. Biolog. Macromolecules*, **71**, pp. 94–102.

23. Dalev, P.G., Patil, R.D., Mark, J.E., Vassileva, E., Fakirov, S. (2000). Biodegradation of Chemically Modified Gelatin Films in Soil, *J. Appl. Polym. Sci.*, **78**, pp. 1341–1347.

24. Xu, J.D., Niu, Y.S., Yue, P.P., Hu, Y.J., Bian, J., Li, M.F., Peng, F., Sun, R.C. (2018). Composite Film Based on Pulping Industry Waste and Chitosan for Food Packaging, *Materials*, **11**, pp. 1–11.

25. Sawaguchi, A., Ono, S., Oomura, M., Inami, K., Kumeta, Y., Honda, K., Sameshima-Saito, R., Sakamoto, K., Ando, A., Saito, A. (2015). Chitosan Degradation and Associated Changes in Bacterial Community Structures in Two Contrasting Soils, *Soil Sci. Plant Nutr.*, **61**, pp. 471–480.

26. Vieira, A.C., Vieira, J.C., Guedes, R.M., Marques, A.T. (2017). Experimental degradation characterization of PLA-PCL, PGA-PCL, PDO, and PGA Fibers, 21st International Conference on Composite Materials (ICCM-21), China: ICCM-21, pp. 1–9.

27. Lunt, J. (1997). Large-Scale Production, Properties and Commercial Applications of Polylactic Acid Polymers, *Polym. Degrad. Stabil.*, **59**, pp. 145–152.

28. Gilding, D.K., Reed, A.M. (1979). Biodegradable Polymers for Use in Surgery - Polyglycolic/Poly (Lactic Acid) Homo and Copolymers, *Polymer*, **20**, pp. 1459–1464.

29. Sudesh, K., Doi, Y. (2000). Molecular Design and Biosynthesis of Biodegradable Polyesters, *Polym. Adv. Technol.*, **11**, pp. 865–872.

30. Sudesh, K., Doi, Y., Magistrali, P., Garcia, S.G. (2005). *Handbook of Biodegradable Polymers*, Chapter 7 "Polyhydroxyalkanoates", Bastioli, C. (ed.), Rapra Technology Limited, United Kingdom, pp. 183–216.

31. Chen, G.Q. (2010). *Plastics from Bacteria: Natural Functions and Applications*, Chapter 1 "Introduction of Bacterial Plastics PHA, PLA, PBS, PE, PTT, and PPP", Chen, G.Q. (ed.), Springer, Berlin Heidelberg, pp. 1–16.

32. Babu, R.P., O'Connor, K., Seeram, R. (2013). Current Progress on Bio-Based Polymers and Their Future Trends, *Prog. Biomater.*, **2**, pp. 8.

33. Liu, H., Xie, F., Yu, L., Chen, L., Li, L. (2009). Thermal Processing of Starch-Based Polymers, *Prog. Polym. Sci.*, **34**, pp. 1348–1368.

34. Cazón, P., Velazquez, G., Ramírez, J.A., Vázquez, M. (2017). Polysaccharide-Based Films and Coatings for Food Packaging: A Review, *Food Hydroc.*, **68**, pp. 136–148.

35. Nguyen, V.L.N., Joly, C., Dantigny, P. (2016). Active Packaging with Antifungal Activities, *Int. J. Food Microbiol.*, **220**, pp. 73–90.

36. Sudesh, K., Iwata, T. (2008). Review Sustainability of Biobased and Biodegradable Plastics, *Clean*, **36**, pp. 433–442.

37. Woolnough, C.A., Yee, L.H., Charlton, T., Fostera, L.J.R. (2010). Environmental Degradation and Biofouling of 'Green' Plastics including Short and Medium Chain Length Polyhydroxyalkanoates, *Polym. Int.*, **59**, pp. 658–667.

38. Sudesh, K., Abe, H., Doi, Y. (2000). Synthesis, Structure and Properties of Polyhydroxyalkanoates: Biological Polyesters, *Prog.Polym. Sci.*, **25**, pp.1503–1555.

39. Tsuge, T. (2002). Metabolic Improvements and Use of Inexpensive Carbon Sources in Microbial Production of Polyhydroxyalkanoates, *J. Biosci. Bioeng.*, **94**, pp. 579–584.

40. Liao, S., Chan, C.K., Ramakrishna, S. (2008). Stems Cells and Biomimetic Materials Strategies for Tissue Engineering, *Mater. Sci. Eng. C.*, **28**, pp. 1189–1202.

41. Shah, A.A., Hasan, F., Hameed, A., Ahmed, S. (2008). Biological Degradation of Plastics: A Comprehensive Review, *Biotechnol. Adv.*, **26**, pp. 246–265.

42. Pathak, V.M., Navneet (2017). Review on the Current Status of Polymer Degradation: A Microbial Approach, *Bioresour. Bioprocess*, **4**, pp. 1–31.

43. Shah, A.A., Hasan, F., Akhter, J.I., Hameed, A., Ahmed, S. (2008). Degradation of Polyurethane by Novel Bacterial Consortium Isolated from Soil, *Ann. Microbiol.*, **58**, pp. 381–386.

44. Kale, S.K., Deshmukh, A.G., Dudhare, M.S., Patil, V.B. (2015). Microbial Degradation of Plastic: A Review, *J. Biochem. Technol.*, **6**, pp. 952–961.

45. Grover, A., Gupta, A., Chandra, S., Kumari, A., Khurana, S.P. (2015). Polythene and Environment, *Int. J. Environ. Sci.*, **5**, pp. 1091–1105.

46. Restrepo-Flórez, J.M., Bassi, A., Thompson, M.R. (2014). Microbial Degradation and Deterioration of Polyethylene: A Review, *Int. Biodeter. Biodegr.*, **88**, pp. 83–90.

47. Bhardwaj, H., Gupta, R., Tiwari, A. (2012). Microbial Population Associated with Plastic Degradation, *Sci. Rep.*, **1**, pp. 1–4.

48. Kim, Y.T., Min, B., Kim, K.W. (2014). *Innovations in Food Packaging*, Han, J.H. (ed.), Academic Press, Cambridge, MA, United States, pp. 13.

49. Gaur, S.S., Ghosh, T., Katiyar, V. (2017). *General Material Properties Required for Food Packaging Applications and Bio-Based Plastics for Food Packaging Applications*, Katiyar, V. (ed.), Smithers Pira, United Kingdom, pp. 97–113.

50. Chandra, R., Rustgi, R. (1996). Biodegradation of Maleated Linear Low-Density Polyethylene and Starch Blends, *Polym. Degrad. Stabil.*, **56**, pp. 185–202.

51. Arvanitoyannisa, I., Biliaderis, C.G., Ogawa, H., Kawasaki, N. (1998). Biodegradable Films Made from Low-Density Polyethylene (LDPE), Rice Starch and Potato Starch for Food Packaging Applications: Part 1, *Carbohydr. Polym.*, **36**, pp. 89–104

52. Araujo, J.R., Waldman, W.R., Paoli, M.A.D. (2008). Thermal Properties of High-Density Polyethylene Composites with Natural Fibres: Coupling Agent Effect, *Polym. Degrad. Stabil.*, **93**, pp. 1770–1775.

53. Zheng, W., Lu, X., Wong, S.C. (2004). Electrical and Mechanical Properties of Expanded Graphite-Reinforced High-Density Polyethylene, *J. Appl. Polym. Sci.*, **91**, pp. 2781–2788.

54. Busch, S.F., Weidenbach, M., Fey, M., Schäfer, F., Probst, T., Koch, M. (2014). Optical Properties of 3D Printable Plastics in the THz Regime and their Application for 3D Printed THz Optics, *J. Infrared Millim. Terahertz Waves*, **35**, pp. 993–997.

55. Manchado, M.A.L., Valentini, L., Biagiotti, J., Kenny, J.M. (2005). Thermal and Mechanical Properties of Single-Walled Carbon Nanotubes–Polypropylene Composites Prepared by Melt Processing, *Carbon*, **43**, pp. 1499–1505.

56. Himma, N.F., Anisah, S., Prasetya, N., Wenten, I.G. (2015). Advances in Preparation, Modification, and Application of Polypropylene Membrane, *J. Polym. Eng.*, **36**, pp. 329–62.

57. Wang, X., Li, Y., Liao, W., Gu, J., Li, D. (2008). A New Intumescent Flame-Retardant: Preparation, Surface Modification, and its Application in Polypropylene, *Polym. Adv. Technol.*, **19**, pp. 1055–1061.

58. Bishay, I.K., Abd-El-Messieh, S.L., Mansour, S.H. (2011). Electrical, Mechanical, and Thermal Properties of Polyvinyl Chloride Composites Filled with Aluminum Powder, *Mater. Des.*, **32**, pp. 62–68.

59. Ramesh, S., Winie, T., Arof, A.K. (2007). Investigation of Mechanical Properties of Polyvinyl Chloride–Polyethylene Oxide (PVC–PEO)-Based Polymer Electrolytes for Lithium Polymer Cells, *Eur. Polym. J.*, **43**, pp. 1963–1968.

60. Shin, C. (2006). Filtration Application from Recycled Expanded Polystyrene, *J. Colloid Interface Sci.*, **302**, pp. 267–271.

61. Hasegawa, N., Okamoto, H., Kawasumi, M., Usuki, A. (1999). Preparation and Mechanical Properties of Polystyrene Clay Hybrids, *J. Appl. Polym. Sci.*, **74**, pp. 3359–3364.

62. Vegt, A.K.V., Govaert, L.E. (2005). *Polymeren: van keten tot kunststof*, Vegt, A.K.V. (ed.), DelftseUitgeversMij, Belfeld, Netherlands, pp. 1–24.

63. Mathew, G., Hong, J.P., Rhee, J.M., Leo, D.J., Nah, C. (2005). Preparation and Anisotropic Mechanical Behavior of Highly-Oriented Electrospun Poly(butylene terephthalate) Fibers, *J. Appl. Polym. Sci.*, **101**, pp. 2017–2021.

64. Ishak, Z.A.M., Leong, Y.W., Steeg, M., Karger-Kocsis, J. (2007). Mechanical Properties of Woven Glass Fabric Reinforced In Situ Polymerized Poly(Butylene Terephthalate) Composites, *Compos. Sci. Technol.*, **67**, pp. 390–398.

65. Liesl, K.M. (ed.) (2003). *Permeability Properties of Plastics and Elastomers: A Guide to Packaging and Barrier Materials*, William Andrew Publishing, New York, pp. 177–178.

66. Tavassoli-Kafrani, E., Shekarchizadeh, H., Masoudpour-Behabadi, M. (2016). Development of Edible Films and Coatings from Alginates and Carrageenan, *Carbohydr. Polym.*, **137**, pp. 360–374.

67. Oishi, Y., Nakaya, M., Matsui, E., Hotta, A. (2015). Structural and Mechanical Properties of Cellulose Composites Made of Isolated Cellulose Nanofibers and Poly(vinyl alcohol), *Compos. A.*, **73**, pp. 72–79.

68. Sun, S., Mitchell, J.R., MacNaughtan, W., Foster, T.J., Harabagiu, V., Song, Y., Zheng, Q. (2010). Comparison of the Mechanical Properties of Cellulose and Starch Films, *Biomacromolecules*, **11**, pp. 126–132.

69. Hassanzadeh, P., Sun, W., deSilva, J.P., Jin, J., Makhnejia, K., Cross, G.L.W., Rolandi, M. (2013). Mechanical Properties of Self-Assembled Chitin Nanofiber Networks, *J. Mater. Chem. B.*, **1**, pp. 1–6.

70. Dutta, P.K., Dutta, J., Tripathi, V.S. (2004). Chitin and Chitosan: Chemistry, Properties, and Applications, *J. Sci. Ind. Res.*, **63**, pp. 20–31.

71. Cho, Y.I., No, H.K., Meyers, S.P. (1998). Physicochemical Characteristics and Functional Properties of Various Commercial Chitin and Chitosan Products, *J. Agric. Food Chem.*, **46**, pp. 3839–3843.

72. Richardson, P.H., Clark, A.H., Russell, A.L., Aymard, P., Norton, I.T. (1999). Galactomannan Gelation: A Thermal and Rheological Investigation Analyzed Using the Cascade Model, *Macromolecules*, **32**, pp. 1519–1527.

73. Noble, O., Turquois, T., Taravel, F.R. (1990). Rheologicai Properties of Galactomannan-Based Gels. Part 1 - Guar and Hydroxypropylguar Gels in Alkaline Media, *Carbohydr. Polym.*, **12**, pp. 203–217.

74. Ahmad, F.B., Williams, P.A. (2001). Effect of Galactomannans on the Thermal and Rheological Properties of Sago Starch, *J. Agric. Food Chem.*, **49**, pp. 1578–1586.

75. Cheng, L.H., Karim, A.A., Norziah, M.H., Seow, C.C. (2002). Modification of the Microstructural and Physical Properties of Konjac Glucomannan-Based Films by Alkali and Sodium Carboxymethylcellulose, *Food Res. Int.*, **35**, pp. 829–836.

76. Xiao, C., Lu, Y., Zhang, L. (2000). Preparation and Physical Properties of Konjac Glucomannan–Polyacrylamide Blend Films, *J. Appl. Polym. Sci.*, **81**, pp. 882–888.

77. Schroeter, J., Hobelsberger, M., Reichenhall, B. (1992). On the Mechanical Properties of Native Starch Granules, *Starch*, **7**, pp. 247–252.

78. Shanks, R., Kong, I. (2012). *Thermoplastic Starch: Thermoplastic Elastomers*, El-Sonbati, A. (ed.), InTech, Croatia, pp. 95–116.

79. Xiao, G., Zhu, Y., Wang, L., You, Q., Huo, P., You, Y. (2011). Production and Storage of Edible Film Using Gellan Gum, *Procedia. Environ. Sci.*, **8**, pp. 756–763.

80. Razali, M.H., Ismail, N.A., Osman, U.M., Amin, K.A.M. (2018). Mechanical and Physical Properties of Gellan Gum (GG) Biofilm: Effect of Glycerol, *ASM Sci. J.*, **1**, pp. 158–165.

81. Teramoto, N., Shibata, M. (2006). Synthesis and Properties of Pullulan Acetate: Thermal Properties, Biodegradability, and a Semi-Clear Gel Formation in Organic Solvents, *Carbohydr. Polym.*, **63**, pp. 476–481.

82. Schut, J.H. (2016). Pioneering Sustainability: Biopolymer Applications Are Expanding in Fits and Starts, *Plast. Eng.*, **72**, pp. 6–13.

83. Han, J.H., Gennadios, A. (2005). *Edible Films and Coating: A Review, Innovations in Food Packaging*, Han, J.H. (ed.), Elsevier Academic Press, New York, pp. 239–262.

84. Gómez-Estaca, J., Gavara, R., Catalá, R., Hernández-Muñoz, P. (2016). The Potential of Proteins for Producing Food Packaging Materials: A Review, *Packag. Technol. Sci.*, **29**, pp. 203–224.

85. Felton, L.A. (2013). Mechanisms of Polymeric Film Formation, *Int. J. Pharm.*, **457**, pp. 423–427.

86. Debeaufort, F., Quezada-Gallo, J.A., Voilley, A. (1998). Edible Films and Coatings: Tomorrow's Packaging: A Review, *Crit. Rev. Food Sci. Nutr.*, **38**, pp. 299–313.

87. Martin, O., Schwach, E., Avrous, L., Couturier, Y. (2001). Properties of Biodegradable Multilayer Films Based on Plasticized Wheat Starch, *Starch-Stärke,*, **53**, pp. 372–380.

88. Sothornvit, R., Olsen, C.W., McHugh, T.H., Krochta, J.M. (2003). Formation Conditions, Water Vapor Permeability, and Solubility of Compression-Molded Whey Protein Films, *J. Food Sci.*, **68**, pp. 1985–1999.

89. Hernandez-Izquierdo, V.M., Krochta, J.M. (2008). Thermoplastic Processing of Proteins for Film Formation: A Review, *J. Food Sci.*, **73**, pp. 30–39.

90. Ullsten, N.H., Cho, S.W., Spencer, G., Gällstedt, M., Johansson, E., Hedenqvist, M.S. (2009). Properties of Extruded Vital Wheat Gluten Sheets with Sodium Hydroxide and Salicylic Acid, *Biomacromolecules*, **10**, pp. 479–488.

91. Lagrain, B., Goderis, B., Brijs, K., Delcour, J.A. (2010). Molecular Basis of Processing Wheat Gluten toward Biobased Materials, *Biomacromolecules*, **11**, pp. 533–541.

92. Yousuf, B., Qadri, O.S., Srivastava, A.K. (2018). Recent Developments in Shelf-Life Extension of Fresh-Cut Fruits and Vegetables by Application of Different Edible Coatings: A Review, *LWT-Food Sci. Technol.*, **89**, pp. 198–209.

93. Guilbert, S. (1986). *Food Packaging and Preservation: Theory and Practice in Technology and Application of Edible Protective Films*, Elsevier Applied Science Publishing Co., London, England, pp. 371–394.

94. Weinbreck, F., Nieuwenhuijse, H., Robijn, G.W., de Kruif, C.G. (2004). Complication of Whey Proteins with Carrageenan, *J. Agric. Food Chem.*, **52**, pp. 3550–3555.

95. Mallakpour, S., Sadaty, M.A. (2016). Thiamine Hydrochloride (Vitamin B1) as Modifier Agent for TiO_2 Nanoparticles and the Optical, Mechanical, and Thermal Properties of Poly(Vinyl Chloride) Composite Films, *RSC Adv.*, **6**, pp. 92596–92604.

96. Sorrentino, A., Gorrasi, G., Vittoria, V. (2007). Potential Perspectives of Bio-Nanocomposites for Food Packaging Applications, *Trends Food Sci. Technol.*, **18**, pp. 84–95.

97. Sinha Ray, S., Okamoto, M. (2003). Polymer/Layered Silicate Nanocomposites: A Review from Preparation to Processing, *Prog. Polym. Sci.*, **28**, pp. 1539–1641.

98. Arora, A., Padua, G.W. (2010). Review: Nanocomposites in Food Packaging, *J. Food Sci.*, **75**, pp. 43–49.

99. Uyama, H., Kuwabara, M., Tsujimoto, T., Nakano, M., Usuki, A., Kobayashi, S. (2003). Green Nanocomposites from Renewable Resources: Plant Oil–Clay Hybrid Materials, *Chem. Mater.*, **15**, pp. 2492–2494.

100. Li, X., Ha, C.S. (2003). Nanostructure of EVA/Organoclay Nanocomposites: Effects of Kinds of Organoclays and Grafting of Maleic Anhydride onto EVA, *J. Appl. Polym. Sci.*, **87**, pp. 1901–1909.

101. Majeed, K., Jawaid, M., Hassan, A., Abu Bakar, A., Abdul Khalil, H.P.S., Salema, A.A., Inuwa, I. (2013). Potential Materials for Food Packaging from Nanoclay/Natural Fibres Filled Hybrid Composites, *Mater. Des.*, **46**, pp. 391–410.

102. Azizi, S.M.A., Alloin, F., Dufresne, A. (2005). Review of Recent Research into Cellulosic Whiskers, their Properties and their Application in Nanocomposite Field, *Biomacromolecules*, **6**, pp. 612–626.

103. Dujardin, E., Blaseby, M., Mann, S. (2003). Synthesis of Mesoporous Silica by Sol–Gel Mineralisation of Cellulose Nanorod Nematic Suspensions, *J. Mater. Chem.*, **13**, pp. 696–699.

104. Shen, Z., Simon, G.P., Cheng, Y.B. (2002). Comparison of Solution Intercalation and Melt Intercalation of Polymer–Clay Nanocomposites, *Polymer*, **43**, pp. 4251–4260.

105. Rhim, J.W., Ng, P.K.W. (2007). Natural Biopolymer-Based Nanocomposite Films for Packaging Applications, *Crit. Rev. Food Sci. Nutr.*, **47**, pp. 411–433.

106. He, C.L., Huang, Z.M., Han, X.J., Liu, L., Zhang, H.S., Chen, L.S. (2006). Coaxial Electrospun Poly(L-Lactic Acid) Ultrafine Fibers for Sustained Drug Delivery, *J. Macromol. Sci. B.*, **45**, pp. 515–524.

107. Bhardwaj, N., Kundu, S.C. (2010). Electrospinning: A Fascinating Fiber Fabrication Technique, *Biotechnol. Adv.*, **28**, pp. 325–347.

108. Luo, X., Xie, C., Wang, H., Liu, C., Yan, S., Li, X. (2012). Antitumor Activities of Emulsion Electrospun Fibers with Core Loading of Hydroxycamptothecin Via Intratumoral Implantation, *Int. J. Pharm.*, **425**, pp. 19–28.

109. Wang, C., Tong, S.N., Tse, Y.H., Wang, M. (2012). Conventional Electrospinning vs. Emulsion Electrospinning: A Comparative Study on the Development of Nanofibrous Drug/Biomolecule Delivery Vehicles, *Adv. Mater. Res.*, **410**, pp. 118–121.

110. Kim, B.S., Oh, J.M., Kim, K.S., Seo, K.S., Cho, J.S., Khang, G., Lee, H.B., Park, K., Kim, M.S. (2009). BSA-FITC-loaded Microcapsules for in Vivo Delivery, *Biomater.*, **30**, pp. 902–909.

111. Su, Y., Li, X., Liu, Y., Su, Q., Qiang, M.L., Mo, X. (2011). Encapsulation and Controlled Release of Heparin from Electrospun Poly(L-Lactide-co-epsilon-Caprolactone) Nanofibers, *J. Biomater. Sci. Polym.* **22**, pp. 165–177.

112. Yu, J.H., Fridrikh, S.V., Rutledge, G.C. (2004). Production of Submicrometer Diameter Fibers by Two-Fluid Electrospinning, *Adv. Mater.*, **16**, pp. 1562–1566.

113. Chakraborty, S., Liao, I.C., Adler, A., Leong, K.W. (2009). Electrohydrodynamics: A Facile Technique to Fabricate Drug Delivery Systems, *Adv. Drug Deliv. Rev.*, **61**, pp. 1043–1054.

114. Huang, C., Chen, S., Lai, C., Reneker, D.H., Qiu, H., Ye, Y., Hou, H. (2006). Electrospun Polymer Nanofibres with Small Diameters, *Nanotechnology*, **17**, pp. 1558–1563.

115. Liao, Y., Zhang, L., Gao, Y., Zhu, Z.T., Fong, H. (2008). Preparation, Characterization, and Encapsulation/Release Studies of a Composite Nanofiber Mat Electrospun from an Emulsion Containing Poly (Lactic-Co-Glycolic Acid), *Polymer*, **49**, pp. 5294–5299.

116. Li, X., Zhang, H., Li, H., Yuan, X. (2010). Encapsulation of Proteinase K in PELA Ultrafine Fibers by Emulsion Electrospinning: Preparation and In Vitro Evaluation, *Colloid Polym. Sci.*, **288**, pp. 1113–1119.

117. Yoo, H.S., Kim, T.G., Park, T.G. (2009). Surface-Functionalized Electrospun Nanofibers for Tissue Engineering and Drug Delivery, *Adv. Drug Deliv. Rev.*, **61**, pp. 1033–1042.

118. Persano, L., Camposeo, A., Tekmen, C., Pisignano, D. (2013). Industrial Upscaling of Electrospinning and Applications of Polymer Nanofibers: A Review, *Macromol. Mater. Eng.*, **298**, pp. 504–520.

119. Bhatia, S.K., Ramadurai, K.W. (2017). *3D Printing and Bio-Based Materials in Global Health*, Springer Briefs in Materials, Springer Nature, Switzerland, pp. 25–30.

120. Park, S.E., Cho, M.H., Lim, J.K., Kim, J.S., Kim, J.H., Kwon, D.Y., Park, C.S. (2007). A New Colorimetric Method for Determining the Isomerization Activity of Sucrose Isomerase, *Biosci. Biotechnol. Biochem.*, **71**, pp. 583–586.

121. Liu, Z., Zhang, M., Bhandari, B., Wang, Y. (2017). 3D Printing: Printing Precision and Application in Food Sector, *Trends Food Sci. Technol.*, **69**, pp. 83–94.

122. Liu, J., Sun, L., Xu, W., Wang, Q., Yu, S., Sun, J. (2019). Current Advances and Future Perspectives of 3D Printing Natural-Derived Biopolymers, *Carbohydr. Polym.*, **207**, pp. 297–316.

123. Schaefer, D., Cheung, W.M. (2018). Smart Packaging: Opportunities and Challenges, *Procedia CIRP*, **72**, pp. 1022–1027.

124. López-Rubio, A., Almenar, E., Hernandez-Muñoz, P., Lagarón, J.M., Catalá, R., Gavara, R. (2004). Overview of Active Polymer-Based Packaging Technologies for Food Applications, *Food Rev. Int.*, **20**, pp. 357–287.

125. Biji, K., Ravishankar, C., Mohan, C., Gopal, T.S. (2015). Smart Packaging Systems for Food Applications: A Review, *J. Food Sci. Technol.*, **52**, pp. 6125–6135.

126. Yildirim, S., Röcker, B., Rüegg, N., Lohwasser, W. (2015). Development of Palladium-Based Oxygen Scavenger: Optimization of Substrate and Palladium Layer Thickness, *Packag. Technol. Sci.*, **28**, pp. 710–718.

127. Bumbudsanpharoke, N., Ko, S. (2015). Nano-Food Packaging: An Overview of Market, Migration Research, and Safety Regulations, *J. Food Sci.*, **80**, pp. 910–923.

128. De Azeredo, H.M. (2009). Nanocomposites for Food Packaging Applications, *Food Res. Int.*, **42**, pp. 1240–1253.

129. Chithrani, B.D., Ghazani, A.A., Chan, W.C. (2006). Determining the Size and Shape Dependence of Gold Nanoparticle Uptake into Mammalian Cells, *Nano Lett.*, **6**, pp. 662–668.

130. Love, S.A., Maurer-Jones, M.A., Thompson, J.W., Lin, Y.S., Haynes, C.L. (2012). Assessing Nanoparticle Toxicity, *Annu. Rev. Anal. Chem.*, **5**, pp. 181–205.

131. Nel, A.E., Mädler, L., Velegol, D., Xia, T., Hoek, E.M.V., Somasundaran, P., Klaessig, F., Castranova, V., Thompsonet, M. (2009). Understanding Biophysicochemical Interactions at the Nano Bio Interface, *Nat. Mater.*, **8**, pp. 543–557.

132. Zhu, X., Lee, D.S., Yam, K.L. (2012). Release Property and Antioxidant Effectiveness of Tocopherol-Incorporated LDPE/PP Blend Films, *Food Addit. Contam.*, **29**, pp. 461–468.

133. Farhoodi, M. (2016). Nanocomposite Materials for Food Packaging Applications: Characterization and Safety Evaluation, *Food Eng. Rev.*, **8**, pp. 35–51.

134. Blasco, C., Picó, Y. (2011). Determining Nanomaterials in Food, *TrAC Trends Anal. Chem.*, **30**, pp. 84–99.

135. Dasgupta, N., Ranjan, S., Mundekkad, D., Ramalingam, C., Shanker, R., Kumar, A. (2015). Nanotechnology in Agro-Food: From Field to Plate, *Food Res. Int.*, **69**, pp. 381–400.

136. Huang, J.Y., Li, X., Zhou, W. (2015). Safety Assessment of Nanocomposite for Food Packaging Application, *Trends Food Sci. Technol.*, **45**, pp. 187–199.

137. Kuorwel, K.K., Cran, M.J., Orbell, J.D., Buddhadasa, S., Bigger, S.W. (2015). Review of Mechanical Properties, Migration, and Potential Applications in Active Food Packaging Systems Containing Nanoclays and Nanosilver, *Compr. Rev. Food Sci. Food Saf.*, **14**, pp. 411–430.

138. Cushen, M., Kerry, J., Morris, M., Cruz-Romero, M., Cummins, E. (2012). Nanotechnologies in the Food Industry: Recent Developments, Risks, and Regulation, *Trends Food Sci. Technol.*, **24**, pp.30–46.

139. Amenta, V., Aschberger, K., Arena, M., Bouwmeester, H., Moniz, F.B., Brandhoff, P., Gottardo, S., Marvin, H.J.P., Mech, A., Pesudo, L.Q., Rauscher, H., Schoonjans, R., Vettori, M.V., Weigel, S., Peters, R.J. (2015). Regulatory Aspects of Nanotechnology in the Agri/Feed/Food Sector in EU and Non-EU Countries, *Regul. Toxicol. Pharmacol.*, **73**, pp.463–476.

140. Neethirajan, S., Jayas, D.S. (2011). Nanotechnology for the Food and Bioprocessing Industries, *Food Bioproc. Technol.*, **4**, pp. 39–47.

141. European Food Safety Authority (EFSA). (2009). The Potential Risks Arising from Nanoscience and Nanotechnologies on Food and Feed Safety, *EFSA J.*, **7**, pp. 958–997.

142. Schipfer, F., Kranzl, L., Leclère, D., Sylvain, L., Forsell, N., Valin, H. (2017). Advanced Biomaterials Scenarios for the EU28 up to 2050 and their Respective Biomass Demand, *Biomass Bioenerg.*, **96**, pp. 19–27.

143. Bhardwaj, U., Dhar, P., Kumar, A., Katiyar, V. (2014). Polyhydroxyalkanoates Cellulose Based Nanobiocomposites for Food Packaging applications, *Food Addit. Packag.*, **1162**, pp. 275–314.

144. Robertson, G.L. (2005). *Food Packaging: Principles and Practice*, 2nd Ed., CRC Press, United States.

145. Tencati, A., Pogutz, S., Moda, B., Brambilla, M., Cacia, C. (2016). Prevention Policies Addressing Packaging and Packaging Waste: Some Emerging Trends, *J. Waste Manag.*, **56**, pp. 35–45.

146. Dentoni, D., Bitzer, V., Pascucci, S. (2016). Cross-Sector Partnerships and Co-Creation of Dynamic Capabilities for Stakeholder Orientation, *J. Bus. Ethics*, **135**, pp. 35–53.

147. Sorensen, J., Sadhu, A., Sampath, G., Sugden, S., Gupta, S.D., Lapworth, D., Marchantet, B.P., Pedley, S. (2016). Are Sanitation Interventions a Threat to Drinking Water Supplies in Rural India? An Application of Tryptophan-Like Fluorescence, *Water Res.*, **88**, pp. 923–932.

148. Klemm, D., Heublein, B., Fink, H.P., Bohn, A. (2005). Cellulose: Fascinating Biopolymer and Sustainable Raw Material, *Angew. Chem. Int. Ed.*, **44**, pp. 3358–3393.

149. Buchanan, C.M., Pearcy, B.G., White, A.W., Wood, M.D. (1997). The Relationship Between Blend Miscibility and Biodegradation of Cellulose Acetate and Poly (Ethylene Succinate) Blends, *J. Environ. Polym. Degrad.*, **5**, pp. 209–217.

150. Buchanan, C.M., Dorschel, D., Gardner, R.M., Komarek, R.J., Matosky, A.J., White, A.W., Wood, M.D. (1996). The Influence of Degree of Substitution on Blend Miscibility and Biodegradation of Cellulose Acetate Blends, *J. Environ. Polym. Degrad.*, **4**, pp. 179–195.

151. Gupta, A., Kumar, V. (2007). New Emerging Trends in Synthetic Biodegradable Polymers–Polylactide: A Critique, *Eur. Polym. J.*, **43**, pp. 4053–4074.

152. Li, W., Tedford, R.A., Thoman, B.J., Christie, T.R. (2013). Thermoformed Article Made from Polybutylene Succinate (PBS) and Modified Polybutylene Succinate (MPBS), Google Patents.

153. Bos, H.L., Sanders, J.P.M. (2103). Raw Materials Demand and Sourcing Options for the Development of a Biobased Chemical Industry in Europe, Part 1: Estimation of Maximum Demand for Biofuel, *Bioprod. Bior.*, **7**, pp. 246–259.

154. Sudesh, K., Bhubalan, K., Chuah, J., Kek, Y.K., Kamilah, H., Sridewi, N., Lee, Y.F. (2011). Synthesis of Polyhydroxyalkanoate from Palm Oil and Some New Applications, *Appl. Microbiol. Biotechnol.*, **89**, pp. 1373–1386.

155. Kamilah, H., Al-Gheethi, A., Yang, T.A., Sudesh, K. (2018). The Use of Palm Oil-Based Waste Cooking Oil to Enhance the Production of Polyhydroxybutyrate [P(3HB). by *Cupriavidusnecator* H16 Strain, *Arab. J. Sci. Eng.*, **43**, pp. 3453–3463.

Chapter 2

Active Ingredients Enabled Edible Coatings and Films Improve Shelf Life of Food: An Overview

Tanvir Arfin
Environmental Materials Division,
CSIR-NEERI (National Environmental Engineering Research Institute),
Nagpur, Maharashtra, India
t_arfin@neeri.res.in

The coatings and films embedded in the edible form to the processed vegetables and fruits enhance their life span, and these even maintain the microbiological, sensory, as well as nutritional qualities of the fruits and vegetables. The different formulations are done and clinically tested for inhibiting the polyphenol oxidase activity and for retarding the movement of browning in fruits and vegetables. The edible coatings and films are responsible for transporting essential materials, and they even provide benefits to the consumer in the form of encapsulating bioactive compounds by offering new innovative products posing functional and nutraceutical effects. These products have various properties, such as solubility and adhesion transparency. After evaluation of such properties, the behavior and composition can be predicted and then the optimization can be

Biomaterials in Food Packaging
Edited by Mohd Yusuf and Shafat Ahmad Khan

ISBN 978-981-4877-98-5 (Hardcover), 978-1-003-25678-6 (eBook)
www.jennystanford.com

done. The coating of the edible products involves different events, namely consumption of oil to be minimized in the deep fried fat products, transportation of bioactive compounds, and life expansion of perishable substances. Hence, the primary emphasis in the study is undertaken on the characterization of newly hydro colloidal films, which are produced by non-conventional sources. Another focus is related to the determination of capability of compounds in discharging the molecules along with features like antioxidants and aromatics. Lastly, the interacting activity is enabling them to show an encapsulating matrix in the particles.

2.1 Introduction

In the current scenario of the research field in the development of society is a significant concern with sustainable development. The ideas of such studies have attracted the interest basically for influence and regeneration aspect, as these are referred to maintain justice on using resources that interact with a human being as well as the ecosystem [1–5]. In the present scientific study, it has been found that integration of a new material in influencive form is regarded as a better option [6–10]. The new materials are employed in various branches of analytical chemistry for the identification of quantitatively and qualitatively with minimum sample available in the very complex form [11–15]. The content is said to be favorable for carrying out research activity in techno-economic processing on large extent [16–20].

Most of the materials applied to the packaging industry are made up of fossil fuels, which are non-degradable. Therefore, these lead to hazardous environmental problems, globally [21]. Significant effects are put into practice to enhance the shelf life and develop the quality of food at the time of packaging by the innovation of new materials like edible and biodegradable films from renewable sources [22]. It is observed that because of the biodegradable nature, the environmental problems could be solved to some extent. The biodegradable packaging material is made up of polymer that is degraded by microorganisms by the process of composting, which on the other hand, liberates the compounds, namely CO_2, H_2O, CH_4, and biomass. Biodegradable polymers are of two types: edible and non-edible [23]. The biodegradable film is produced by the derived food components, employing the wet and dry manufacturing processes. The edible film (EF) resulting from the processes is a free-standing sheet kept in between the elements of food. EF is a layer of material employed for wrapping and covering food products to extend their life. The EF can be consumed with the food also. The edible coating (EC) is a thin layer of material that can be eaten and is applied to the outer surface of the food by the processes of panning, dipping, and spraying. The edible packaging material can be fed with the meal or as a part of the product. It even accomplishes the function such as acting like a carrier for additives, namely antioxidants and antimicrobial agents. The EF and EC can be used even for various other functions such as inhibiting moisture, CO_2 or O_2 migration, and enhancing mechanical integrity of food products [24]. There are some differences between EC and EF, such as EC is applied in the form of liquid to the food products, which can be possible by immersing the products in a solution obtained from the structural matrix, namely carbohydrate and lipid. Whereas, EF is firstly molded into solid sheets and then wrapped over the food products. It is preliminary essential that EF needs to be safe, according to GRAS (generally recognized as safe), and should fulfill the limits according to the Food and Drug Administration (FDA). The material applied for food contact will have to go through regulatory scrutiny, where biopolymers act as additives that migrate toward the food products.

The EC and EF applied to food, such as horticultural products, are affected by features, such as optical and mechanical characteristics. Such features are affected by various factors like pH, solvent, temperature, etc. [25].

2.2 Background

EC or EF has been used for many years for food products to safeguard the materials. Cellulose coatings are most popular in the meat industry and waxing in the vegetable and fruit industry. During the mid-20th century, a layer is used for preventing loss of water and providing shining texture to vegetables and fruits. The EF and EC are shaping the barrier for the changes in term of chemical, biological, and bodily means. During the purchase of vegetables and fruits, buyers mainly think about healthy food items, depending on freshness and appearance. The main issue regarding this is to manage pathogenic microorganisms and spoilage growth in the fruit market and godowns. The first approach to get rid of this problem is to make use of EC [26]. The EC provides an extra protecting coating to get sparkle and influence the storage atmosphere to modify inner gasoline composition. Nowadays, the EC is employed intensively to preserve vegetables and fruits, including capsicum, cucumber, apple, and others. Application of EC is a successful process and does not depend on the manipulation of the inner composition of gasoline.

2.3 Functions and Advantages

The functions and advantages of EF and EF are exhibited in Figure 2.1. The main benefits of such food packaging are its eco-friendly nature. The fresh product involves enhanced shelf life. The EC or EF is the transporter of antimicrobial substances, and it is finally considered for food products. It is an active component that is united in the matrix of polymer and can be eaten along with the food, improving the essential requirements, such as safety, nutritional, and sensory needs. It also fulfills some other functions, such as protecting the products from physical, microbial, and chemical actions.

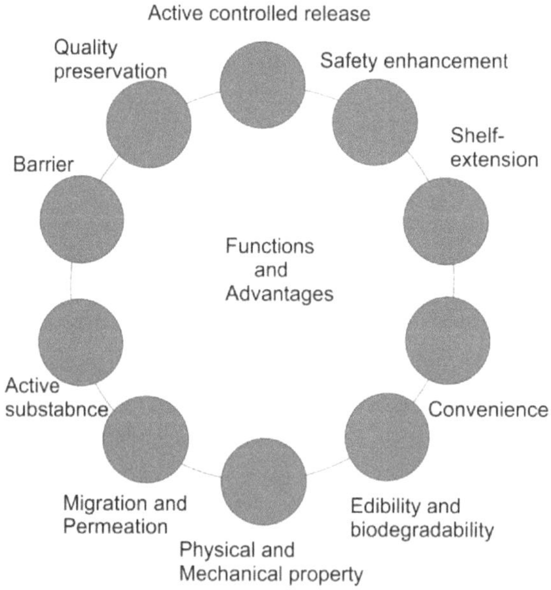

Figure 2.1 Functions and advantages of EC and EF.

The biodegradable films are produced by making the use of two basic techniques. Firstly, using wet solvent processing called solution casting. This was developed 100 years ago. The solution was spread on the plates such as acrylic, Teflon, or silicon plates, which were later following with the process of drying under controlled conditions, such microwave energy and hot air, at the ambient situation. The extrusion was used in manufacturing of thermoplastic polymers after 1950, and it was regarded as the central production process applied for the manufacture of plastics. The method of extrusion employs the elevated temperature, and it gets soften by trimming, and finally, the polymer is melted as a result of which cohesive film matrix is formed.

2.4 Importance of EC and EF

The film is employed for improving the organoleptic features of packaged foods. The film/coating behaves as a carrier of various substances, such as antimicrobials and coloration [27]. The film is implemented in the form of an interior heterogeneous component at interfaces on a specific sheet of food products. It is suitable for saving the deteriorative component moisture as well as the movement of solute in food products, such as pizzas and candies [28]. The EF or EC is also applied as a multilayer food packaging material. The functional activity and permeability of the EC and EF are minor in comparison to the artificial film, which is causing pollution [29]. Along with the function as a selective barrier for moisture, gas, and solute migration, the EC helps in minimizing the growth of microorganisms in solid and semi-solid food. It can be possible by reducing the diffusion rate of antimicrobial agents to coat substances instead of food [30].

2.5 Structural Matrix

The EC and EF are categorized based on the structural material. The primary sources of molecular groups are lipids, polysaccharides, and proteins. The origins of EF and EC are shown in Figure 2.2.

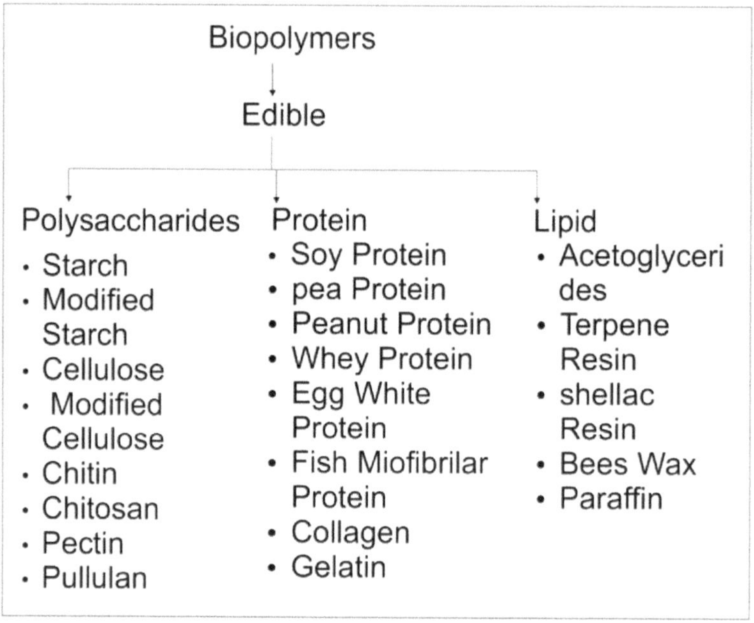

Figure 2.2 Various types of biopolymer origin.

The edible and biodegradable films are manufactured with the combination of different polysaccharides, lipids, and proteins that have useful features of a specific compound. The barrier and mechanical features of the films depend on the compound employed during the matrix of polymer and the compatibility [31]. The main compounds used in the EF and EC structural matrices are summarized in Table 2.1.

Table 2.1 Summary of various materials applied in EC and EF

S. No.	Material	Reference
1.	Chitosan	[32]
2.	Alginate	[33]
3.	Lipid	[34]
4.	Cellulose	[35]
5.	Starch	[36]

2.6 Synthesis of EF

EF is prepared by the dispersion of film-material using alcohol and water as solvents. For producing EF, colors, plasticizers, and flavors are added by adjusting pH. Some of the polymeric film solutions need to be heated for facilitating the dispersion. Later, the solution is cast and then dried at ambient temperature for attaining standalone films. Food application can be accessible by adopting various methods, such as spraying, brushing, and dipping.

2.7 Barrier Feature

The barrier features of EF are volatile permeability, gas permeability, and so on, which are the significant factors in food application. The effectiveness of the film is decided through the investigation of permeability of water vapor and permeability of oxygen (O_2). For determining the rate of oxidation and respiration of food products, the O_2 and CO_2 permeability is regarded as an essential parameter. The EF is composed of polar polymer that forms cracks because of the high sensitivity of changing humidity. The lipid material exhibits least water vapor, and it also performs as a distinguished moisture barrier. The demerit of the lipid material is that it is considered as a less effective gas barrier and cannot form a film without any help of agents, namely proteins and lipids.

2.8 Active Biopackaging

The active biopackaging is considered as the creative prospect, as it possesses the inherent ability to control the transmission of active agents from packaging substances. It is observed that the active agents on directly combining with the packaging substances can be recognized as favorable and reliable, but their quantity should be limited as per the use, which is reflected in Figure 2.3.

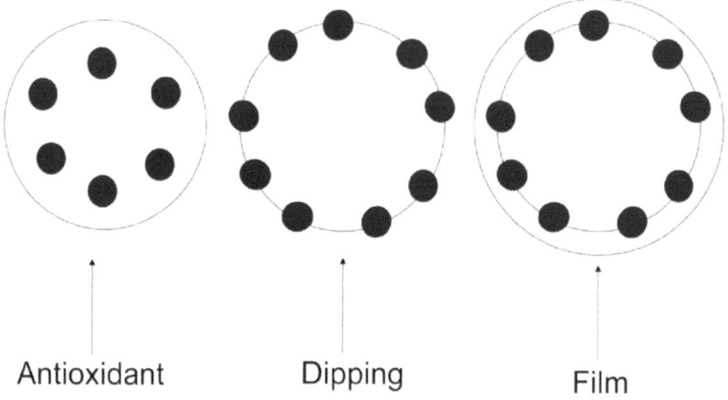

Figure 2.3 Different adaptation processes of additives.

For attaining active packaging and coating functions, the emulsifiers, antioxidants, and antimicrobial agents are assimilated within the film-forming solution. There are various antimicrobials, which are integrated into the EF and EC, which involve polypeptides, organic acids, and so on. The primary function of EF and EC is to protect food products from oxidation and spoilage that results in improving the quality and safeguard the products [37].

The most significant approach for food production is the optimization of EF composition because it needs to be framed as per the features of vegetables and fruits, on which it has to be applied [38]. Therefore, it is necessary to characterize and test different coating solutions on the food for maintaining and enhancing its quality [39].

The multicomponent and composite EF is optimized due to the mechanical features, transparency of food, and handling of food products. Hence, to fulfill such target, suitable designing of the surface methodology is to be done for determining the optimal mixture of components that could be beneficial [40]. As such, on adding lipid to enhance the features of the moisture barrier, the transparency features get affected.

The coating, which is dependent on polysaccharides, has enhanced application, and it has even improved the shelf life of vegetables and fruits because of its ability to be selective permeability of polymers toward O_2 and CO_2. Summary of some compounds with their effects is shown in Table 2.2. The polysaccharide-based coating is applied for modifying the internal environment of fruits, which delays senescence [38]. The EC develops passive modified atmosphere, influencing the changes in fresh and least processed food products, such as color and volatile compound [39].

Table 2.2 Summary of various biomaterials behavior

S. No.	Component	Behavior	Reference
1.	Chitosan	Antimicrobial	[41]
2.	Chitosan	Tissue conservation	[42]
3.	Chitosan	Antifungal	[43]
4.	Chitosan	Shelf life	[44]
5.	Chitosan	Mammalian cell	[45]

The demand of the consumers in accumulating in term of preservation of food naturally is leading to improve the alternative approach. The biopolymer is used widely, as it is generated from renewable sources or their by-products. The EF or EC is the main topic of interest because of its feature to increase the shelf life of food products. The shelf life of the food component can be enlarged by a decrease in respiration. The EC is a biodegradable material and eco-friendly substance, which is employed for reducing the application of plastic packaging. The enrichment of shelf life is significant, as it increases the economic benefits for food processing companies.

2.9 Chemistry of EF and EC

The quantity of material applied in the food industry is confined so, it should be capable of forming a film. The material should be capable of dispersing in a favorable food safe solvent and should be in harmony to plasticizers, antimicrobial agents, etc. The material is classified as composite, lipid, and hydrocolloid [46–48].

2.9.1 Polysaccharide

Polysaccharide is a natural polymer employed for the preparation of EF or EC, involving the cellulose and pullulan [49]. It is an essential coating considered as an effective blocker for O_2 because of its organized H-bond network. Since it exhibits hydrophilic nature, it is not capable of acting as a moisture barrier [50]. The coating comprising polysaccharide polymer shows fragile water vapor barrier feature, so it enables in the delay of moisture loss from meals [51]. Its coating has different characteristics, such as it is colorless, oil-free coating, and has minor caloric content. Such layer is employed to extend the shelf life of vegetables, fruits, and meat by the reduction of dehydration, surface darkening, and so on.

2.9.1.1 Cellulose

Cellulose is a renewable polymer resource found in large amount in the world. It is predicted that through the photosynthesis process, 10^{11}–10^{12} tons of cellulose are synthesized every year in the pure form, but sometimes it is combined with lignin or different polysaccharides of woody plants [52–55]. It is the polymeric raw material employed for the two specific functions. For instance, it has been used in the form of construction material as textile fibers, and intact wood. The stable cellulose derivatives are applied in the various technical aspect and domestic life [56–58]. Multiple features of cellulose were obtained, which could be noticed from the information based on dyeing the cellulose fibers. It holds a unique position in the anecdote of the polymer. It also possesses hydrophilic as well as hydrophobic feature that accomplishes self-association of the respective polymer in water [59–61]. The structure of cellulose is shown in Figure 2.4.

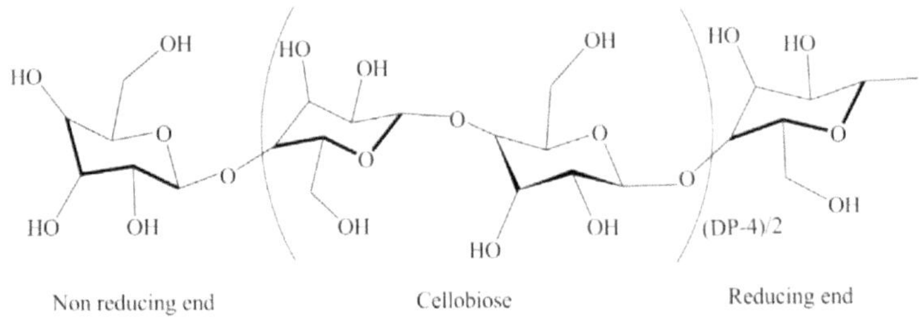

Figure 2.4 Structure of cellulose.

The cellulose, as well as its derivatives, has been used for manufacturing the processed food from many decades. The cellulose that is physically modified provides a useful strategy for many products in bulking features such as minimized or low-calorie diet, and flow of the products posing an ability of natural sweeteners and favor packets, as well as imbiber the oil flavor. The wide use of cellulose is mainly due to its rheology and textural characteristics, controlled water interaction, and chemical features. Microcrystalline cellulose (MCC) and the soft cellulose play a significant role to provide bulking low-calorie food products. The chemically modified cellulose derivatives perform five vital functions such as organizing flow features, stability of foam, emulsification, strength as well as development for water binding, and ice crystal production. The physical and chemical characteristics can devise the use of cellulose in food application [62, 63].

The cellulose derivatives are chosen as the best option for developing baked products because of their beneficial properties. Bread requires a specific size value for the development of the fine texture and cell structure. The selection of various qualities as well as the quantity of diverse wheat flour decides the qualification of the food product. Therefore, additives are used for correcting such circumstances. There is a requirement of a particular readjusting for supplying nut structure and also the baked quality that could be compared with products comprised wheat flour. The high-fiber bread applies different derivatives of cellulose. The MCC or cellulose derivatives can be substituted with the wheat flour, even the modified cellulose performs with least ability in the baked products.

However, the specific component needs to be changed for attaining the appropriately qualified products.

2.9.1.2 Chitin

Chitin, abundantly found in nature, derives the poly-cationic polysaccharide called chitosan. The chitosan has more than 5000 units of glucosamine, which can be collected from shrimps shell; crab comprised chitin and N-acetyl glucosamine unit. It is obtained through the process of partial chemical de-acetylation of chitin that is extracted from the crustacean group, including animals such as crab, shrimp through deproteination, decolorization, and decalcification, and also from fungi through the process of enzymatic extraction [64, 65]. The structure of chitosan is shown in Figure 2.5. It has features, such as biodegradable, antimicrobial, and biocompatible, which enable it to be used for different applications. Till date, very restricted consideration is given to the use of chitosan in the food industry. It is a cost-effective polymer because it can be easily obtained from nature. The main aim of the current review is the transition of marine, as well as terrestrial processing, which dumps into the profitable by-products making the material adequate for research in term of food and its development. The extensive application involves the recovery of waste substance by various approaches, such as processing castoffs, protection, development, amplification, cleansing, and deacidification. In this support, multiple researchers have suggested by carrying out antifungal and antimicrobial performance [66, 67].

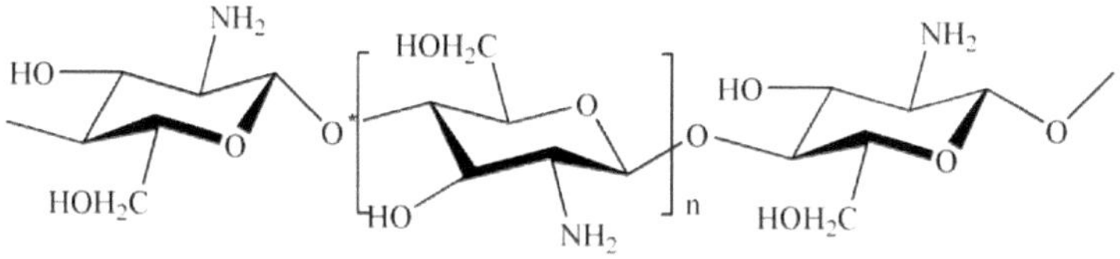

Figure 2.5 Structure of chitosan.

The introduction of green extraction approach based on the green chemistry theory includes addition of higher concentration, supporting the application of microorganism enzymes for extraction of chitin. The result of the approach clarified that the use of microorganisms in comparison to that of enzyme is much more superior, as it conserves the structure of chitin. The biological method of extracting chitin is high reproducible, even in a short period by pure exploitation, and it uses a small solvent with consuming low energy [68]. But yet, the biological approach faces some limitation in the laboratory-scale production, as shown in Figure 2.6.

Figure 2.6 Development of chitosan from materials.

The physiochemical and biological features of chitosan oligomer (CO) attained by the marine diversity depend on the molecular weight, where order length is bounded. The naturally obtained CO offers more significant advantages in comparison to the synthesized chitosan. The antioxidant potential of CO and its derivatives can be improved by converting the chitosan nanobiofilm (CN) to reinforce the application of CN in the food industry. The chitosan nanoparticles released tend to raise free radical scavenging behavior of biofilm for the extended time of food storage in the industry. Therefore, the brief physiological and sensory estimation are required for determining the molecular mechanism of antifungal and antimicrobial characteristics. This can provide little knowledge of the influence of chitosan for the exertion of high-quality food products.

2.9.1.3 Starch

Starch is made up of two glucose polymers, such as amylose and amylopectin, as exhibited in Figure 2.7. It is a polymer of glucose residue connected by the glycoside bonds. Such a relationship is delicate where the hydrolysis process is attained at a low pH value [69]. It is the carbohydrate stored in very high amount in plants and also regulated as a critical factor to determine the quality of food products. It is the only polymer that can be applied daily in large amount for food and well as non-food application [70]. The starch has various features such as naturally occurring polymer, present in large quantity, cost effective, and consumed and eaten by all living creatures. The main biopolymer used in large amount is the starch, as it can be available very quickly and easily, because it is produced from various sources, namely corn and rice [71].

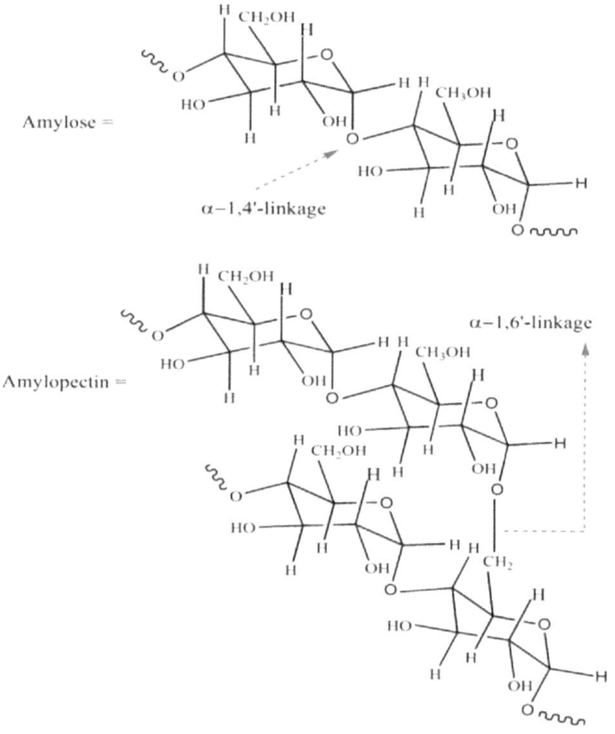

Figure 2.7 Structure of protein.

Corn starch is the most valuable starch in comparison to other starches for industrial application. The starch segregated from the corn is intermixed in the baked products, namely cake and cookies, as it possesses the ability for preserving the moisture content of the products, preventing the development of crystal from sugar, and finally improving the juiciness of the goods [72]. The corn starch helps in providing consistency and offering efficient gelling features at the time of heating corn-banana [73].

Wheat with the botanical name "Triticumaestivum" is the highly consumed crop and it is used in a different ways like bread, as it provides energy and different exclusive nutrients [74]. The baked products obtained from wheat, such as cake and bread, depend on molecular structure and characteristics of the wheat [75]. Corn is even considered as an excellent source to provide thickening in formula milk of infants [76].

2.9.1.4 Pectin

Pectin, the structural hetero-polysaccharide, is available in terrestrial plants. It bears rhamnose and methyl-esterified galacturonic acid units. Its structure is shown in Figure 2.8. The proportion of the rhamnose units and the availability of other essential units are mainly dependent on the plants, from which pectin gets detached. Since pectin can be produced from various plants, but yet the industrial pectin gets extracted from citrus peel as well as apple pomace, at the mild acidic circumstance. In such condition, pectin leads to form gels, and such feature has enabled pectin to be used as an additive in products, such as jam and confectionaries. It is even the significant food component that is present in high amount in the agricultural waste as well. It holds nutritional advantages for health-related issues of human beings and also has pharmaceutical behavior, which makes it clear that it can be applied in various food components. Its coatings are further studied to specify its capability in retarding the migration of lipid and loss of moisture and also to enhance the appearance as well as better handling of food products.

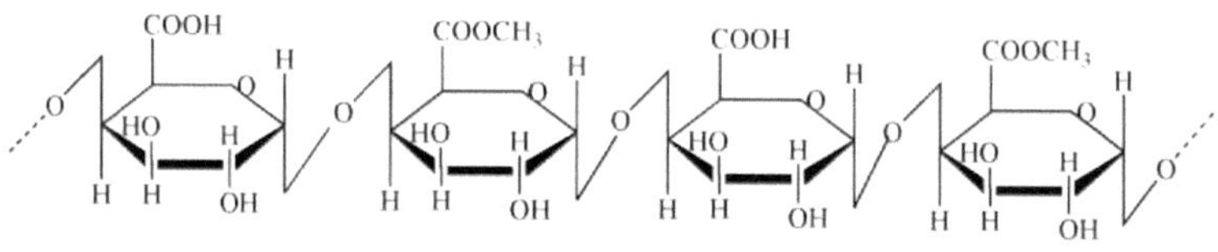

Figure 2.8 Structure of pectin.

Maftoonazad and Ramaswamy [77] employed pectin-based composite coating on the fruit avocado to test the limits of change in quality at different temperature of storage and kinetic parameters, and to determine the loss of quality of the avocado that was stored. The results signified that such coating is responsible for reducing the rate of various parameter changes in the avocado at the time of storage, and it even prolonges the shelf life of the product at 10 °C for one month.

2.9.1.5 Pullulan

Pullulan, a neutral linear polysaccharide, is made up of α-1, 6-linked maltotriose residue as well water soluble products. It is a fungal exopolysaccharide obtained from the starch through the fungus *Aureobasidium pullulans*. Bauer, in 1938, had given his observation about this exopolymer by the name pullulan. The structure of pullulan is shown in Figure 2.9.

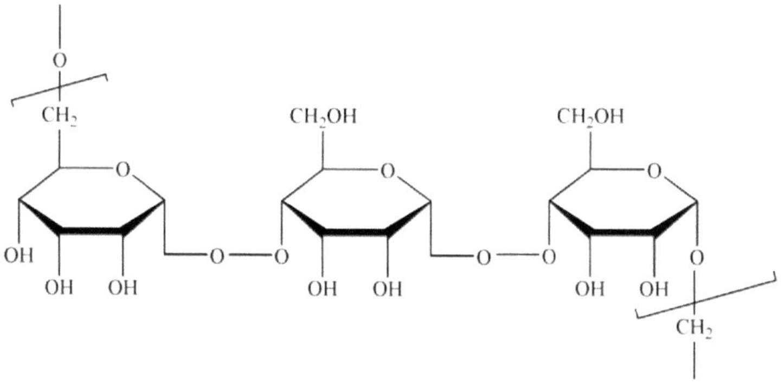

Figure 2.9 Structure of pullulan.

It is used on a large scale in the food processing industry. The pullulan slowly digested macromolecule posing the ability of tasteless and odorless, so it is considered to be applied on low-calorie food that provides bulk as well as texture to the products. It is an oxygen barrier polysaccharide, posing excellent moisture retention ability and also inhibiting fungal growth. All such properties enable pullulan to be an excellent material used in the preservation of food and thus it is employed in large scale in the food industry [78].

2.9.1.6 Alginate

Alginate is a polymer posing features such as biocompatible, hydrophilic, biodegradable, and economical. It is found within Phaeophyceae brown algae in the marine as a structural component, and also obtained from some bacteria in the form of capsular polysaccharides. The structure of alginate is shown in Figure 2.10.

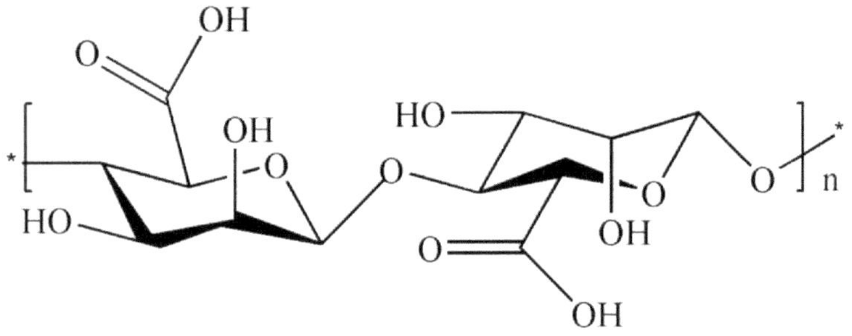

Figure 2.10 Structure of alginate.

The alginate is a versatile biopolymer with specific chemical and physical features. It possesses the selective binding ability of multivalent cations, which are the basis of gel formation. Also, it has hydroxyl and carboxylic acid features that can be able to accept and release a proton with the change in pH, due to which alginate is considered as pH-sensitive [79].

2.9.2 Protein

The material-based protein is obtained from silk and wool of animals [80–83]. The polypeptide chains are linked with each other to frame a fibrous structure. The fibrous protein is featured by both primary and secondary structures, in which the basic material is made up of amino acid sequencing, and secondary structure is comprised H-bond in between the polypeptides chain, as shown in Figure 2.11. The features of the protein are exhibited in Figure 2.12.

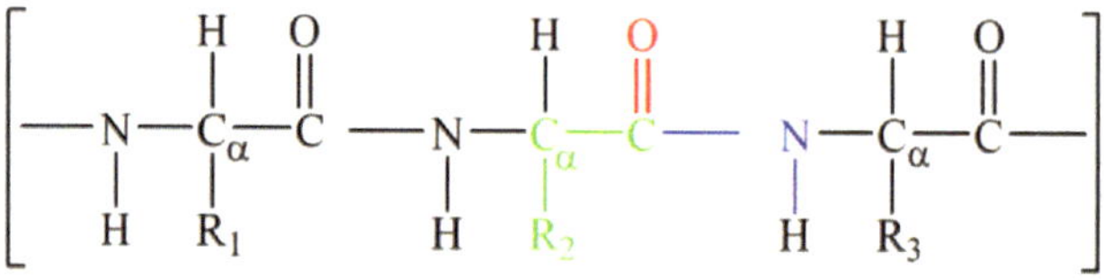

Figure 2.11 Structure of protein.

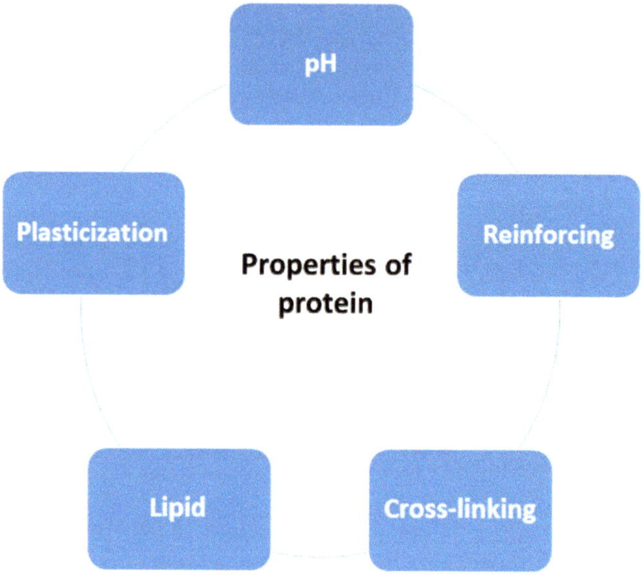

Figure 2.12 Various properties of protein.

Bacteriocin is an excellent antimicrobial agent for obtaining an active biopolymer film to be used in the food packaging industry. But it is found that the use of such bacteriocin is limited in food preservation because of the decrease in activity of antimicrobial peptides as well as protein [84]. As such bacteriocins are manufactured from various microorganisms, the direct assimilation of strain in the film can be considered as the promising substitutes.

The production of protein film comes with various problems based on costing and performance. The cost of the film can be decreased by suitable exploitation of by-products of the food industry for the extraction of protein, and the performance problem can be overcame by reinforcement or through modification method of the protein matrix. Therefore, more elaborate and precise studies are required to be carried out for putting the features of films in large-scale processing and application in the food industry.

2.9.2.1 Gelatin

The skin of pig has been applied in the form of raw material for manufacturing gelatin since 1930, and it even considered as the essential material in large-scale production of food, but for pharmaceutical work. Protein is also obtained from fish and mammal. It is also produced by the bones of cattle, which is very complicated, and the extraction process is costly. The structure of gelation is shown in Figure 2.13.

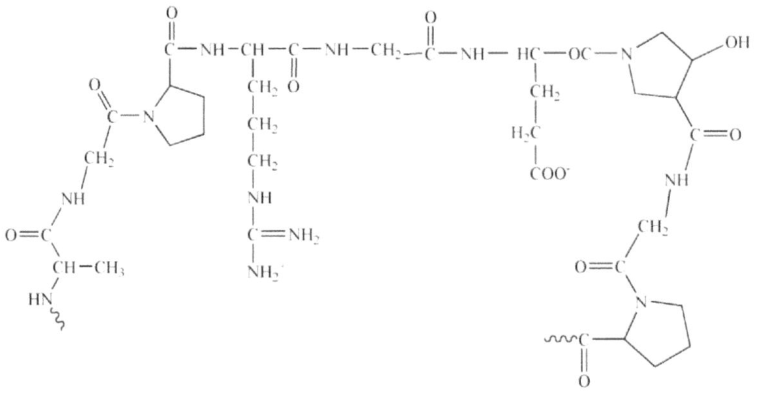

Figure 2.13 Structure of gelatin.

It is a biocompatible and biodegradable polymer obtained by chemical, physical, and thermal degradation of collagen [85]. Depending on the origin and varieties of tissue, the various processes for the extraction of gelatin can be used, as exhibited in Figure 2.14. Collagen is a fiber that is biopolymeric in nature, and it is also available in high abundant. It is the protein found in connective tissue serving as the vital component for stress bearing in connective tissue and it is also rich in residues such as proline and hydroxyproline.

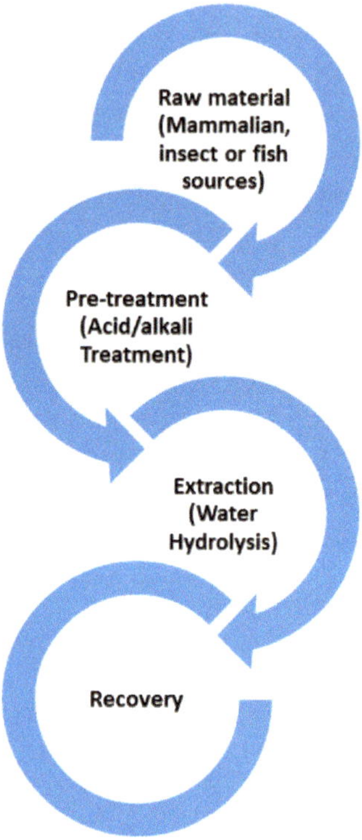

Figure 2.14 Gelatin production process.

The melting temperature of gelatin gel is less in comparison with other starches, which makes it favorable to be used in the food industry in term of flavor release [86].

In upcoming years, the agro-industrial by-products have gathered interest for the production of gelatin, and even it is at the pace of development. Therefore, the revalorization is increasing, as it is a capable material of exploration. The application of gelatin residue liberated from capsules produced by nutraceutical area is increasing day by day, for the disposal and treatment processes of waste that are expensive and have different environmental problems.

2.9.3 Composite

Research activities on edible components have mainly intensive on the composite, multi-component films for developing significant benefits of every element and also for minimizing demerits [87]. The films are incorporated in various forms such as suspension, dispersion, and so on. The successive layer film or coating has a significant demerit, such as the method of preparation includes four stages including two casting stages and two drying stages. The laminated film is least employed in the food industry, in spite of exhibiting sound barrier feature, as compared to water vapor [88]. The research activities carried out in such field show that successive layer films delaminate after some time, and it establishes cracks or pinholes showing the non-uniform surface [89]. The functional characteristics of such components are dependent on the process of preparation, amount, and type of the components [90].

2.10 Economic and Nutritional Influences

The vegetables and fruits are the essential ingredients of diet for every human being, and these are in demand in each part of the world. Both vegetables and fruits are the origins of vitamins, antioxidants, minerals, fibers, and etc. It is also responsible for providing flavor to the compounds that are affected by biotic and abiotic adversities. The use of vegetables and fruits in the diet leads to decrease danger caused by various chronic diseases, namely dementia, stroke, and cancer [91]. The healthy diet advantages on the consumption of the stored vegetables and fruits get reduced because of the decrease in Vitamin C and polyphenols, which is mainly due to physiological changes taking place at the time of storage [92]. Protein is recognized as an essential component required for better survival of human beings and animals. Its main characteristic is to provide the proper amount of amino acids. The first class protein that is subscribed to add dietary as well as nutritive value to the food is dependent on the amount of amino acid and its use after the process of digestion and absorption. The demand for plant-based protein, in spite of the animal protein, has increased due to economic issues, health anxiety, and emerging drift toward vegetarianism [93]. Therefore, it is evident that the film formed from plant protein enables it to be an excellent substance for healthy EF and plastic opportunity biodegradable packaging materials [94].

2.11 Protection of Food Product: Mechanism Phenomena

The primary role of the fruit industry is to reduce the browning of fresh fruits that occurs due to polyphenol oxidase mechanism taking place after cutting and peeling [95].The browning activity can be treated by inhibitors, which impart the growth of brown discoloration [96]. The film forming substance can be either hydrophobic or hydrophilic, but to maintain useful feature only H_2O or C_2H_5OH can be employed as a solvent at the time of processing. Various other materials can be further added to the matrix for improving the functional features.

The main benefit of the dry process as the eco-friendly method is that it does not make any use of solvents. There are different approaches that are utilized to prolong the life of fruits [97].

2.12 Healthy Foods: Public Demands

The demand for vegetables and fruits are increasing continuously in the market, as these are functional food components, and these contain nutrients, antioxidant, and other important ingredients, which prevent various types of cancers and degenerative diseases. The vegetables and fruits have a short shelf life, as they are decomposable. The main challenge for the whole world is the preservation of vegetables and fruits. Such trouble can be solved by EC and EF, which help in preserving the products in a safe and healthy way and are beneficial to the consumer as well as the environment. Nowadays, herbal EC is used in the fitness and nutraceutical field. It possesses an extraordinary barrier residence for CO_2, O_2, and others, and it is recognized as an outstanding transporter of nutrients. These are the main factors that contribute to its market value.

2.13 Conclusion

In the developing world of science, the edible coatings and films are considered as an essential part of output forward for various applications [98–102]. The different functions of the materials said to be commercially applied, showing protuberant results and positive aspects [103–107]. The independent studies carried out have made it clear that EF and EC are excellent healthy agents for the protection of food products from different means, since they occur naturally, and are cheap and renewable as well. The EF is a superb transporter of nutrients that the human beings take through food. The research activities performed till date have offered outstanding results, with least drawbacks, and also provided the very particular concept of essential oils as well as herbal coatings. More explosive researchers are needed to increase the beneficial outcomes of EC and EF for human beings involving production and applications. Today, there is a trend of employing active envelopes, which involves reducing the oil consumption of deep-fried products. The application of EC and EF in the form of appropriate packaging for the food industry has gathered attraction, due to the promising attempt to prolong the shelf life of the food products. Such film and coating show different functional aspects when used in lipids to reduce moisture and other issues. The EC and EF perform the same function as synthetic packaging film, but yet these are chosen as the best option for packaging based on a specific application. The EF is capable enough for minimizing a few traditional polymeric packaging substances for correct use. Therefore, to carry out such activities, the bio-based packaging should work as a standard packaging by providing the critical function of preservation and cost-effective features. In the current chapter, the extensive study related to chemistry, merits, and requirements of healthful EF and EC are explained, keeping in minds the nutritional as well as economic aspects, followed by the increasing demand for food products from the human beings.

Acknowledgement

Authors would like to acknowledge the Knowledge Resource Centre, CSIR-NEERI, (CSIR-NEERI/KRC/2019/MAY/EMD/1), for their support.

References

1. Bushra, R., Arfin, T., Oves, M., Raza, W., Mohammad, F., Khan, M.A., Ahmad, A., Azam, A., Muneer, M. (2016). Development of PANI/MWCNTs decorated with cobalt oxide nanoparticles towards multiple electrochemical, photocatalytic and biomedical application sites, New J. Chem., 40, pp. 9448–9459.

2. Arfin, T., Bushra, R., Mohammad, F. (2016). Electrochemical sensor for the sensitive detection of o-nitrophenol using graphene oxide-poly(ethyleneimine) dendrimer-modified glassy carbon electrode, *Graphene Technol.*, **1**, pp. 1–15

3. Arfin, T., Rafiuddin (2009). Transport studies of nickel arsenate membrane, J. Electroanal. Chem., 636, pp. 113–122.

4. Arfin, T., Rafiuddin (2009). Electrochemical properties of titanium arsenate membrane, *Electrochim. Acta*, **54**, pp. 6928–6934.

5. Arfin, T., Mohammad, F. (2014). Electrochemical, dielectric behaviour and in vitro antimicrobial activity of polystyrene-calcium phosphate, *Adv. Ind. Eng. Manag.*, **3**, pp. 25–38.

6. Arfin, T., Fatma, S. (2014). Synthesis, influence of electrolyte solutions on impedance properties and in vitro antibacterial studies of organic-inorganic composite membrane, *Adv. Ind. Eng. Manag.*, **3**, pp. 19–30.

7. Arfin, T., Fatima, S. (2014). Conductometric studies with polystyrene calcium phosphate membrane, *Asian J. Adv. Basic Sci.*, **2**, pp. 1–14.

8. Arfin, T., Mohammad, F. (2013). DC electrical conductivity of nano-composite polystyrene-titanium-arsenate membrane, *J. Ind. Eng. Chem.*, **19**, pp. 2046–2051.

9. Mohammad, F., Arfin, T. (2013). Cytotoxic effects of polystyrene-titanium-arsenate composite in cultured H9c2 cardiomyoblasts, *Bull. Environ. contam. Toxicol.*, **91**, pp. 689–696.

10. Arfin, T., Yadav, N. (2012). Impedance characteristics and electrical double-layer capacitance of composite polystyrene-cobalt-arsenate membrane. *J. Ind. Eng. Chem.*, **19**, pp. 256–262.

11. Arfin, T., Rafiuddin (2012). Metal ion transport through a polystyrene-based cobalt arsenate membrane: Application of irreversible thermodynamics and theory of absolute reaction rates, *Desalination*, **284**, pp. 100–105.

12. Arfin, T., Yadav, N. (2012). Impedance characteristics and electrical double layer capacitance of polystyrene-based nickel arsenate membrane, *Anal. Bioanal. Electrochem.*, **4**, pp. 135–152.

13. Arfin, T., Rafiuddin (2011). An electrochemical and theoretical comparison of ionic transport through a polystyrene-based cobalt arsenate membrane, *Electrochim. Acta*, **56**, pp. 7476–7483.

14. Arfin, T., Jabeen, F., Kriek, R.J. (2011). An electrochemical and theoretical comparison of ionic transport through a polystyrene based titanium-vanadium (1:2) phosphate membrane, *Desalination*, **274**, pp. 206–211.

15. Arfin, T., Rafiuddin (2010). Thermodynamics of ion conductivity of alkali halide across a polystyrene-based titanium arsenate membrane, *Electrochim. Acta*, **55**, pp. 8628–8631.

16. Mohammad, F., Arfin, T., Al-Lohedan, H.A. (2019). Biocompatible polylactic acid-reinforced nickel-arsenate composite: Studies of electrochemical conductivity, mechanical stability, and cell viability, *Mater. Sci. Eng. C*, **102**, pp. 142–149.

17. Arfin, T., Falch, A., Kriek, R.J. (2012). Evaluation of charge density and the theory for calculating membrane potential for a nano-composite nylon-6,6 nickel phosphate membrane, *Phys. Chem. Chem. Phys.*, **14**, pp. 16760–16769.

18. Arfin, T., Bushra R., Kriek, R.J. (2013). Ionic conductivity of alkali halides across a polyaniline-zirconium (IV)-arsenate membrane, *Anal. Bioanal. Electrochem.*, **5**, pp. 206–221

19. Arfin, T., Rangari, S.N. (2018). Graphene-oxide-ZnO nanocomposite modified electrode for the detection of phenol, *Anal Methods*, **10**, pp. 347–358.

20. Mohammad, F., Arfin, T., Al-lohedan, H.A. (2019). Enhanced biosorption and electrochemical performance of sugarcane bagasse derived a polylactic acid-graphene oxide-CeO_2 composite, *Mater. Chem. Phys.*, **229**, pp. 117–123.

21. Arfin, T., Sonawane, K. (2018). Bio-based materials for food packaging: Green and sustainable advanced packaging, Chapter 1 "Bio-based materials: Past to future", Ahmed, S. (ed.), Springer Nature, Singapore, pp. 1–32.

22. Tharanathan, R.N. (2003). Biodegradable films and composite coatings: Past, present and future, Trends Food Sci. Technol.,14, pp. 71–78

23. NurHanani, Z.A., Roos, Y.H., Kerry, J.P. (2014). Use and application of gelatin as potential biodegradable packaging materials for food products, Int. J. Biol. Macromol., 71, pp. 94–102.

24. O'Sullivan, A., Shaw, N.B., Murphy, S.C., van de Vis, J.W., van Pelt-Heerschap, H., Kerry, J.P. (2006). Extraction of collagen from fish skins and its use in the manufacture of biopolymer films, J. Aquat. Food Prod. T., 15, pp. 21–32.

25. Guilbert, S., Gontard, N., Gorris, G.M. (1996). Prolongation of the shelf-life of perishable food products using biodegradable films and coatings, LWT-Food Sci. Technol., 29, pp.10–17.

26. Rojas-Grau, M.A., Raybaudi-Massilia, R.M., Soliva-Fortuny, R.C., Avena-Bustillos, R.J., McHugh, T.H., Martin-Belloso, O. (2007). Apple puree-alginate edible coatings as carrier of antimicrobial agents to prolong shelf-life of fresh-cut apples, *Postharvest Biol. Technol.*,**45**, pp. 254–264.

27. Vasconez, M.B., Flores, S.K., Campos, C.A., Alvarado, J., Lia, N. (2009). Antimicrobial activity and physical properties of chitosan-tapioca starch based edible films and coatings, *Food Res. Int.*, **42**, pp. 762–769.

28. Bourtoom, T. (2008). Edible films and coatings: Characteristics and properties, *Int. Food Res. J.*, **15**, pp. 237–248.

29. Kester, J.J., Fennema, O.R. (1986). Edible films and coatings: A review, *Food Technol.*, **40**, pp. 47–59.

30. Aloui, H., Khwaldia, K. (2016). Natural antimicrobial edible coatings for microbial safety and food quality enhancement, *Rev. Food Sci. Food Saf.*, **15**, pp. 1080–1103.

31. Altenhofen, M., Krause, A.C., Guenter, T. (2009). Alginate and pectin composite films cross linked with Ca^{2+} ions: Effect of the plasticizer concentration, Carbohydr. Polym., 77, pp. 736–742.

32. Hajji, S., Younes, I., Affes, S., Boufi, S., Nasri, M. (2018). Optimization of the formulation of chitosan edible coatings supplemented with carotenoproteins and their use for extending strawberries postharvest life, *Food Hydrocoll.*, **83**, pp. 375–392.

33. Abdallah, M.R., Mohamed, M.A., Mohamed, H., Talaat, M., Emara (2018). Application of alginate and gelatin-based edible coating materials as alternatives to traditional coating for improving the quality of pastirma, *Food Sci. Biotechnol.*, **27**, pp. 1589–1597.

34. Dhall, R.K. (2013). Advances in edible coatings for fresh fruits and vegetables: A review, *Crit. Rev. Food Sci. Nutr.*, **53**, pp. 435–450.

35. Garcia, M.A., Ferrero, C., Bertola, N., Martino, M., Zaritzky, N. (2002). Edible coatings from cellulose derivatives to reduce oil uptake in fried products, *Innov. Food Sci. Emerg. Technol.*, **3**, pp. 391–397.

36. Pagella, C., Spigno, G., De Faveri, D.M. (2002). Characterization of starch-based edible coatings, *Food Bioprod. Process*, **80**, pp. 193–198.

37. Kim, I.H., Yang, H.J., Noh, B.S., Chung, S.J., Min, S.C. (2012). Development of a defatted mustard meal-based composite film and its application to smoked salmon to retard lipid oxidation, Food Chem., 133, pp. 1501–1509

38. Rojas-Grau, M.A., Soliva-Fortuny, R., Martın-Belloso, O. (2009). Edible coatings to incorporate active ingredients to fresh cut fruits: A review, Trends Food Sci. Technol., 20, pp. 438–447

39. Oms-Oliu, G., Soliva-Fortuny, R., Martin-Belloso, O. (2008). Using polysaccharide-based coatings to enhance quality and antioxidant properties of fresh melon, LWT- Food Sci. Technol., 41, pp. 1862–1870

40. Ozdemir, M., Floros, J.D. (2008). Optimization of edible whey protein films containing preservatives for mechanical and optical properties, J. Food Eng., 84, pp. 116–123.

41. Maqbool, M., Ali, A., Ramachandran, S., Smith, D. R., Alderson, P.G. (2010). Control of postharvest anthracnose of banana using a new edible composite coating, Crop Prot., 29, pp. 1136–1141

42. El Gaouth, A., Arul, J., Wilson, C., Benhamou, N. (1997). Biochemical and cytochemical aspects of the interactions of chitosan and Botrytis cinerea in bell pepper fruit, Postharvest Biol. Technol., 12, pp. 183–194

43. Martınez-Camacho, A.P., Cortez-Rocha, M.O., Ezquerra-Brauer, J.M., Graciano-Verdugo, A.Z., Rodrıguez-Felix, F., Castillo-Ortega, M.M., Yepiz-Gomez, M.S., Plascencia-Jatomea, M. (2010). Chitosan composite films: thermal, structural, mechanical an antifungal property, Carbohydr. Polym., 82, pp. 305–315

44. Kanatt, S.R., Rao, M.S., Chawla, S.P., Sharma, A. (2013). Effects of chitosan coatings on shelf-life of ready-to-cook meat products during chilled storage, LWT- Food Sci. Technol., 53, pp. 321–326

45. Zielinski, B.A., Aebischer, P. (1994). Chitosan as a matrix for mammalian cell encapsulation, Biomaterials, 15, pp. 1049–1056

46. Arfin, T. (2019). Marine polysaccharides: Advances and multifaceted application, Chapter 1 "Marine polysaccharides: An overview", Ahmed, S., Soundarajan, A. (eds.), Pan Stanford Publishing, Singapore, pp. 1–11

47. Arfin, T., Tarannum, A. (2017). Green polymeric materials: Advanced and sustainable development, Chapter 2 "Polymer materials: From the past to the future", Ahmed, S., Annu, S., Ikram, S. (eds.), Nova Science Publishers, New York, pp. 35–52.

48. Arfin, T. (2018). Handbook of bionanocomposite: Green and sustainable materials, Chapter 9 "MWCNT polymer composites: Environmental applications", Ahmed, S., Kanchi, S. (eds.), Pan Stanford Publishing, Singapore, pp. 235–246.

49. Krochta, J.M., De Mulder-Johnston, C. (1997). Edible and biodegradable polymer films: Challenges and opportunities, Food Technol., 51, pp. 61–74.

50. Yang, L., Paulson, A.T. (2000). Effects of lipids on mechanical and moisture barrier properties of edible gellan film, Food Res. Int., 33, pp. 571–578.

51. Kester, J.J., Fennema, O.R. (1986). Edible films and coatings: A review, Food Technol., 40, pp. 47–59.

52. Arfin, T, Tarannum, A., Sonawane K. (2018). Green and sustainable advanced materials: Processing and characterization, Vol. I, Chapter 1 "Green and sustainable advanced materials: An overview", Ahmed, S., Hussain C.M. (eds.), Scrivener Publishing LLC, USA, pp. 1–34.

53. Arfin, T., Tarannum, A. (2018). *Handbook of nanomaterials for industrial applications,* Chapter 6 "Engineered nanomaterials for industrial application: An overview", Hussain, C.M. (ed.), Elsevier, the Netherlands, pp. 127–134.

54. Athar, S., Bushra, R., Arfin, T. (2017). *Nanocellulose and nanohydrogel matrices: Biotechnological and biomedical applications,* Chapter 7 "Cellulose nanocrystals and PEO/PET hydrogel material in biotechnology and biomedicine: Current status and future prospects", Jawaid, M., Mohammad, F. (eds.), Wiley-VCH, Germany, pp. 139–173.

55. Borkar, R., Waghmare, S.S., Arfin, T. (2017). *Nanocellulose and nanohydrogel matrices: Biotechnological and biomedical applications,* Chapter 2 "Bacterial cellulose and polyester hydrogel matrices in biotechnology and biomedicine: Current status and future prospects", Jawaid, M., Mohammad, F. (eds.), Wiley-VCH, Germany, pp. 21–46.

56. Khan, A.U., Malik, N., Arfin, T. (2017). *Nanocellulose and nanohydrogel matrices: Biotechnological and biomedical applications,* Chapter 12, "Nanofibrillated cellulose and copoly (amino acid) hydrogel matrices in biotechnology and biomedicine", Jawaid, M., Mohammad, F. (eds.), Wiley-VCH, Germany, pp. 331–352.

57. Arfin, T., Mohammad, F. (2016). Electrochemical, antimicrobial and anticancer effects of ethyl cellulose-nickel (II) hydrogen phosphate, Innovations Corrosion Mater. Sci., 6, pp. 10–18.

58. Arfin, T., Mohammad, F. (2015). Electrical conductivity, mechanical stability, antibacterial and anticancer activities of ethyl cellulose-tin (II) hydrogen phosphate, Adv. Mater. Lett., 6, pp. 1058–1065.

59. Arfin, T., Kumar, C. (2014). Synthesis, characterization, conductivity and antibacterial activity of ethyl cellulose manganese (II) hydrogen phosphate, *Anal. Bioanal. Chem.*, **6**, pp. 403–421.

60. Mohammad, F., Arfin, T., Al-Lohedan, H.A. (2018). Synthesis, characterization and applications of ethyl cellulose-based polymeric calcium (II) hydrogen phosphate composite, J. Electron. Mater., 47, pp. 2954–2963.

61. Mohammad, F., Arfin, T., Al-Lohedan, H.A. (2017). Sustained drug release and electrochemical performance of ethyl cellulose-magnesium hydrogen phosphate composite, Mater. Sci. Eng. C, 71, pp. 735–743.

62. Mohammad, F., Arfin, T., Saba, N., Jawaid, M., Al-Lohedan, H.A. (2018). Electrically conductive polymers and polymer composites: From synthesis to biomedical applications, Chapter 9 "Electrical conductivity and biological efficacy of ethyl cellulose and polyaniline-based composites", Khan, A., Asiri, A.M. (eds.), Wiley-VCH, Germany, pp. 181–197.

63. Waghmare, S.S., Arfin, T. (2015). Defluoridation by adsorption with chitin-chitosan-alginate-polymer-cellulose-resins-algae and fungi: A review, *Int. Res. J. Eng. Tech.*, **2**, pp. 1179–1197.

64. Arfin, T. (2018). Handbook of biopolymers: Advances and multifaceted applications, Chapter 7 "Current innovative chitosan-based water treatment of heavy metals: A sustainable approach", Ahmed, S., Kanchi, S., Kumar, G. (eds.), CRC Press, the U.K., pp. 167–183.

65. Mohammad, F., Arfin, T., Al-Lohedan, H.A. (2017). Enhanced biological activity and biosorption performance of trimethyl chitosan-loaded cerium oxide particles, J. Ind. Eng. Chem., 45, pp. 33–43.

66. Arfin, T. (2017). Chitosan: Derivatives, composites and applications, Chapter 5 "Chitosan and its derivatives: Overlook of commercial application in diverse field", Ahmed, S., Ikram, S. (eds.), Scrivener Publishing LLC, USA, pp. 115–150.

67. Arfin, T., Mohammad, F. (2016). Natural polymers: Derivatives, blends and composites, Vol. 1, Chapter 13 "Chemistry and structural aspects of chitosan towards biomedical, Ikram, S., Ahmed, S. (eds.), Nova Science Publishers, USA, pp. 265–280.

68. Aytekin, O., Elibol, M. (2010). Cocultivation of lactococcuslactis and teredinobacterturnirae for biological chitin extraction from prawn waste, *Bioprocess Biosyst. Eng.*, **33**, pp. 393–399

69. Mohammad, F., Arfin, T., Bwatanglang, I.B., Al-Lohedan, H.A. (2019). *Bio-based polymers and nanocomposites: Preparation, processing, properties & performance*, Chapter 8 "Starch-based nanocomposite: Types and industrial applications, Sanyang, M.L., Jawaid, M. (eds.), Springer Nature, Switzerland, pp. 157–181.

70. Arfin, T., Sonawane, K. (2018). *Green and sustainable advance materials: Application*, Vol. 2, Chapter 8 "An excellence method on starch-based materials: A promising stage for environmental application", Ahmed, S., Hussain, C.M. (eds.), Scrivener Publishing LLC, USA, pp. 177–208.

71. Mogarkar, P.R., Arfin, T. (2017). *Natural polymers: Derivatives, blends and composites*, Vol. 2, Chapter 5 "Chemical and structural importance of starch-based derivative and its applications, Ikram, S., Ahmed, S. (eds.), Nova Science Publishers, USA, pp. 73–87.

72. Ali, A., Wani, T.A., Wani, I.A., Madoodi, F.A. (2016). Comparative study of the physic-chemical properties of rice and corn starches grown in Indian temperate climate, *Journal of the Saudi Society of agricultural Sciences*, **15**, pp. 75–82.

73. Alimi, B.A., Workneh, T.S., Oyeyinka, S.A. (2017). Structural, rheological and in vitro digestibility properties of composite corn-banana starch custard paste, LWT- Food Sci. Technol., 79, pp. 84–91.

74. Lu, Y., Luthria, D., Fuerst, E.P., Kiszonas, A.M., Yu, L., Morris, C.F. (2014). Effect of processing on phenolic composition of dough and bread fractions made from refined and whole wheat flour of three wheat varieties, J. Agric Food Chem., 62, pp. 10431–10436

75. Zhou, Q., Li, X., Yang, J., Zhou, L., Cai, J., Wang, X., Dai, T., Cao, W., Jiang, D. (2018). Spatial distribution patterns of protein and starch in wheat grain affect baking quality of bread and biscuit, J. Cereal. Sci., 79, pp. 362–369.

76. Sitohang, K.A.K., Lubis, Z., Lubis, L.M. (2015). The effect of ratio of wheat starch and breadfruit flours with kinds of stabilizer on the quality of breadfruit cookies, Journal Rekayasa Pangandan Pertanian, 3, pp. 308–315.

77. Maftoonazad, N., Ramaswamy, H.S. (2008). Effect of pectin-based coating on the kinetics of quality change associated with stored avocados, J. Food Proc. Preserv., 32, pp. 621–643.

78. Leathers, T.D. (2003). Biotechnological production and applications of pullulan, *Appl. Microbiol. Biotechnol.*, **62**, pp. 468–473.

79. Arfin, T., Sonawane, S. (2018). *Alginates: Applications in the biomedical and food industries*, Chapter 3 "Alginate: recent progress and technological prospects", Ahmed S. (ed.), Scrivener Publishing LLC, USA, pp. 45–58.

80. Arfin, T., Athar, S., Rangari, S. (2018). *Handbook of biopolymers: Advances and multifaceted applications*, Chapter 4" Proteins and their novel applications", Ahmed, S., Kanchi, S., Kumar, G. (eds.), CRC Press, the U.K., pp. 75–93

81. Arfin, T., Mogarkar, P.R. (2018). *Biocomposites: Biomedical and environmental applications*, Chapter 15 "Bio-based material protein and its novel applications", Ahmed, S., Ikram, S., Kanchi, S., Bisetty, K. (eds.), Pan Stanford Publishing, Singapore, pp. 405–432.

82. Malik, N., Khan, A.U., Naqvi, S., Arfin, T. (2016). Ultrasonic investigation of α-amino acids with aqueous solution of urea at different temperatures: A physicochemical study, *J. Appl. Sol. Chem. Model.*, **5**, pp. 168–177.

83. Malik, N., Khan, A.U., Naqvi, S., Arfin, T. (2016). Ultrasonic studies of different saccharides in α-amino acids at various temperatures and concentrations, *J. Mol. Liq.*, **221**, pp. 12–18

84. Kristo, E., Koutsoumains, K.P., Biliaderis, C.G. (2008). Thermal, mechanical and water vapour barrier properties of sodium caseinate films containing antimicrobials and their inhibitory action on listeria monocytogenes, *Food Hydrocoll.*, **22**, pp. 373–386.

85. Athar, S., Arfin, T. (2017). *Natural polymers: Derivatives, blends and composite*, Vol. 2, Chapter 11 "Commercial and prospective applications of gelatin", Ikram, S., Ahmed, S. (eds.), Nova Science Publishers, USA, pp. 199–216.

86. Djagny, V.B., Wang, Z., Xu, S. (2001). Gelatin: A valuable protein for food and pharmaceutical industries, *Crit. Rev. Food Sci. Nutr.*, **41**, pp. 481–492.

87. Galus, S., Lenart, A., Voilley, A., Debeaufort, F. (2013). Effect of oxidized potato starch on the physicochemical properties of soy protein isolate-based edible films, *Food Technol. Biotechnol.*, **51**, pp. 403–409.

88. Debeaufort, F., Voilley, A. (1995). Effect of surfactants and drying rate on barrier properties of emulsified edible films, *Int. J. Food Sci. Technol.*, **30**, pp. 183–190

89. Gallo, J.A.Q., Debeaufort, F., Callegarin, F., Voilley, A. (2000). Lipid hydrophobicity, physical state and distribution effects on the properties of emulsion-based edible films, *J. Membr. Sci.*,**180**, pp. 37–46

90. Fabra, M.J., Perez-Masia, R., Talens, P., Chiralt, A. (2011). Influence of the homogenization conditions and lipids self-association on properties of sodium caseinate based films containing oleic and stearic acids, *Food Hydrocoll.*, **25**, pp. 1112–1121.

91. Boeing, H., Bechthold, A., Bub, A., Ellinger, S., Haller, D., Kroke, A., Leschik-Bonnet, E., Muller, M.J., Oberritter, H., Schulze, M., Stehle, P., Walzi, B. (2012). Critical review: Vegetables and fruits in the prevention of chronic diseases, *Eur. J. Nutr.*, **51**, pp. 637–663.

92. Lin, D., Zhao, Y. (2007). Innovations in the development and application of edible coatings for fresh and minimally processed fruits and vegetables, *Compr. Rev. Food Sci. F.*, **6**, pp. 60–75

93. Karim, A.A., Bhat, R. (2009). Fish gelatin: Properties, challenges, and prospects as an alternative to mammalian gelatins, *Food Hydrocoll.*, **23**, pp. 563–576

94. Brauer, S., Meister, F., Gottlober, R.P., Nechwatal, A. (2007). Preparation and thermoplastic processing of modified plant proteins, *Macromol. Mater. Eng.*, **292**, pp. 176–183

95. Toivonen, P.M.A. (2008). Influence of harvest maturity on cut-edge browning of 'granny smith' fresh apple slices treated with anti-browning solution after cutting, *LWT-Food Sci. Technol.*, **41**, pp. 1607–1609.

96. Gaozalez-Aguilar, G.A., Celis, J., Sotelo-Mundo, R.R., de la Rosa, L.A., Rodrigo-Garcia, J., Alvarez-Parrilla, E. (2008). Physiological and biochemical changes of different fresh-cut mango cultivars stored at 5 °C, *Int. J. Food Sci. Technol.*, **43**, pp. 91–101.

97. Rico, D., Martin-Diana, A.B., Barat, J.M., Barry-Ryan, C. (2007). Extending and measuring the quality of fresh-cut fruit and vegetables: A review, *Trends Food Sci. Technol.*, **18**, pp. 373–386.

98. Arfin, T., Mohammad, F. (2015). *Advances in materials science research*, Vol. 21, Chapter 7 "Dendrimer and its role for the advancement of nanotechnology and bioengineering", Wythers, M.C. (ed.), Nova Science Publishers, New York, pp. 157–174.

99. Arfin, T., Mohammad, F., Yusof, N.A. (2015). *Polystyrene: Synthesis, characteristics and application*, Chapter 10 "Applications of polystyrene and its role as a base in industrial chemistry", Lynwood, C. (ed.), Nova Science Publishers, New York, pp. 269–280.

100. Arfin, T., Mohammad, F. (2013). *Halides: Chemistry, physical properties and structural effects*, Chapter 2 "Synthesis, characterization and influence of electrolyte solutions towards the electrical properties of nylon-6,6 nickel carbonate membrane: Test for the theory of uni-ionic potential based on thermodynamics of irreversible processes", Lefebure, J. (ed.), Nova Science Publishers, New York, pp. 39–66.

101. Onwudiwe, D.C., Arfin, T., Strydom, C.A. (2014). Synthesis, characterization, and dielectric properties of N-butyl aniline capped CdS nanoparticles, *Electrochim. Acta*, **116**, pp. 217–223.

102. Arfin, T. (2018). *Handbook of bionanocomposite: Green and sustainable materials*, Chapter 9 "MWCNT polymer composites: environmental applications", Ahmed, S., Kanchi, S. (eds.), Pan Stanford Publishing, Singapore, pp. 235–246.

103. Sophia, A.C., Arfin, T., Lima, E.C. (2019). *A new generation material graphene: Applications in water technology*, Chapter 18 "Recent developments in adsorption of dyes using graphene based nanomaterials", Naushad, Mu. (ed.), Springer Nature, Switzerland, pp. 439–471.

104. Arfin, T., Sonawane, K., Saidankar, P., Sharma, S. (2019). *Integrating green chemistry and sustainable engineering*, Chapter 14 "Role of microbes in the bioremediation of toxic dyes", Shahid-ul-Islam (ed.), Scrivener Publishing LLC, MA, pp. 443–472.

105. Arfin, T., Sonawane, K., Saidankar, P., Sharma, S. (2019). *Integrating green chemistry and sustainable engineering*, Chapter 18 "Biotechnology past-to-future", Shahid-ul-Islam (ed.), Scrivener Publishing LLC, MA pp. 617–645.

106. Mohammad, F., Arfin, T. (2014). Thermodynamics and electrochemical characterization of core-shell type gold-coated super paramagnetic iron oxide nanoparticles, *Adv. Mater. Lett.*, **5**, pp. 315–324.

107. Waghmare, S., Arfin, T., Manwar, N., Lataye, D., Labhsetwar, N., Rayalu, S. (2015). Preparation and characterization of polyalthia longifolia based adsorbent for removing fluoride from drinking water, *Asian J. Adv. Basic Sci.*, **4**, pp. 12–24.

Chapter 3

Surface Chemistry for Intelligent Food Packaging

Sivapriya Veeraiyan,[a] Abdul Azeez Nazeer,[a] Saravanan Dhandapani,[b] and Sudarshana Deepa Vijaykumar[c]

[a] *Nano-Bio Translational Research Laboratory, Department of Biotechnology, Bannari Amman Institute of Technology, Sathyamangalam, India*
[b] *Department of Textile Technology, Kumaraguru College of Technology, Coimbatore, India*
[c] *Department of Biotechnology, National Institute of Technology, Tadepalligudem, Andhra Pradesh, India*
sudarshanadeepa@nitandhra.ac.in

Food packaging is the major factor that determines the quality of finished food products, which have to undergo transportation, storage, and end application. The four thumb rules of food packaging are protection of the content, communicate the details about the product, convenient to transport and open, and containment. In recent years, food packaging has witnessed a new era of development with two major concepts: (i) active packaging to extend the shelf life and perpetuate the quality of the food, and (ii) intelligent packaging to monitor the condition of the food products and their surroundings (indicators). The intelligent packaging system has three central technologies: (i) indicators to communicate the food quality to users, (ii) data carriers to track the storage and

Biomaterials in Food Packaging
Edited by Mohd Yusuf and Shafat Ahmad Khan

ISBN 978-981-4877-98-5 (Hardcover), 978-1-003-25678-6 (eBook)
www.jennystanford.com

distribution, and (iii) sensors for quick and precise quantification of analytes in the food. Understanding the surface chemistry of these sensors and indicators can conduce to bestow the biomaterials in these applications. The use of biomaterials in intelligent food packaging is advantageous over conventional materials, in the aspect of environment friendliness. Hence, in this chapter, the surface chemistry of the intelligent packaging to monitor or indicate the food quality that performs the convenience and communicative functions of packaging is discussed.

3.1 Introduction

Food packaging technology is repetitively fluctuating to encounter the mounting challenges of the contemporary civilization. Packaging is a solitary process to retain the quality of the food products throughout the distribution, stocking, and application. It decelerates the rate of decay, enabling the supply and promotion more competent. Package guard products from peripheral milieu interconnect with the patron over extolled scripts, brand emblem, and graphics; billet the existence of the patron or assemble the management of packaged food easier for the patron; and turn as ampoules for differently molded products with the aim of enhancing the logistic proficiency. Preservation and protected delivery of packaged food products are the foremost goals all over the food industry [1]. Nevertheless, the loss of food traits, due to physical, chemical, and biological ruins, arises during the storage and distribution of the food products. Safeguarding the food quality is an imperative research goal since it is closely associated with the aim all over the world to improve the effective and healthy food supply chain. Food quality control is essential, as it aids to shield patrons against food-borne disease and to exploit the productivity of the food commerce. Different dynamic constituents can be merged into the packaging material to develop the functionality of food packages. Novel ideas have significantly subsidized to accomplish an advanced perception for effective food packaging, i.e. the active packaging and intelligent packaging ideas.

The active packaging materials (APMs) are the substances envisioned to prolong the shelf life or to extend the pristineness of the packaged food. The purpose of APMs is to integrate with the constituents that would absorb or release the elements from the packaged food or into the atmosphere around the food. Numerous APMs, such as oxygen (O_2) and ethylene scavengers, moisture curbers, carbon dioxide (CO_2) scavengers and emitters, controlled release of antioxidant and antimicrobial agents, and sensory devices, to regulate the liberation or adsorption of flavors and odors, have been merged to progress the quality of food packages.

The intelligent packaging materials (IPMs) are the substances that surveil the changes and communicate the condition in the environment around the packaged food. Intelligent packaging intends to carry the information associated with the food quality to investors of the food supply chain. For instance, an intelligent packaging system can display the freshness and shelf life of the food products; it can display temperature of the food product using thermochromic inks; and it can even show the temperature history via time–temperature indicators. This chapter summarizes the intelligent devices integrated with food packages, as there is a growing interest in intelligent developments in the food sector.

3.2 Simple Food Package and Intelligent Devices

The conventional/simple food packaging techniques, such as wrapper and cold storage, have helped to preserve food materials for an extended duration by guarding the products. But they do not provide the information dealing with the quality regarding gas concentration, temperature change, and growth of microorganisms within the packaged environment. Hence, there is a need for an intelligent device system that continuously supervises and conveys the information concerning the food quality whenever necessary, since simple packaging is not sufficient for keeping up the quality of the food products. Therefore, various intelligent components should be combined with the simple packaging system to convey the customers concerning the condition of the food package.

3.3 Intelligent Packaging Technique

Unlike the conventional packaging techniques, the intelligent packaging techniques (IPTs) are used to record the variations in both the internal and external environments and to convey the status of the packaged food products to users [2]. In general, they sense and record the quality of the products and their environments, and indicate the information about the packaged food quality. This approach is highly beneficial, as it provides timely indication to the users concerning the food quality inside the packaging system. The IPT further displays the detailed information on ingredients, expiry date, manufacturing place, batch and process, and storage stipulations [3]. Study has been conducted on developing the improvised transparent labels or seals that remain unchanged until the package is damaged. Once the package is damaged, the chemically reactive sensors change its color indicating the damage [4]. This can be accomplished by integrating sensors or indicators that potentially sense the changes in the packing system. Therefore, IPT is a prodigious accomplishment for the continuous monitoring and maintenance of the packaged food items for the prolonged period to the consumers [5].

3.4 Devices Used in IPT

There are various approaches including the microbial and chemical techniques to regularly supervise the quality of the food materials throughout processing. Nevertheless, there are certain parameters that need continuous assessment during the entire supply chain. Unfortunately, the quality parameters vary recurrently, which could affect the quality of the food items, making it difficult for the patrons to appraise the quality inside the package. The IPT sorts out this limitation and enables to know the condition of the food inside the package. There are devices fabricated for this purpose that are either placed inside or outside the package to screen the quality of the food products or to convey the consumers regarding the safety issues associated with the products.

There are three main technologies of intelligent packaging systems that are broadly acknowledged: (i) indicators, which aim to deliver the information on food quality to the users, (ii) data carriers, which are precisely envisioned for recording the information about packaging, storage time and place, distribution method, and delivery data, and (iii) sensors, which precisely quantify analytes in the packaged food [6].

3.5 Indicators of IPT

The indicators are used to convey information related to the status or any characteristic changes in the food materials contained within the package or the environment surrounding the package by perceiving visual changes. The information regarding changes can be immediately displayed visually, e.g. different color concentrations [7]. The unique feature of the indicators is that they provide various direct visual signals depending on the food environment. They do not have any specific receptors to provide information to the users regarding food quality. The indicators that synced with the food packaging materials include time-associated temperature indicators, freshness indicators, and gas indicators [8].

The time-allied temperature indicator is used to measure variations in specific temperature of the food package. It will also provide information regarding the presence of any microbial contamination and the structural anomaly if any happened while processing and packing the food [9]. The freshness indicator tells about the deterioration of food products due to chemical changes or microbial growth. Gas indicators are usually produced as labels or badges to indicate the emission of various gases from the food products within the packaging system [10, 11].

3.5.1 Temperature Indicators

Temperature is an important parameter that determines the storage period of a particular food product. Immediate rise or fall in temperature is responsible for maintaining the quality of the processed food products. It is mandatory to check the temperature at every phase of the delivery from packing to consumption of the food products for the purpose of quality. There are intelligent tools used for evaluating the temperature of the food materials at different times during the delivery of the package. The temperature indicators are classified in two different types: simple temperature indicator (STI) and time temperature indicator (TTI).

The STI shows the fluctuations in temperature and warns the consumers about the rise and fall of temperature with respect to the critical point. It signals the users about the protein denaturation and possible existence of microorganisms in the processed food package [12].

The TTI is the first-generation temperature indicator, which is an integrated pointer of time and temperature. It conveys the information about the history of both time and temperature of the packaged food. It also displays any unfavorable condition such as temperature change over time [13]. TTIs are used in the form of small self-adhesive labels attached to the food packages or containers.

Chemistry involved: The simple principle behind TTI depends on mechanical, enzymatic, microbiological, chemical, and electrochemical reactions, which describe the result in the form of mechanical distortion or color change [14]. Kim et al. (2016) [15] developed a sample isopropyl palmitate (IPP), which is a diffusion-based TTI used for examining the microbial quality of unpasteurized juice. The different temperature ranges have been calculated by the diffusion of IPP, which is placed inside the indicator. The diffusion of IPP up to 7 mm from the indicator is proved to be the accurate indicator level that represents the evaluation of spoilage by microorganisms in the packed unpasteurized juice. However, the technologically advanced TTI could be used for the successful indication of microbial spoilage at 13 °C or even higher. A precise and cost-effective TTI fabricated from the natural components to use as IPT device was developed by Pereira et al. [13]. The TTI used as labeling film was developed using isolated anthocyanin pigment from red cabbage combined with chitosan and polyvinyl alcohol. Anthocyanin can be used as a natural indicator, as it undergoes chemical changes during pH variations. Hence, this pigment can be used for analyzing quality of the food products by assessing the pH variations in the packed food. The developed system can able to detect the variations in temperature circuitously by assisting the pH change of the food products. In last few years, there has been an increasing development and application of several categories of TTIs, such as photochrome-driven, polymer-based, microbe-based, diffusion-based, and enzyme-based TTIs for the assessment of time and temperature of different unpreserved food items.

TTIs have received growing attention for their advantage in providing the data on the time vs temperature history of the packed food, which is very instrumental in instigating the physical and chemical worsening kinetics. These are readily operational devices and user friendly. The digital signals obtained by the indicators are easily understandable by the users, as they display direct information related to the quality of the packed content at a particular temperature [13].

3.5.2 Gas Indicators

To preserve the fresh, unprocessed, and organic food products, such as fruits and vegetables, are challenging to pack and store for a long time, due to various reasons including their respiration, leakage of gas from and to the package, and variation in gas concentration due to the microbial growth and metabolism inside the packaged system [16]. The gas indicators are used as a solution for such problems. These indicators provide information on the concentration of O_2 and CO_2 inside the packaging system by varying the color of the package, due to the particular enzymatic or chemical reaction. These indicators provide information regarding the presence or absence of gases, as these are directly attached to the packaged food materials [1]. As the gas indicators are placed within the systems, some parameters such as their toxicity, reactivity, and hydrophobic nature of the substances are needed to be taken serious care of while designing the devices. The components integrated while designing should have food contact approval. There is an increasing interest in the development of O_2 and CO_2 indicators among researchers, as there is a remarkable importance of these gases in food packaging applications [17, 18]. An IPT system to surveil the intermediate moisture level through the assessment of CO_2 gas concentration inside the food package was developed by Nopwinyuwong et al [19]. The microorganism growth is detected by the respective color indication, as an outcome that represents the emission of CO_2 gas from the spoiled food.

Chemistry involved: The gas indicator devices have three major components, namely the Redox dyes such as methylene blue or 2,6-dichloroindophenol, an alkaline compound, and a reducing agent such as proton donor. These indicators suffer a drawback of dye leakage, as they are in contact with the environmental moisture around the package. The gas indicators are commercially presented in the form of labels, printed layers, and tablets attached to the package or as monolayer coatings over the polymer sheets.

Vu and Won [18] developed O_2 gas indicator, which detects the emission of O_2 and inhibits the leaching out of dyes printed on the packaging medium. The recent advancement in this indicator, which uses UV-stimulated colorimeter to sense the O_2 leakage, has addressed this issue and reduced the dye leaching by encapsulating the device. The UV-activated O_2 indicators have a redox dye Zein, photosensitive Thionine, P25 TiO_2, dispersive medium glycerol, and polymer to encapsulate these components. The dye leakage was stretched virtually to 80% when the Zein coated film was placed in an aqueous environment for 24 h. The use of alginates to encapsulate the indicator system can decrease the leakage of dye to approximately 6%, since the ion binding ability can inhibit the cation dye from leakage into water. This indicator system is highly sensitive and changes color drastically upon contact with the gas present inside the package.

Hong and Park [20] developed an IPT device, which detects the presence of CO_2 gas without engendering any damage to the packaging material. Bromocresol purple or methyl red is a basic dye, unified to the polymeric surface (polypropylene resin and calcium hydroxide) of the food package. This consequently helps detect the changes in the quality of the food during transportation and storage. This IPT system is based on calorimetric change with respect to pH regardless of temperature variation. The change in concentration of CO_2 inside the food package gives rise to alteration in the pH that is detected by the indicator and changes the color.

3.5.3 Freshness Indicators

Freshness indicators are developed to detect the microbial contamination and notify the consumers regarding the freshness of food materials in the packaging system [21]. Due to the growing demand for fresh and healthy foods in the last two decades, the research and development of freshness indicators have increased massively. This indicator has to be proposed as smart device that could monitor the grade of the packaged food products during transportation and storage.

Chemistry involved: Freshness deterioration may be due to the revelation of food products to unfavorable state and increased shelf life. The organic substances emit nitrogen, amine, and other chemicals, when they start decaying due to various factors, including microbes. Hence, the freshness indicators are generally fabricated with the ingredients that are associated to the microbial growth, such as starch source, ethanol, nitrogen compounds (trimethyl amines), different organic acids (lactic acid or acetic acid), CO_2, sulfuric compounds, and biogenic amines, to evaluate the freshness of packaged food products [22]. They provide unswerving information to the users regarding the integrity of the products concerning the chemical changes or microbial growth within the package.

The quality of meat products is determined by hydrogen sulfide (H_2S) indicators. The H_2S is a gas released during aging of meat matrix and is associated with the color of the protein myoglobin, which acts as the quality indicator for fresh meat products. Smolander et al. introduced a freshness indicator with the principle similar to the pre-existed modified-atmosphere technique to monitor the packed poultry meat [23]. Other freshness indicators are developed based on the microbial sensitivity using chemicals such as ethanol, diacetyl, and CO_2.

Rukchon et al. [24] came up with an indicator to monitor the freshness of the packed chicken breast meat. The system is developed based on the pH variation of the meat and the pH sensitive dye used for the detection. This system works by determining the resulting color produced at different levels of CO_2 emitted from the packaged chicken breast meat, which would vary the pH.

Rokka et al. [25] built a freshness indicator to surveil the packed whole broiler chicken quality. It was developed based on the thermo-sensitive biogenic amines, such as putrescine, tyramine, and cadaverine. The levels of different amines vary with respect to the varying temperatures in the broiler chicken package, thereby representing the freshness of the meat at different times.

Zhai et al. [26] developed a freshness indicator to monitor the freshness of packed fish using colorimetric principle. They mended the starch/polyvinyl alcohol with roselle (*Hibiscus sabdariffa L.*) anthocyanins that would change the color in the presence of total volatile nitrogen amines leaked from the spoiled seafood. These indicators work even at the freezing temperatures. The colorimetric film assesses the visible color changes. Hence, this film is used to detect the freshness of fish inside the packaging device.

Ripe Sense is an aroma testing IPT device developed by a New Zealand-based group of labs, Jenkins Group. This indicator is used to alert the ripening of fruits at various stages, deliberating their phase of ripening. Initially, the indicator displays red color that turns into orange, and further into yellow, with the passage of time of maturity levels of fruits. This device can be used for monitoring the freshness of aromatic fruits, such as pears, kiwi, melon, mango, and stone fruits including peach, apricot, and plum, and others [27].

3.5.4 Sensors

Sensors are the utmost integral, pioneering, and propitious technologies for the upcoming intellectual packaging systems [3]. A sensor is a system that has a processing and control unit made up of electronic components, software, and an interconnection network. A sensor is used to detect a signal for monitoring the physicochemical properties to which the system responds.

Generally, the sensors are designed with four major functional units: (i) Receptor, which is a signal/analyte receiving part of the sensor that is made of a sampling unit, where the reaction between the analyte and the sensor surface molecules take place. The analytes are adsorbed on the surface. Upon the adsorption, the reaction between the analyte and the surface chemicals alter the energy, which is closely related to detect the analyte transforms or disorients, and the physicochemical characteristics of the receptor such as redox potential, pH, and temperature. (ii) Transduction element, which is a measuring part of the sensor such as an electrode. This unit is able to transform the variations in energy and carry the data into an analytical signal, which is optical, electrical, chemical, or thermal. (iii) Signal processing electronics, which translate the processed data of the transducer into the readable output language. (iv) A signal display unit, which shows the result of the analysis [28].

The ideal sensor should possess the ensuing features: (i) specificity for the target species (i.e. selectivity); (ii) sensitivity to changes in target species concentrations; (iii) fast response time; (iv) extended lifetime of at least several months; and (v) small size (miniaturization), with the possibility of low-cost manufacture. There are various types of sensors that have been developed for food applications, such as electrochemical sensors and luminescence sensors.

Electrochemical sensor is a significant section of the sensors, where an electrode is used as the transducer [29]. The electrochemical sensor is driven by the redox reaction at the periphery of the electrode with the analyte medium; thereby stimulates the voltage of a potentiostat. The electrons transferred between the electroactive species in the analyte solution and the electrodes generate a current, which is directly proportional to the analyte concentration. The emission of fluorescence or chemiluminescence signals in luminescence sensor is measured only after the subsequent analyte adsorbed on an appropriate solid support, which gives origin to the expression solid-phase luminescence. The luminescent signals produced by the sensors are associated with the concentration of the analytes in the food packages under certain conditions [30]. Most of the developments are concerned on the sensing and monitoring the food package for pollutants or contaminants that could spoil or intoxicate the food. Even though the most advanced flexible printed chemical sensors are amalgamated into the food packaging system, the IPT still has a wide scope to witness more advancement, in terms of most progressive sensor technologies that can integrate with the intelligent devices into the packaging systems. The future scope of IPT would be on developing the biosensors and gas sensors.

3.5.4.1 Biosensors

The major difference between the chemical receptor and biological sensor is the receptor molecules that are made up of synthetic chemical compounds in chemical sensors, while the biocomponents, such as antigens, antibodies, enzymes, nucleic acids, and phages, constitute the receptor surface of the latter [31]. Currently, the biosensors are employed only in few food packaging applications. For instance, Food Sentinel System was developed by SIRA Technologies (USA); it is a packaging barcode technology that provides complete details on the history of the product that went through adverse conditions. This device is provided with a membrane of antibody corresponding to the certain pathogen, and the membrane is attached to the barcode. An ink is attached to the biosensor will turn red in the presence of contaminated bacteria and the data transmission through the barcode will be rendered when scanned [32]. Similarly, a visual diagnostic tool, Toxin Guard (Toxin Alert Inc., USA) is used to detect pathogens that may deteriorate food. This type of sensor basically works on the immunoassay of pathogenic antigen presents in analyte with the antibody attached on the polymer packaging film. In this system, the toxic metabolite from the pathogen reacts with the antibodies decorated on a thin film of flexible polymer layer exhibiting a clear color change in the device that indicates the presence of pathogens in the food package [33]. A system that interpolates biochemistry and electronics, Bioett, is used to track the changes in the temperature of the packed food materials during the course of refrigerated transportation. This system includes a biosensor incorporated into the food package, a detector system made up of radio frequency (RF) chip to read the data sent by the biosensor, and a database to store the processed information about the products. In this IPT device, the data can be read using a handheld scanner at any point in the food supply chain [8]. A commercially available biosensor, which is flexible to detect the toxins in food, was developed by Flex Alert; it specifically detects the pathogens, including *Escherichia coli* O157, *Listeria spp.*, and *Salmonella spp.*, and aflatoxins [4].

3.5.4.2 Gas sensors

Studies have been carried out recently for the advancement of sensors that can analyze quantitatively the gaseous analytes with satisfactory results. Recognized systems for gas analysis include piezo electric crystal sensors, metal oxide semiconductor field-effect transistors (MOSFETs), organic conducting polymers, potentiometric CO_2 sensors, and amperometric O_2 sensors. Though these systems possess several limitations, these IPTs encompass various critical analyses of packages in most of the cases. The current developments mainly engrossed in O_2 and CO_2 sensors, which help overcome the limitations. The progress of smart sensors to detect O_2 concentration that can pervade in and around the packaging system has attracted significant attention over the past few decades. The general O_2 sensors are based on sensing the luminescence and the progressive advancements when associated with the systems based on absorption or reflectance. These signify an alternative method to evaluate the O_2 concentration and visually indicate the results with increased sensitivity and quantitative accuracy measurements [34]. Diverse structures of these IPTs provide the opportunity to quantify the analytes in three-dimensional (3-D) samples with non-destructive analysis. Additionally, they provide faster response without the need to produce any electrical connection or consume analyte.

Huber et al. developed a novel non-destructive, non-invasive fiber-optic O_2 meter that is used to detect O_2 permeability of plastic bottles and containers [35]. The basic principle of this is the reduction of luminescence produced by the collision between molecular O_2 and luminescent dye molecules, when they are in the excited state in both gaseous and liquid phases. The disadvantage of cross-sensitivity among H_2S, CO_2, ammonia (NH_3), pH, or any ionic species such as sulfide (S^{2-}), sulfate (SO_4^{2-}), or chloride (Cl^-), is eliminated in this IPT and the salinity factor does not impact the measurement. Even the turbidity of the medium and the variations in the stirring rates during the food processing do not influence the measurement. These IPTs can be used with the alcohol and the combination of alcohols. Fitzgerald et al. invented a competent and more sensitive phosphorescence-based O_2 sensor using the complexes of fluorescent ruthenium(II) and platinum(II)-octaethylporphiryne-ketone (PtOEPK) dye [36]. They have also tested the real sample, such as smoked fish, raw and cooked meat, and packaged sliced ham, using this sensor. This indicates that the sensor provides accurate and dependable outcomes on real-time applications in the market.

Baleizao et al. [37] developed a rapid and sensitive optical dual sensor based on the luminescence measurements for the gaseous O_2 and temperature monitoring. It consists of polymer films prepared using two different luminescent compounds, one sensitive to temperature, while other to O_2. Ruthenium tris-1,10-phenanthroline, due to its effective temperature-dependent luminescence, is used as a temperature-sensitive dye and is conjugated with poly(acrylonitrile) to evade cross-sensitivity with O_2. For detecting the O_2, Fullerene C70 probe is used, due to its stability, thermally activated, sustained fluorescence at elevated temperature, and exceptional O_2 sensitivity.

The dual sensor exhibits an operational temperature range 0–120 °C, and detects O_2 limits in the range up to 50 ppm O_2 concentrations.

The progress of CO_2 sensors is only the second preference with respect to O_2 sensors for food packaging applications, as O_2 plays an important role in degradation of several food products. The application of modified atmosphere packaging (MAP) system has been recognized, as it controls the volume of CO_2 in food packages [38]. Conservative methods for a qualitative and quantitative analysis of CO_2 consist of infrared (IR) spectroscopy, mass spectroscopy (MS), and gas chromatography (GC). However, these techniques proposed brilliant results, but are expensive due to sophisticated equipment utilization [39].

There has been a great effort from the past two decades to manufacture sensitive, rapid, economic, tuneable, and easily contracted sensors to analyze CO_2 in the food package. Von Bultzingslowen et al. [40] fabricated an optical sensor based on the pH-driven fluorescent indicator to monitor CO_2 in MAP practices using 1-hydroxypyrene-3,6,8-trisulfonate (HPTS), immobilized in a hydrophobic, organically modified silica (ormosil) matrix. The reference luminophore is encapsulated in polymer nanobeads to inhibit the cross-sensitivity of O_2, Cl^-, and varying pH.

Borisov et al. [41] settled a silicone matrix-based optical CO_2 sensor emulsion at room temperature using the ionic liquids and (RTILs)–1-butyl-3-methylimidazolium salts. This fluorimetric sensor employs 8-hydroxypyrene-1,3,6-trisulfonate (HPTS) for the quantitative analysis of CO_2, which is applicable in various fields. Borchert et al. developed yet another opto-chemical CO_2 sensor with an indicator dye, such as a phosphorescent, a reporter dye (pentafluorophenyl)porphyrin (PtTFPP), and a colorimetric pH indicator α-naphtholphthalein, combined with the plastic matrix organized with a phase transfer agent tetraoctyl- or cetyltrimethylammonium hydroxide [42]. Experiments carried out to test the efficiency of these sensors revealed that at 4 °C, the sensor acquired its high sensitivity to CO_2 for 21 days in the food and peripheral atmosphere environments. This efficiency is adequate for several packaged food products.

3.5.5 Data Carriers

The notification on the quality control methods and information about the time-dependent parameters of the packed food at every particular time through the data carrier devices during the food supply chain is more competent for the benefits of food safety and quality. The data carrier device does not afford any data related to the status of food quality, instead it is proposed for traceability, automation, counterfeit protection, and prevention from theft [43]. The data carriers are usually used on large containers, such as multi-box containers that contain a number of small packed units, and shipping crates that are used to transport a large number of packed containers, pallets, and large paperboard packages. Barcodes and radio frequency identification (RFID) tags are the most significant data carrier devices in the food packaging industry. These data carrier devices belong to the chief category of convenience-enhancing intelligent systems [44].

3.5.5.1 Barcodes

The first ever Universal Product Code (UPC) barcodes originated in the market application in the 1970s. These barcodes are majorly applied in the high-scale retail trade and in stores to enable inventory control, since these are economical [45]. A barcode is a sequential pattern of parallel lines organized to denote 12 digits of the data. An optical barcode data scanner reads the encoded data and transmits the information to an operating system, such as a computer, where it is translated, displayed, and stored [1]. Initially, the barcodes developed were one dimensional (1-D), and the primary operating principle was closely similar to that of laser beam from the vertical code bars cutting a horizontal slice. The first-generation barcode had very limited storage capacity such as to store the item number and manufacturer's number [46]. Hence, the file size of the data was reduced and the reduced space symbology (RSS) barcode was developed to encode huge amount of data in a limited space. The RSS-14 stacked omni-directional barcode and the RSS-expanded barcode are the most commonly used RSS [47]. The two-dimensional (2-D) barcode can able to encode more information to be stored, as compared to 1-D barcode, which can be linked by spaces and dots organized in a matrix. This allows storing an increased density of data within a reduced space. The quick response (QR) barcode is the most recently used technology that can encode myriad volume of data and then store using various encoding approaches, such as numeric (numbers alone), alphanumeric (number and letters), byte/binary, and kanji (logographic Chinese characters). Reading the 2-D barcode symbologies require a scanning device that is potential to scan the data in both vertical and horizontal matrix simultaneously.

3.5.5.2 RFID systems

The RFID tags are one of the most sophisticated technologies for the data carrier applications. The three basic functional units of the RFID system are a tiny antenna made up of a microchip connected to a tag to store the data, a reader to send and receive radio signals, and a middleware [48]. The long-distance communication up to 100 m with the help of RF technology and the space of 1 MB to store enormous data are the two significant advantages of the RFID system. These help in improvising automatic identification of products and traceability operations [49]. There are two types of tags used in the RFID, namely the active tags that work on the power from a battery source, and the passive tags that do not need any such power from external source.

Despite the RFID technology being very popular, it took long time to find its spot in the food packaging market, as it is not economical [50]. However, it should not be considered as an alternative for barcodes for their advantages on distant communication and storage space. Currently, the RFID tags are used only to control the road traffic, identification of marked pallets, security alarms, parking guidance for automobiles, and even the tracing and identification of wild animals in forest for the census studies. Their application in the food industry is limited to a few purposes such as product identification and tracing the supply path, cold chain monitoring, livestock management, and shelf life prediction [51].

3.6 Other IPT Systems

Some of the useful intelligent devices, such as doneness indicators that are used to inform the proper completion of processed food and thermochromatic inks, are found to have fewer applications when comparedtoothersystems[44].Thethermosensitivethermochromatic inks that change color with the change in temperature are printed on the food packages. There will be an indication of color change, as the temperature within the food packages deviates from the optimal consumption temperature. Sometimes, the color change is convoyed by a synchronized display of a brief message, such as "ready to serve". Majorly these thermochromatic inks are synthesized in companies based in the U.S. (LCR Hallcrest, CTI Inks, QCR Solutions Corporations, and others) and the U.K. (Siltech Limited, B&H Colour Change, and others). The difficulty in spotting the color change is one of the main limitations of this indicator [44]. The IPT device tackling theft, counterfeiting, and tampering are trapping huge interest, although it is not common in the food industry. Electronic article surveillance (EAS) is an example of the system that is in contradiction with theft control equipment, which is in the form of electronic-tagging system, while anti-counterfeiting and anti-tampering devices are in the form of holograms, thermochromatic inks, micro-tags, tear labels, and tapes [1, 52].

3.7 Summary and Future Prospects of IPT

The IPT has modernized the food packaging sector. This technology has enabled the producers and distributors to provide food products to the consumers with the best possible quality and safety. The consumers are majorly benefitted from the technology, as they are provided with the fresh and better-quality food products, with the help of TTIs, freshness indicators, gas concentration indicators, and others. The distributors and retailers, on the other hand, are benefitted by the IPT devices, including barcodes, RFID tags, thermochromatic inks, holograms, and tear taps, to protect the food products from counterfeiting and misplacements. The future of IPT is very optimistic with the progression of the technology. Electronic labeling through embedding electronic chips on the packages with the ink technology is one of the most sophisticated technologies that may substitute the existing IPT systems. With the advancing technologies, such as artificial intelligence (AI), there are several developments in interactive IPT devices for food packages. These systems could guide the consumers to improve the storage parameters, thereby increasing the shelf life of the food products. Simple IPT systems, such as thermochromatic inks, are used to prevent the food package from the sunlight by varying the color in response to the temperature without any need of electronic assistance. There are various single-purpose-IPT devices, such as gas concentration indicators, freshness indicators, and TTIs, used in the current food packaging industry. A technology to unify all these sensors together to fabricate an integrated multi-purpose device would be more prevailing and useful for providing data concerning the quality feature of the food products. This demands the collaboration of various fields of research, such as the food nutrition, mechanical engineering, and electronics and instrumentation engineering, into the intelligent food packaging domain. Though the intelligent packaging plays a vital role in improving the safety and quality of delicate food products, progress has to be made in this technology to minimize the cost of production and processing. According to an educated approximate calculation, the cost of an intelligently packed food is at least double the cost of the original food product. Further, food administrative agencies have raised concerns and legal issues related to the safety of the reactive labels coming in proximity to the eatables. Hence, besides working on research for cost reduction of IPTs, there is a stringent need to work on safe implementation of the intelligent packaging system.

References

1. Han, J.H. (2014). Innovations in food packaging, 2 Ed., Chapter 8 "Intelligent packaging for food products", Han, J.H. (ed.), Academic Press, Amsterdam, pp. 171–209.

2. Yam, K.L. (2012). Emerging food packaging technologies: Principles and practice, Chapter 8 "Intelligent packaging to enhance food safety and quality", Yam, K.L., Lee, D.S. (eds.), Woodhead Publishing, Philadelphia, pp.139–155.

3. Bagchi, A. (2012). Intelligent sensing and packaging of foods for enhancement of shelf life: Concepts and applications, *Int. J. Scientific Eng. Res.*, **3(10)**, pp. 1–13.

4. Vanderroost, M., Ragaert, P., Devlieghere, F., Meulenaer, B.D. (2014). Intelligent food packaging: The next generation, *Trends Food Sci. Technol.*, **39**, pp. 47–62.

5. Realini, C.E., Marcos, B. (2014). Active and intelligent packaging systems for a modern society, *Meat Sci.*, **98(3)**, pp. 404–419.

6. Kerry, J.P., O'Grady, M.N., Hogan, S.A., (2006). Past, current and potential utilization of active and intelligent packaging systems for meat and muscle-based products: A review, *Meat Sci.*, **74**, pp. 113–130.

7. O'Grady, M.N., Kerry, J.P. (2008). Meat biotechnology, Chapter 19 "Smart packaging technology", Toldrà, F. (ed.), Springer, New York, pp. 425–451.

8. Hogan, S.A., Kerry, J. (2008). Smart packaging technologies for fast moving consumer goods, Chapter 2 "Smart Packaging of Meat and Poultry Products", Kerry, J., Butler, P. (eds.), John Wiley & Sons, Chichester England, pp. 33–60.

9. Siro, I. (2012). Progress in food preservation, Chapter 2 "Active and intelligent packaging of food", Bhat, R., Alias, A.K., Paliyath, G. (eds.), John Wiley & Sons, New York, pp. 23–48.

10. Lee, K., Ko, S. (2014). Proof-of-concept study of a whey protein isolate based carbon dioxide indicator to measure the shelf-life of packaged foods, *Food Sci. Biotechnol.*, **23**, pp. 115–120.

11. Vu, C.H.T., Won, K. (2014). Leaching-resistant carrageenan-based colorimetric oxygen indicator films for intelligent food packaging, *J. Agric. Food Chem.*, **62**, pp. 7263–7267.

12. Pault, H. (1995). Brain boxes or simply packed, *Food Processing UK*, **64(7)**, pp. 23–24.

13. Pereira Jr, V.A., de Arruda, I.N.Q., Stefani, R. (2015). Active chitosan/PVA films with anthocyanins from Brassica oleraceae (Red Cabbage) as time-temperature indicators for application in intelligent food packaging, *Food Hydrocoll.*, **43**, pp. 180–188.

14. Vaikousi, H., Biliaderis, C.G., Koutsoumanis, K.P. (2008). Development of a microbial time/temperature indicator prototype for monitoring the microbiological quality of chilled foods, *Appl. Environ. Microbiol.*, **74(10)**, pp. 3242–3250.

15. Kim, J.U., Ghafoor, K., Ahn, J., Shin, S., Lee, S.H., Shahbaz, H.M., Shin, H.H., Kim, S., Park, J. (2016). Kinetic modeling and characterization

of a diffusion-based time-temperature indicator (TTI) for monitoring microbial quality of non-pasteurized angelica juice, *LWT-Food Sci. Technol.*, **67**, pp. 143–150.

16. Brody, A.L. (2001). What's active about intelligent packaging?, *Food Technol.*, **55**, pp. 75–78.
17. Roberts, L., Lines, R., Reddy, S., Hay, J. (2011). Investigation of polyviologens as oxygen indicators in food packaging, *Sensors Actuators B: Chem.*, **152**, pp. 63–67.
18. Vu, C.H.T., Won, K. (2013). Novel water-resistant UV-activated oxygen indicator for intelligent food packaging, *Food Chem.*, **140**, pp. 52–56.
19. Nopwinyuwong, A., Trevanich, S., Suppakul, P. (2010). Development of a novel colorimetric indicator label for monitoring freshness of intermediate-moisture dessert spoilage, *Talanta*, **81**, pp. 1126–1132
20. Hong, S.I., Park, W.S. (2000). Use of color indicators as an active packaging system for evaluating kimchi fermentation, *J. Food Eng.*, **46**, pp. 67–72.
21. Lund, B.M., Baird-Parker, A.C., Gould, G.W. (2000). The microbiological safety and quality of foods, Springer, Gaithersburg, the U.S.
22. Heising, J.K., van Boekel, M.A.J.S., Dekker, M. (2015). Simulations on the prediction of cod (Gadusmorhua) freshness from an intelligent packaging sensor concept, *Food Pack. Shelf Life*, **3**, pp. 47–55.
23. Smolander, M., Hurme, E., Latva-Kala, K., Luoma, T., Alakomi, H.L., Ahvenainen, R. (2002). Myoglobin-based indicators for the evaluation of freshness of unmarinated broiler cuts, *Innov. Food Sci. Emerg. Technol.*, **3**, pp. 279–288.
24. Rukchon, C., Nopwinyuwong, A., Trevanich, S., Jinkarn, T., Suppakul, P. (2014). Development of a food spoilage indicator for monitoring freshness of skinless chicken breast, *Talanta*, **130**, pp. 547–554.
25. Rokka, M., Eerola, S., Smolander, M., Alakomi, H.L., Ahvenainen, R., (2004). Monitoring of the quality of modified atmosphere packaged broiler chicken cuts in different temperature conditions: B. Biogenic amines as quality-indicating metabolites, *Food Control*, **15**, pp. 601–607.
26. Zhai, X., Shi, J., Zou, X., Wang, S., Jiang, C., Zhang, J., Huang, X., Zhang, W., Holmes, M. (2017). Novel colorimetric films based on starch/ polyvinyl alcohol incorporated with roselle anthocyanins for fish freshness monitoring, *Food Hydrocoll.*, **69**, pp. 308–317.
27. Pocas, M.F.F., Delgado, T.F., Oliveira, F.A.R. (2008). Smart Packaging Technologies for Fast Moving Consumer Goods, Chapter 9 "Smart packaging technologies for fruits and vegetables", Kerry, J., Butler, P. (eds.), John Wiley & Sons, Chichester England, pp. 151–166.

28. Neethirajan, S., Jayas, D.S., Sadistap, S. (2009). Carbon dioxide (CO_2) sensors for the agri-food industry: A review, *Food Bioprocess Technol.*, **2**, pp. 115–121.

29. Goulart, L.A., Cruz de Moraes, F., Mascaro, L.H. (2016). Influence of the different carbon nanotubes on the development of electrochemical sensors for bisphenol A, *Mater. Sci. Eng.: C*, **58**, pp. 768–773.

30. Ibañez, G.A., Escandar, G.M. (2011). Luminescence sensors applied to water analysis of organic pollutants: An update, *Sensors*, **11**, pp. 11081–11102.

31. Wang, J. (2006). Analytical electrochemistry 3, Chapter 2 "Study of electrode reactions and interfacial properties", John Wiley & Sons, Hoboken, New Jersey, pp. 29–66.

32. Yam, K.L., Takhistov, P.T., Miltz, J. (2005). Intelligent packaging: Concepts and applications, *J. Food Sci.*, **70**, pp. R1–R10.

33. Han, J.H. (2014). Innovations in food packaging 2 Ed., Chapter 1 "A Review of Food Packaging Technologies and Innovations", Academic Press, Amsterdam, pp. 3–12.

34. MacCraith, B., McDonagh, C., O'Keefe, G., Keyes, E., Vos, J., O'Kelly, B., McGilp, J.F. (1993). Fibre optic oxygen sensor based on fluorescence quenching of evanescent-wave excited ruthenium complexes in sol–gel derived porous coatings, *Analyst*, **118**, pp. 385–388.

35. Huber, C., Nguyen, T.A., Krause, C., Humele, H., Stangelmayer, A. (2006). Oxygen ingress measurement into PET bottles using optical-chemical sensor technology, *Monatsschrift fur Brauwissenschaft*, **59**, pp. 5–15.

36. Fitzgerald, M., Papkovsky, D.B., Smiddy, M., Kerry, J.P., O'Sullivan, C.K., Buckley, D.J., Guilbault, G.G. (2001). Nondestructive monitoring of oxygen profiles in packaged foods using phase-fluorimetric oxygen sensor, *J. Food Sci.*, **66**, pp. 105–110.

37. Baleizao, C., Nagl, S., Schaferling, M., Berberan-Santos, M.N., Wolfbeis, O.S. (2008). Dual fluorescence sensor for trace oxygen and temperature with unmatched range and sensitivity, *Anal. Chem.*, **80**, pp. 6449–6457.

38. Fu, A.H., Molins, R.A., Sebranek, J.G. (1992). Storage quality characteristics of beef rib eye steaks packaged in modified atmospheres, *J. Food Sci.*, **57**, pp. 283– 287.

39. Schulz, K., Jensen, M.L., Balsley, B.B., Davis, K., Birks, J.W. (2004). Tedlar bag sampling technique for vertical profiling of carbon dioxide through the atmospheric boundary layer with high precision and accuracy, *Environ. Sci. Technol.*, **38(13)**, pp. 3683–3688.

40. Bultzingslowen, C., McEvoy, A.K., McDonagh, C., MacCraith, B.D., Klimant, I., Krause, C., Wolfbeis, O.S. (2002). Sol-gel based optical

carbon dioxide sensor employing dual luminophore referencing for application in food packaging technology, *Analyst*, **127**, pp. 1478–1483.

41. Borisov, S.M., Waldhier, M.C., Klimant, I., Wolfbeis, O.S. (2007). Optical carbon dioxide sensors based on silicone-encapsulated room-temperature ionic liquids, *Chem. Mater.*, **19**, pp. 6187–6194.

42. Borchert, N.B., Kerry, J.P., Papkovsky, D.B. (2013). A CO_2 sensor based on Pt porphyrin dye and FRET scheme for food packaging applications, *Sensors Actuators B: Chem.*, **176**, pp. 157–165.

43. McFarlane, D., Sheffi, Y. (2003). The impact of automatic identification on supply chain operations, *Int. J. Logistics Managem.*, **14**, pp. 1–17.

44. Robertson, G.L. (2012). Food packaging: Principles and practice, 3 Ed., CRC Press, Boca Raton, the U.S.

45. Manthou, V., Vlachopoulou, M. (2001). Bar-code technology for inventory and marketing management systems: A model for its development and implementation, *Int. J. Prod. Econom.*, **71**, pp. 157–164.

46. Drobnik, O. (2015). Barcodes with IOS: Bringing together the digital and physical worlds, IOS Press, Manning, the U.S.

47. Yam, K.L., Takhistov, P.T.W., Miltz, J.W. (2009). The Wiley Encyclopedia of Packaging Technology, 3 Ed., Chapter (I) "Intelligent packaging", Yam, K. (ed.), John Wiley & Sons, New York, pp. 605–616.

48. Sarac, A., Absi, N., Dauzère-Pérès, S. (2010). A literature review on the impact of RFID technologies on supply chain management, *Int. J. Prod. Econom.*, **128**, pp. 77–95.

49. Plessky, V.P. and Reindl, L.M. (2010). Review on SAW RFID tags. *IEEE Trans. Ultrason. Ferroelectrics Frequency Control*, 57(3), pp. 654–668.

50. Preradovic, S., Karmakar, N.C. (2012). Multiresonator-based chipless RFID: Barcode of the future, Springer, New York.

51. Uysal, I., Emond, J., Bennett, G. (2011). Tag testing methodology for RFID enabled temperature tracking and shelf life estimation, RFID-Technologies and Applications (RFID-TA), *2011 IEEE International Conference*, pp. 8–15.

52. Rodrigues, E.T., Han, J.H. (2011). Encyclopedia of agricultural, food, and biological engineering, 2 Ed., Chapter "Packaging: Intelligent", Heldman, D.R., Moraru, C.I. (eds.), CRC Press, Boca Raton, pp. 1199–1204.

Chapter 4

Brief Overview of Aspiring Properties of Functional Biopolymers for Food Packaging Applications

Pooja Agarwal, Anjali Gupta, and Divya Tripathy
Division of Chemistry, Galgotias University, Greater Noida, Uttar Pradesh, India
poojaocm5@gmail.com

Increased use of plastics has created serious environmental issues because of their resistance to biodegradation. In last few years, interest has been developed for biodegradable polymeric materials in packaging because of their biodegradable nature and also having desirable properties like conventional polymers. Biopolymers have been categorized into three classes based on their origins. Biopolymers can be produced from biomass, can be synthesized from monomers that are obtained from biological sources, and can be produced from microorganisms. Functions of biopolymers depend upon different features like physical, mechanical, thermal, and barrier properties. Besides, the recent trend of lifestyle also poses challenges toward innovative modes of food packaging. Functionality in packaging material can be involved through active packaging concept. Modern

Biomaterials in Food Packaging
Edited by Mohd Yusuf and Shafat Ahmad Khan

ISBN 978-981-4877-98-5 (Hardcover), 978-1-003-25678-6 (eBook)
www.jennystanford.com

food-packaging techniques involve hygienic processing and as a result extend the shelf-life of food without causing any bad effect on the health of the consumer. This chapter highlights different kinds of biopolymers; their preferable characteristics for food packaging application, with special emphasis on biopolymeric materials, which appear as a promising solution for food packaging; and new advancement in terms of functionalities, with the replacement of conventional plastic packaging materials.

4.1 Introduction

Packaging term refers to the protection, safety, secure handling, and improved usability of products. The commonly used packaging materials are wood, plastic, paper, metal, and glass. Presently, plastic is the most frequently used packaging material because of its unique features like lightweight, non-permeable, flexible, stable (inactive to environment), durable, and inexpensive. Due to these properties, it persists for the long time in the environment and accumulated as solid waste [1, 2]. Plastic also contains additives, plasticizers, and colorants, which are responsible for serious environmental problems during its disposal [1, 3]. Therefore, environmental concerns and petroleum resource limitations have increased the attention for products that are safer, environmentally friendly, non-toxic, and cost effective. Over the past two decades, it has been the major concern of researchers to develop polymers from renewable resources in the area of plastic productions. Use of biodegradable polymers as bioplastic is a promising approach to decrease the amount of plastic waste. In addition to biodegradability, biopolymers have other features like air penetrability, sealability at low temperature, ease availability, and economically favorable. Starch, cellulose, protein, polylactic acid (PLA), and polyhydroxybutyrate (PHB) are the latest developments designed for packaging purposes. These have been conversed in this chapter, along with the development in functionality for active packaging.

The concept of active packaging has good effect on marketing of food, due to its performance and fitness for food packaging. In the last decade, many developments in packaging technology have been done, such as active packaging, which can behave as an antimicrobial agent, an oxygen scavenger, a respiration regulator, and an odor absorber [4]. The packaging industry will stimulate due to these evolving developments for better quality and safety products that help in decreasing losses of food [5], along with the environmental sustainability.

4.2 Biopolymer

It is an organic polymer that can degrade into simpler molecules like carbon dioxide, methane, and water. It is a good alternative to petroleum-based polymers (traditional plastics). The most familiar biopolymers are starch, protein, peptide, DNA, and RNA. The biopolymers are derived from renewable natural resources, which are eco-friendly and mostly non-toxic. These are of biological origin such as animals, plants, and microorganisms, or can be synthesized chemically from materials that are of biological origin such as sugar, starch, oil, and natural fat [6–12].

Origin and description of bio-based polymers: Biologically origin polymers can be classified into three groups based on their source and synthesis:

Group 1: The biopolymers in this category are obtained by biomass directly. Several polysaccharides like starch, cellulose, and proteins (casein and gluten) are some example of biopolymers in this group. All the polymers of this category are hydrophilic in nature rather crystalline. These polymers have excellent barrier property, still poor in performance particularly in context of packaging of moist food products.

Group 2: The biopolymers in this category are manufactured by a conventional polymerization method from monomers of biological origin. One of the excellent examples of biopolyester is PLA, which can be produced by polymerization of bio-based monomer lactic acid.

Group 3: The biopolymers in this group are obtained from microorganisms. The most common example of this class is polyhydroxyalkanoate (PHA).

4.2.1 Polysaccharide-Based Biopolymer

4.2.1.1 Starch

Starch is the most abundant polysaccharide, and it is inexpensive and widely available. Due to these factors, starch-based biomaterial has received great attention. Starch is a hydrocolloid biopolymer belongs to the class of carbohydrate. It consists of linear chain amylose and branched amylopectin, made up of glucose molecule joined together by glycosidic linkage. The variation of contents of amylase and amylopectin affects the end properties of the starch. Elongation strength of the starch rises with the increase of amylose content, when there is no plasticizer. High amylopectin is favorable when plasticizer is used. For the preparation of biopolymers, several varieties of starches like potato, corn, rice, cassava, and tapioca are used [13]. Starch is generally used due to its thermoplastic property. Its plasticization occurs by the process of destructuration at high temperature in presence of water or plasticizer. However, poor moisture barrier and mechanical properties limit its uses. These shortcomings are improved by preparation of its bio-composites with several other biopolymers and additives.

4.2.1.2 Cellulose

Cellulose is biological origin polymer, which is available in abundance on the planet. It is a major component of plant cell walls. Cellulose-rich materials are wood, cotton, and paper. The cellulose is found in plants in abundance and works as a strengthening material [7]. It is a linear homopolymer of glucose, which is linked by β-glycosidic linkage. It is a tougher fiber because hydrogen bonding occurs between glucose units, within a chain, and even between the neighboring chains. Building and arrangement of hydroxyl group in cellulose are likely to create strong hydrogen-bonded crystal structure. This allows the crystallization of chains in linear conformation. Micro-fibrils in the range of sub-micron diameter are associated along the cellulose fibers and are called single fiber. Commonly produced cellulose derivatives are methyl cellulose, carboxymethyl cellulose (CMC), ethyl cellulose, hydroxyl ethyl/propyl cellulose, and cellulose acetate [14].

4.2.1.3 Protein

Proteins are polymeric structures made by repeating units of amino acids with peptide linkages. There are total 22 amino acids that are usually found in protein [15]. The proteins can be synthesized by the chemical route or fermentation process [16]. These are used for coating and film preparation, which can be attained from animals e.g. casein, whey from milk protein, as well as from plants e.g. corn zein protein, soy protein, etc. These have different physicochemical characters according to sources from which they are obtained. The proteins in wheat gluten and corn zein are occurred in globular form, while casein and gelatin are example of fibrous protein [17–19]. Protein-based films are superior to polysaccharide-based films, in respect to mechanical strength and barrier properties, along with the better nutritional value [20]. Although, there are restrictions in packaging with protein films and coatings because of their weak mechanical strength related to synthetic polymeric material and their low-moisture barrier property due to hydrophilicity [21].

4.2.2 Synthetic Biopolymer

4.2.2.1 PLA

Application of PLA has increased in recent years because of its economical property and commercial feasibility. It is a biopolyester of 2-hydroxypropanoic acid commonly known as lactic acid. Microbial fermentation of carbohydrate is major source of lactic acid, and it also synthesizes chemically from petrochemicals. The lactic acid can exists in two stereo-isomeric forms: L(+) and D(-) lactic acids. Suitable microbe is used for the synthesis of required isomer L-lactic acid or D-lactic acid. There are three common methods to synthesize PLA viz: direct condensation of D- or L-lactic acid, dehydration polymerization method, and the most prominent method is ring opening polymerization of lactide dimer. The PLA obtained can be classified as poly(D-lactide) (PDLA), poly(L-lactide) (PLLA), and poly(D, L-lactide) (PDLLA), according to the isomer of lactide dimer selected, and all these isomeric forms of polymer have different properties [22–24].

The properties of PLA, like high-molecular weight, good process ability, water-solubility resistance, and biodegradability, make it a suitable food packaging material. These are the key reasons that PLA has become a key substitute as a green packaging material [24, 25]. It has superior properties related to several polymers like better mechanical strength, flavor perseverance, and odor barrier properties than non-degradable polymers like polyethylene and poly vinyl chloride. Also, the PLA has better thermal resistance and processing capability than polystyrene. It also has more grease resistance and printability to polyethylene. Thus, it is the future biodegradable polymer that can be employed in packaging of different food items. Although it has some restrictions in unchanged form, it degrades easily at extensive temperature and also brittle in nature.

4.2.2.2 PHA

PHA belongs to polyester family produced by various microorganisms. Bacterial fermentation of sugar and lipid under unbalanced growth condition is the major production method of this polymer. The PHA has macromolecular structure made up of 3-hydroxy fatty acid monomer units. It is a biodegradable and biocompatible polymer that can be acquired from renewable resources. Thermo-mechanical property of PHA is similar to synthetic polymers like polypropylene [25, 26]. Due to the biodegradable and biocompatible nature of this biopolyester, it is used in several applications as degradable packaging materials like bottles, films, sheets, laminates, coatings, and fibers. The characteristics of the PHA depend on monomer arrangement. Poly (3-hydroxybutyrate) is naturally occurring β-hydroxy acid thermoplastic linear polyester and biomimetic analog of PHA [27–29]. More than 100 monomers and copolymers can be derived from PHAs. Recent researches have also showed that, with PHB, many varieties of PHAs can be produced by microbial fermentation. Nature of the carbon source and microorganism involved decides the monomer composition of PHAs. The PHAs with middle chain lengths show elastomeric property with low-melting point and have comparatively lower crystallinity; whereas, PHB is crystalline thermoplastic substance. Water vapor permeability of PHA is very low and closer to that of LDPE, which is a very substantial property for food packaging application [30].

The homopolymer chain of PHB makes it brittle and shows glass thermal degradation in the region of melting point [31]. It is similar to isotactic polypropylene (iPP) in the aspect of melting temperature (175–180 °C) and mechanical transition temperature (55 °C). It is also similar in mechanical property as of synthetic degradable polyesters. The PHB can be decomposed by several microorganisms in numerous environmental conditions [32]. Its hydrolytic degradation produces 3-hydroxy butyric acid. Properties of PHA like good mechanical strength, flavoring agent, printability, heat locking, temperature stability, resistance to lubricants, and high dye efficacy, increase its food packaging application [15, 33]. "Metabolix PHA" is a blend of PHB and poly (3-hydroxyoctanoate), produced by a U.S.-based company Metabolix. It has been approved by the FDA for making packaging material that preserves all the desired features of non-degradable plastics [34]. Further, the PHB and associated materials can be applied in various areas like covering, agricultural, hygiene-related, and biomedical products. Some other recently end-use applications are paints, pressure-sensitive adhesives, biodegradable rubbers, and cheese coverings. Therefore, the future prospects of PHAs are very promising.

4.3 Desirable Properties of Biopolymers for Food Packaging Application

4.3.1 Gas Barrier

Specific gas pressure and atmospheric conditions are required in the food packaging industry to sustain the freshness and shelf life of food products and to retain quality during their storage. There is increasing requirements of food that is packed in presence of shielding atmosphere with a particular combination of gases, including oxygen, nitrogen, and carbon dioxide, or combination of these [33]. Packaging material requires certain gas barrier properties to hold up specific gas composition in the packaging. The gas barrier feature of bioplastics is very much associated to their penetration capability [14]. The gas penetration behavior of biopolymeric materials can be enhanced with humidity, due to hydrophilic behavior of the materials. Multilayer structure employs better barrier property for food packaging materials. Multilayers can be produced by various methods such as electron beam evaporation, atomic deposition technique, magnetron sputtering, and sol/gel process [14]. One of the major requirements in the food packaging industry is to maintain water and oxygen penetrability throughout the shelf life of thermally treated food items.

Researches showed that gas barrier characteristics of PLA and PLA-based substances are strongly related with the humidity [33]. As reported by many researches, crystallinity also influences mechanical and gas barrier characteristics of PLA sheets for food packaging. The gas barrier property is also enhanced with the crystallinity of the materials. According to a study, the barrier property of plasticized edible film is related to the mechanical strength of the materials with the decrease in tensile strength and increase in percentage elongation (25–45%) [12, 33]. Percentage of solubility in water increases with plasticization effect; however, elastic modulus and glass transition temperature decrease.

4.3.2 Thermal and Mechanical

Thermal and mechanical performances are significant since they determine fitness of biopolymers. Therefore, next to the barrier property, the thermal and mechanical behaviors of packaging substances are significant for processing. Good mechanical and thermal properties are also important for the safety of products against thermal or mechanical damage during storage period. Most of the biopolymer materials show similar properties to conventional polymers, such as polystyrene, which is comparatively rigid material; polyethylene, which is relatively flexible polymer; and PET, which is comparatively stiff material. The mechanical properties in aspect to modulus and stiffness are very much similar as compared to traditional polymers. The modulus range of bio-based materials can vary from 2500–3000 MPa and it is low for stiff polymers like thermoplastic starches, i.e. up to 50 MPa, and also lower for rubbery materials like medium chain PHAs. Various techniques like plasticization, blending with polymeror fillers, cross-linking method, or addition of fibers, can be applied for the processing of biopolymers to achieve the required mechanical properties [35]. Thermal and mechanical properties of bioplastics can be varied through plasticization process, by increasing thermal degradation temperature [33, 36]. Like bacterial cellulose can be used to provide particular mechanical properties in the materials. Speculation says that the materials from biological origin can be synthesized having comparable strength as compared to the materials that are in use currently [36].

4.3.3 Moisture Barrier

The water resistance capability of bioplastic to prevent the entry of undesired vapors is termed as moisture barrier property. This property can be defined as the diffusivity, permeability, and solubility across the packaging film, and also the attraction of the packaging substance toward moisture, which is calculated by water vapor transmission rate. Hydrophobic nature of bio-based materials reduces moisture barrier property of natural polymers [14, 30]. Water absorbance of packaging materials can create unwanted effects like moisture recovery of dry food or surface drying of frozen food. These concerns can be overcome with materials having good moisture barrier property.

There are several methods to prepare moisture-resistant films, e.g. polyester, wax, and fatty acid, which can be used as hydrophobic materials for external coating. The cross-linking of bio-materials can be done with inorganic fillers, blending with moisture-resistant substances of polymers. This is also a promising method, strengthening natural fibers such as jute, coir, and sisal [37]. Morphological properties like crystallinity and chain conformation also characterize the barrier property. This property increases with the crystallinity of the materials [35]. Improved moisture barrier property of packaging material is a significant factor in the food packaging industry because it enhances shelf life of the food. Relative study on degree of water vapor transmission of biopolymeric substance with petroleum-based synthetic polymers showed that bio-based materials have equivalent properties over the conventional petroleum-based materials.

4.3.4 Biodegradability

Biodegradability of a polymer can be defined as the ability of constituent, product, or waste of material to be broken down into simpler constituents by the action of naturally occurring microorganisms like bacteria, fungi, or algae in a biodegrading environment. The process of biodegradation is a subjective to environmental factors like temperature, moisture, existing nutrients, and pH, in which the materials and microbes are exposed [14]. The fragmentation takes place in the first step of biodegradation, which is initiated by moisture, heat, and microbial enzymes. It is followed by the second step in which transformation of longer molecular material into smaller molecules takes place, which is started by enzymes and naturally occurring acids. When the molecules are broken into appropriate size, these are absorbed by the cell walls of the microorganisms and metabolized for energy [38]. In this whole procedure, carbon dioxide and methane are produced. The polymer degradation can take place either due to photo degradation, microbial action, or by chemical action. Degradability of a polymer can be accelerated by the use of oxidation and additive technology in oxo-degradation. Degradation is generally introduced by light, heat, or microorganisms in the control degradation process.

4.4 Functional/Active Packaging

As discussed, bio-based functional packaging materials are obtained from renewable resources and can be degraded by natural compositing method with least environmental impact. Other than this, recent development is made for active packaging. The active packaging enhances storage time and preserves food quality (Figure 4.1). This can be used as an alternate to traditional food processing procedures like high-thermal treatment and preservation by additives, which can bring acidification and dehydration to the food [39]. Functionality can be created by the incorporation of particular components in biopolymers to produce active packaging materials. Incorporation of the polymer substances with additives having anti-microbial properties is a new development in the field of active packaging [40]. The benefits of using such polymeric materials are that these can release functional agents like antioxidants and antimicrobials, and also maintain undesirable food components like ethylene, oxygen, and water [41]. Direct incorporation of additives into the food creates undesirable odor and taste, while controlled release of functional materials into the food by packaging cover for prolonged duration of storage and supply, and controls the growth of undesirable properties in the food [42]. Synthetic additives produce toxicity due to their movement into the food products. In place of these additives, essential oils have been used nowadays. The essential oils as natural extracts can be obtained from species of herbs and spices, tocopherol, and extracts from plants, which are considered safe and facilitate the chemical stability of oxygen-sensitive food [43]. Selection of an antioxidant for functional packaging system is an important step, which needs suitable considerations. Type of food and its quality efficiency are the base for the selection of an antioxidant. The antioxidant selected should be capable to make homogenous spreading to the food and also compatible with the packaging material. To reduce or slow down the growth of microorganisms in the packaged food is the main purpose of using antimicrobial substances in food packaging [44]. Antimicrobial agents can be either directly added in packaging materials for slow dispersion through food surface or can be filled in vapor form. Oxygen scavengers also can be introduced in food packaging to prevent oxidation by eradicating the oxygen. Most commonly used oxygen scavengers are ferrous oxide, which reacts with oxygen to reduce its extent [15–18]. Carbon dioxide scavengers and mediators can be used inside packaging system to reduce the frequency of respiration of food and these can also minimize difference in pressure, thus these avoid collapsing of package due to oxygen absorption [41]. Absorbent pads or moisture-mediated bicarbonate chemicals can be used to add carbon dioxide in packets. Moisture controlling substances like natural clays, calcium oxide, and silica gel can also be employed for active packaging [43]. These act as desiccants for dry food while as internal humidity regulators for the food with large moisture extent. These substances can be applied as porous sachets kept inside or can be applied as permeable water vapor barrier plastic holders with desiccants [38, 39]. Humidity controllers can decrease the moisture loss, preserve necessary relative moisture, and reduce additional moisture content within package [40]. Treatment of packaging surfaces or films through micro pores or perforations can behave as gas adsorbents [44]. Bicarbonate of soda and activated charcoal as flavor binding agents and food-derived colorants and pigments as color providing agents are also used in packaging [42, 45]. Anti-foaming, light-absorbing, and temperature-regulating agents are required in some particular packaging systems [46–48].

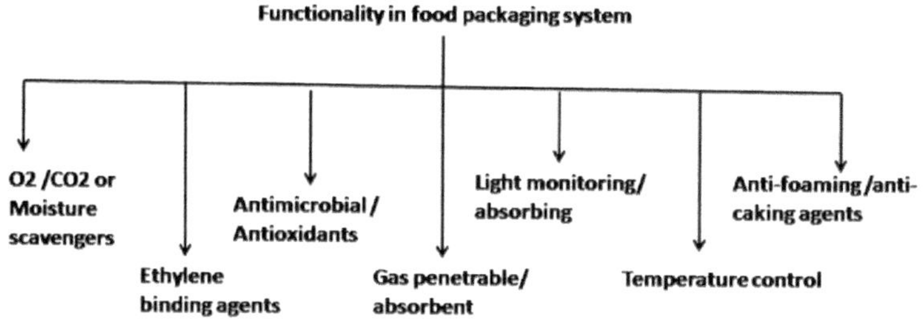

Figure 4.1 Active/functional packaging system.

4.5 Conclusion

Non-biodegradable petroleum-based polymers can be replaced by the bio-based renewable polymeric materials in food packaging, which will help create sustainable and green environment, and will also reduce the waste disposal cost. Biopolymers for packaging application obtained from renewable resources show comparable properties with the conventional plastics. These biopolymer materials can be obtained from natural feedstock, microorganisms, or chemical synthetic procedures. Moreover, research and development (R&D) in this field offers advancement in innovative functional food packaging techniques. These advanced packaging approaches deliver enhanced food quality, protection, feasibility, and bioactivity of functional components. Usefulness of biopolymers and innovative packaging techniques is increasing extensively because of their agreeable health and environmental impact and also they help in reducing the waste.

References

1. Andrady, A.L., Neal, M.A. (2009). Applications and societal benefits of plastics, *Philosophical Trans. Royal Soc. B: Biol. Sci.*, **364(1526)**, pp. 1977–1984.

2. Ashok, A., Mathew, M., Rejeesh, C.R. (2016). Innovative value chain development of modified starch for a sustainable environment: A review, *Int. J. Polym. Sci. Eng.*, **2(1)**, pp. 20–32.

3. Thompson, R.C., Swan, S.H., Moore, C. J., Vom Saal, F.S. (2009). Our plastic age. *Phil. Trans. R. Soc. B*, **364**, pp. 1973–1976. doi:10.1098/rstb.2009.0054.

4. Brody, A.L., Bugusu, B., Han, J.H., Sand, C.K., Mchugh, T.H. (2008). Innovative food packaging solutions, *J. Food Sci.*, **73(8)**, pp. 107–116.

5. Vanderroost, M., Ragaert, P., Devlieghere, F., De Meulenaer, B. (2014). Intelligent food packaging: The next generation, *Trends Food Sci. Technol.*, **39(1)**, pp. 47–62.

6. Thompson, R.C., Swan, S.H., Moore, C.J., Vom Saal, F.S. (2009). Our plastic age, *Philosophical Trans. Royal Soc. B: Biol. Sci.*, **364(1526)**, pp. 1973–1976.

7. Pandey, J.K., Takagi, H., Nakagaito, A.N., Saini, D.R., Ahn, S.H. (2012). An overview on the cellulose based conducting composites, *Comp. Part B: Eng.*, **43(7)**, pp. 2822–2826.

8. Prathipa, R., Sivakumar, C., Shanmugasundaram, B. (1999). Biodegradable polymers for sustainable packaging applications: A review, *J Polym. Environ.*, **23(4)**, pp. 449–458.

9. Ruban, S.W. (2009). Biobased packaging-application in meat industry, *Vet. World*, **2(2)**, pp. 79–82.

10. Jacobsen, S., Fritz, H.G. (1999). Plasticizing polylactide: The effect of different plasticizers on the mechanical properties, *Polym. Eng. Sci.*, **39(7)**, pp. 1303–1310.

11. Yalcin, B., Cakmak, M., Arkın, A.H., Hazer, B., Erman, B. (2006). Control of optical anisotropy at large deformations in PMMA/chlorinated-PHB (PHB-Cl) blends: Mechano-optical behavior, *Polymer*, **47(24)**, pp. 8183–8193.

12. Modi, S., Koelling, K., Vodovotz, Y. (2011). Assessment of PHB with varying hydroxyvalerate content for potential packaging applications, *Eur. Polym. J.*, **47(2)**, pp. 179–186.

13. Ezeoha, S.L., Ezenwanne, J.N. (2013). Production of biodegradable plastic packaging film from cassava starch, *IOSR J. Eng.*, **3(10)**, pp. 14–20.

14. Kumar, M.S., Yaakob, Z. (2011). Handbook of bioplastics and biocomposites engineering applications, Chapter 5 "Biobased materials in food packaging applications", Pilla, S. (ed.), John Wiley & Sons, Hoboken New Jersey, pp. 121–159.

15. Ambrogelly, A., Palioura, S., Söll, D. (2007). Natural expansion of the genetic code, *Nature Chem. Biol.*, **3(1)**, pp. 29–35.

16. Aung, S.P.S., Nu, W.L., Win, K.C. (2017). Chitosan: Hydrogels and composites for their applications, Chapter 3 "Enzymes for production of chitin, chitosan and chitooligosaccharide and determination of activities of enzymes using chitinous substrates", Ahmed, S. Ikram, S. (eds.), Scrivener Publishing LLC, Beverly MA, pp. 71–88.

17. Coltelli, M.B., Wild, F., Bugnicourt, E., Cinelli, P., Lindner, M., Schmid, M., Rodríguez-Turienzo, L. (2016). State of the art in the development and properties of protein-based films and coatings and their applicability to cellulose based products: An extensive review, *Coatings*, **6(1)**, pp. 1–59. doi:10.3390/coatings6010001

18. Lacroix, M., Vu, K.D. (2014). Innovations in food packaging, Chapter 11 "Edible coating and film materials: proteins", Han, J. (ed.), Academic Press, London, pp. 277–304.

19. Dangaran, K., Tomasula, P.M., Qi, P. (2009). Edible films and coatings for food applications, Chapter 2 "Structure and function of protein-based edible films and coatings", Embuscado, M.E., Huber, K.C. (eds.), Springer, New York, pp. 25–56.

20. Zink, J., Wyrobnik, T., Prinz, T., Schmid, M. (2016). Physical, chemical and biochemical modifications of protein-based films and coatings: An extensive review, *Int. J. Molecular Sci.*, **17(9)**, pp. 1376.

21. Wittaya, T. (2012). Structure and function of food engineering, Chapter 3 "Protein-based edible films: Characteristics and improvement of properties", Eissa, A.A. (ed.), InTech Open, London, pp. 43–70.

22. Byun, Y., Kim, Y.T. (2014). Innovations in food packaging, Chapter 14 "Bioplastics for food packaging: Chemistry and physics", Han, J. (ed.), Academic Press, London, pp. 353–368.

23. Hu, B. (2014). Biopolymer-based lightweight materials for packaging applications, *Lightweight Mater. Biopolym. Biofibers*, **1175**, pp. 239–255.

24. Avérous, L., Pollet, E. (2012). Environmental silicate nano-biocomposites, Chapter 2 "Biodegradable polymers", Avérous, L., Pollet, E. (eds.), Springer, London, pp. 13–39.

25. Galego, N., Rozsa, C., Sánchez, R., Fung, J., Vázquez, A., Santo Tomas, J. (2000). Characterization and application of poly (β-hydroxyalkanoates) family as composite biomaterials, *Polym. Testing*, **19(5)**, pp. 485–492.

26. Weber, C.J., Haugaard, V., Festersen, R., Bertelsen, G. (2002). Production and applications of biobased packaging materials for the food industry, *Food Additives Contam.*, **19(S1)**, pp. 172–177.

27. Kim, M.N., Lee, A.R., Yoon, J.S., Chin, I.J. (2000). Biodegradation of poly (3-hydroxybutyrate), Sky-Green and Mater-Bi by fungi isolated from soils, *Eur. Polym. J.*, **36(8)**, pp. 1677–1685.

28. Weber, C.J. (2000). Biobased packaging materials for the food industry: Status and perspectives, a European concerted action, KVL Publishing, the U.S.

29. Anandharamakrishnan, C., Kolli, U.K. (2014). Polymers for Packaging Applications, Chapter 10 "Bionanocomposites and their potential applications in food packaging", Alvi, S., Thomas, S., Sandeep, K.P., Kalarikkal, N., Yargalla, S. (eds.), Apple Academic Press, Oakville, Canada, pp. 229–262.

30. Iguchi, M., Yamanaka, S., Budhiono, A. (2000). Bacterial cellulose: A masterpiece of nature's arts, *J. Mater. Sci.*, **35(2)**, pp. 261–270.

31. Burgueno, R., Quagliata, M.J., Mehta, G.M., Mohanty, A.K., Misra, M., Drzal, L.T. (2005). Sustainable cellular biocomposites from natural fibers and unsaturated polyester resin for housing panel applications, *J. Polym. Environ.*, **13(2)**, pp. 139–149.

32. Lopez-Rubio, A., Gavara, R., Lagaron, J.M. (2006). Bioactive packaging: Turning foods into healthier foods through biomaterials, *Trends Food Sci. Technol.*, **17(10)**, pp. 567–575.

33. Suppakul, P., Miltz, J., Sonneveld, K., Bigger, S.W. (2003). Active packaging technologies with an emphasis on antimicrobial packaging and its applications, *J. Food Sci.*, **68(2)**, pp. 408–420.

34. Flores, S., Conte, A., Campos, C., Gerschenson, L., Del Nobile, M. (2007). Mass transport properties of tapioca-based active edible films, *J. Food Eng.*, **81(3)**, pp. 580–586.

35. Peltzer, M., Wagner, J., Jiménez, A. (2009). Migration study of carvacrol as a natural antioxidant in high-density polyethylene for active packaging, *Food Additives Contam.*, **26(6)**, pp. 938–946.

36. Gómez-Estaca, J., Lopez-de-Dicastillo, C., Hernández-Muñoz, P., Catalá, R., Gavara, R. (2014). Advances in antioxidant active food packaging, *Trends Food Sci. Technol.*, **35(1)**, pp. 42–51.

37. Appendini, P., Hotchkiss, J.H. (2002). Review of antimicrobial food packaging, *Innov. Food Sci. Emerg. Technol.*, **3(2)**, pp. 113–126.

38. Kerry, J.P., O'grady, M.N., Hogan, S.A. (2006). Past, current and potential utilisation of active and intelligent packaging systems for meat and muscle-based products: A review, *Meat Sci.*, **74(1)**, pp. 113–130.

39. Mohan, C.O., Ravishankar, C.N., Gopal, T.S. (2010). Active packaging of fishery products: A review, *Fishery Technol.*, **47(1)**, pp. 1–18.

40. Bhardwaj, A., Alam, T., Talwar, N. (2019). Recent advances in active packaging of agri-food products: A review, *J. Postharvest Technol.*, **7(1)**, pp. 33–62.

41. Vermeiren, L., Devlieghere, F., van Beest, M., de Kruijf, N., Debevere, J. (1999). Developments in the active packaging of foods, *Trends Food Sci. Technol.*, **10(3)**, pp. 77–86.

42. Kenyó, C., Renner, K., Móczó, J., Fekete, E., Kröhnke, C., Pukánszky, B. (2014). Effect of desiccant characteristics on the properties of PS/zeolite functional packaging materials, *Polym. Comp.*, **35(11)**, pp. 2112–2120.

43. Otoni, C.G., Espitia, P.J., Avena-Bustillos, R.J., McHugh, T.H. (2016). Trends in antimicrobial food packaging systems: Emitting sachets and absorbent pads, *Food Res. Int.*, **83**, pp. 60–73.

44. Korotcenkov, G. (2014). Handbook of gas sensor materials: Properties, advantages and shortcomings for applications, Vol 2, Chapter 12 "Nanocomposites in gas sensors: Promising approach to gas sensor optimization", Korotcenkov, G. (ed.), Springer, New York, pp. 181–184.

45. Cadwallader, K. (2015). Flavor challenges and solutions for high protein functional foods and beverages, Protein Trends and Technologies Seminar, May 5–6, Oak Brook, Illinois, USA.

46. Gherezgihier, B.A., Mahmud, A., Admassu, H., Shui, X.W., Fang, Y., Tsighe, N., Mohammed, J.K. (2017). Food additives: Functions, effects, regulations, approval and safety evaluation, *J. Acad. Ind. Res.*, 6, pp. 62–68.

47. Coates, J. (2008). Sentelligence Inc, assignee. Low-cost on-line and in-line spectral sensors based on solid-state source and detectors combinations for monitoring lubricants and functional fluids, US Patent No. 7,339,657.

48. Singh, S., Gaikwad, K.K., Lee, M., Lee, Y.S. (2018). Temperature sensitive smart packaging for monitoring the shelf life of fresh beef, *J. Food Eng.*, **234**, pp. 41–49.

Chapter 5

Recent Research and Development in Food Packaging Technologies: A Review

Syed Dilshad Alam,[a] Rupak Raja,[a,b] Vikas Shirsath,[a] Arvind Kumar Jain,[b] Shafat Ahmad Khan,[b] and Imran Ali[c,d]

[a] *Jubilant Biosys Limited, Noida, Uttar Pradesh, India*
[b] *Division of Chemistry, School of Basic and Applied Sciences, Galgotias University, Greater Noida, Uttar Pradesh, India*
[c] *Department of Chemistry, Jamia Millia Islamia, New Delhi, India*
[d] *Department of Chemistry, College of Sciences, Taibah University, Al-Madinah Al-Munawarah, Kingdom of Saudi Arabia*
dilshad826@gmail.com

In the current scenario, packaging for food items is a great concern. The main focus of the food industry is to provide safe, hygienic, and testy packed food products. There is a lot of packed food items exist in the market for consumption. The main objective of packaging is to envisage the quality of food items packed inside it for a long period. Novel packaging ideas in practice ensure that the products are properly packaged with date and code, and entertain food quality through the food labeling and auto coding. This chapter reviews the latest advancement, research and development (R&D) in food packaging and technologies, and their applications. Various methods for incorporating food packaging technologies, issues related to packaging designs and different materials, and solutions employed are also reviewed in the chapter.

Biomaterials in Food Packaging
Edited by Mohd Yusuf and Shafat Ahmad Khan

ISBN 978-981-4877-98-5 (Hardcover), 978-1-003-25678-6 (eBook)
www.jennystanford.com

5.1 Introduction

In the present era, uses of packed food products are very common among families from villages to cities. In large cities, packed food is the first choice of people to eat. Commonly, packed food attracts the young generation. Due to the short time, the packed food products are very easy to cook. It has also been noticed that most of the families prefer to take lunch and dinner, outside in different types of restaurant. Some packed food items are ready to eat and easily available in stores. The packing materials that are being used should be tested and certified by certain regulatory bodies. Due to degradation nature of the packing materials, it is necessary.

Innovations in food packaging improve, combine, or extending the four key modalities [1] such as:

(i) *Protection*: To prevent food products from external damage (leak or break-up) and internal fouling;

(ii) *Communication*: Communication through proper way with regulated information showing the ingredients and nutritional contents and best use time;

(iii) *Convenience*: Food packaging devoted to consumers for their best convenience; and

(iv) *Containment*: Fundamental information about the packaged items with easy handling, uses, and transportation [2].

The literature dictates that most of the industries are opting smart indicators and their demand is continuously increasing. At present, various types of indicators available and are being used with different benefits for producers, retailers, and consumers [3].

5.2 Key Functions and Requirements of Food Packaging

In general, shelf-life of food has increased through the packaging process, so that packaged food products could be stored for a certain timespan and consumed whenever needed. However, it is an integral way to choose appropriate packaging materials with the balance in food protection with other issues, including material and energy consumption, environmental concern, and strict regulations to minimize pollution. There are many key functions for radiant food packaging to be undersigned, including preservation, convenience, communication, protection, temper indication, containment, and food waste reduction [1, 2]. Advances are made on the latest trends concerning the goal of food packaging to continue food safety with minimization of adverse impacts on the environment.

5.3 Packaging Technologies Prospects

Packaging of food products is a key factor to prevent them from deterioration and extend their shelf-life. Simultaneously, the packaging also protects the products against external as well as internal damage, which can be caused by biological, chemical, or physical interaction. Importantly, the packaging is also functioning as a physical barrier to moisture, O_2, volatile entities, and microbes. It is well documented that the package is an integral part of the preserving system that provides a shield between the food items and the external environment [4]. It is the fundamental requirement of food packaging as a preservation system, which is directly related to the safety of customers. The performance of packaging is based on various factors, such as distribution method, processing operation, initial food quality, and size and shape of the package [5]. The benefits of food processing include marketing and distribution tasks, preservation, toxin removal, and improvement of overall quality of food. By using food processing technology, the manufacturers and suppliers of processed food products are getting more benefits. In general, widely used tools for food preservation are modified atmosphere packaging (MAP) and vacuum packaging (VP) [6]. By adopting the above mentioned strategies of packaging, the food can be protected against deteriorative effects, such as discoloration, flavor change, odor, texture change, nutrient loss, pathogenicity, and other measurable factors. It is also noticed that MAP and VP, which seem to be the ideal preservative methods applicable for many food types, are cheap and easy to handle. Their continuous demand for preserving fresh and natural products without adding any hazardous chemical is a big choice of interest. Nowadays, bio-plastic materials (polymer-based) are used widely in food packaging, without much effect on the food quality [7]. A pathway of food packaging technology is shown in Figure 5.1. Besides, the different classes of food packaging technologies are given below.

5.3.1 Conventional Technology

The conventional packaging technology has been used since the past six decades. Conventional systems of food packaging have been utilized to provide physical support from external damage and other kinds of stimuli during the distribution, storage, and transportation [7, 8]. To the latest regulations and designing of food packaging, four key functionalities should be met in order to maintain shelf-life of the item that satisfies human environment properly. The conventional technology is known due to its availability in the open market. It means that the presence of an infrastructure supports it uses [8].

5.3.2 Modern Technology

The developments of advanced technologies are increasing day by day. Nowadays, consumers are being in practice to use processed food more than staples. Interestingly, the current advances in food processing and technology are fulfilling the demand by opting monitoring methods and techniques with control as well as refined automation. The packed food that the consumers are using is an example of hard work in food research, and thorough investigation of a variety of food properties and actual compositions. Food can be defined in many ways, in food events and food conferences. For instance, it is defined as: food is a fuel that provides energy to perform daily functions and maintain the metabolic processes in the body [9]. Therefore, the ideas of modern packaging systems for food packing support the following:

5.3.2.1 Active packaging

The very first Labuza [10] had used the term active packing. The active packaging refers to the packaging that maintains the quality of products in terms of external environment, self-life, safety, and sensory properties. The term "active packaging" is entirely different from "intelligent packaging", which gives information to the consumer regarding current properties of the food, records, or its history. Rooney [11] enhanced the R&D of active packaging with its scope followed by Gontard and Brody et al. [12, 13]. It contains the interaction of environment to the packaging and product. On the other hand, it is a system in which physical, chemical, and biological activities are responsible for the condition of the packaged food, which cause an extension of its shelf-life, sustainability, and sensory properties, and microbial safety as well, while maintaining its quality [14, 15]. The packaging production industry is a big area, globally, characterized by its internal diversity. Its each sector affects the overall packaging system's demand in the market and is systematically growing. Contrarily, it is a big challenge for manufacturers and R&D units associated with food packaging materials to improve active packaging [16, 17]. In this context, the manufacturers are looking for the best solutions and being compelled to develop improved packaging materials, which are capable of protecting food, extending its storage period, and improving transportation, gas barrier, and environmental performances [18], during the bio-chemical reactions with the internal atmosphere and the product [19–22]. High-impact processes are emerged using active packaging, including physiological processes, chemical processes, physical processes, microbiological changes, and infections [23–25].

It is a well-known fact that inside packaging, atmosphere can be curbed using scavengers or emitters for gases/moisture to remove the harsh elements from the inside environment and enhance shelf-life of food materials [26]. Another group of packaging is based on emitters, which releases friendly substances to the packaging system to impart improved shelf-life [27]. Emitters can be preservatives (sodium benzoate, benzalchonium chloride, mehylparaben, propylparaben, etc.), humidity regulators, fragrant substances, food additives, and biologically active compounds that inhibit the growth of spoilage microbes. Mostly, ethanol, sulfur dioxide, and carbon dioxide are being used as antimicrobials to avoid the growth of microorganisms [26].

The selection of choosing packaging substitutes to impart active properties for any food or beverages is a scientific art, based on the nature of the item. Besides, other things considered for this include process engineering limitations, convenience, economic advantage, usefulness, environmental impacts, and secondary effects resulting from some other changes in the processing or packaging. By using a new technology, a new product introduced and can remove others, due to the lifestyle changes and the limitation of old experience [28]. Figure 5.1 depicts the pictorial representation of processing of packaging to end user supply.

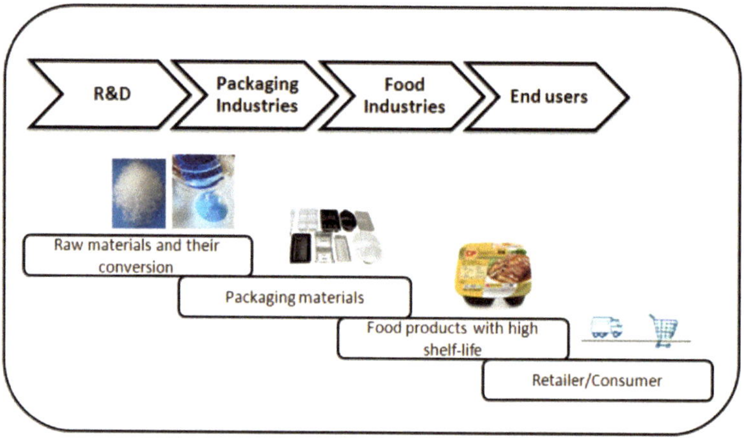

Figure 5.1 Schematic pathway of food packaging technology. Reproduced with minor modification from ref. [29], under the Creative Commons Attribution 4.0 International (CC BY 4.0) License, 2019.

5.3.2.2 Intelligent packaging

As per the European Union, an intelligent packaging referred as the packing system that gives the exact condition of packaged food during transportation and storage [30]. It is an advanced form of traditional food packaging, and for the consumers, it is a medium of information about the product, in terms of its properties, records of changes, and quality. It may also useful to improve Hazard Analysis and Critical Control Points (HACCP) and Quality Analysis and Critical Control Points (QACCP) systems [31]. The given two control checkpoints examine the following considerations [32]:

(i) to sense unsafe food within time;

(ii) to recognize possible hazards to health and make strategies for preventing, reduction, and elimination thereof; and

(iii) to identify most effective methodologies that could improve the food quality.

5.3.2.2.1 *Intelligent packaging technology: Future perspective*

The advanced intelligent packaging has been reached the new phases of technologies. The emerging organic or inorganic-based photonics and electronics, which are being studied about optical and electrical circuits, can be integrated with organic or inorganic materials like polymers instead of silicon. Researches currently based on the generation of new advanced materials with suitable functionalities are to be a part of intelligent packaging [31, 32]. The researchers are continuously working in this area for the development of advanced intelligent packaging, which is in the early stage and will take time to be commercialized in future.

5.3.2.2.2 *Carbon photonics*

Of late, a carbon nanomaterial (CN) proves that it has superior electrical and mechanical properties. At the same time, it also demonstrates a unique optical property, which might be used for the futuristic fabrication of optical sensors, as an alternate or balance to silicon photonics; for example, optical biosensors based on CN were produced successfully with considerable functionality [33]. Similarly, carbon dots and emission-color-tuning CNs have been developed with great analytical and bioanalytical potentials, as a class of strongly fluorescent photonics [34–36].

5.3.2.3 Smart packaging

Earlier it was a big challenge to develop smart packaging; for instance, to combine and integrate the active and intelligent packaging conceptual technologies. Such type of packaging system offers several angles to monitor changes in the products and their surrounding environment by getting feedbacks [37]. Natural preservatives and antioxidants inside the food items give high potential of protection and improved shelf-life. Also, the active packaging having antioxidant properties maintains the quality of food and it is one of the best-gifted substitutes to traditional packaging. The antioxidants possess the potential to reduce oxidation of the food, which is responsible for food spoilage [38].

5.3.2.4 Time–temperature indicator (TTI)

The TTIs play critical responsibility in optimizing temperature in the food product, which is a specific quality parameter. Generally, they composed of small, self-adhesive labels attached to single package or larger configuration, for Fresh-Check from Lifeline technologies. The TTIs display the temperature used during distribution with different colors concerning mechanical, chemical, electrochemical, enzymatic, or microbial changes. They are considered as user-friendly, specialized to provide regularity of food safety and quality, and also applied to monitor the temperature changes in the distribution and storage processes [39].

5.3.2.5 Freshness indicator

It is the direct and quick source of product quality information after the microbial growth in the food product and gives indication even at just starting phase. The quality standard is identified by the chemical reaction between integrated indicators within the package and growth metabolites of spoilage microbes [40]. The main role of freshness detector is to determine amine, ammonia, carbohydrate diacetyl, and ethanol; all of these are in the packaging. This can also be applied to provide a rough estimation of the remaining shelf-life of consumable products.

5.3.2.6 Nanotechnology in food packaging

At present, nanotechnology enhances and is capable to transform food packaging materials continuously. It is applied to monitor the matter of very small size of 1–100 nanometers [41]. Due to nanoscale innovation, much interesting and amazing food packaging with maintaining quality and food safety is achieved. Aluminum-based nanolayer coated inside many snacks food packages is a simple innovation and an example of nanotechnology that is already using in food packaging [42]. Nanomaterials are being developed due to their value-addition nature. They can be prepared top to down from huge structures through many methodologies, including self-assembly, layer-by-layer deposition, lasers, chemical reduction, vaporization, and biomass. The alternate common bottom-up method for the synthesis of complex nanoparticles is being in practice [43].

Nanosensor is a very small device that binds and responsible for detection and displaying a signal. Contrarily, these sensors have the strength to detect and respond to physicochemical signal (sensors) and biological signal (biosensors), resulting to transfer that response into a signal that can be used by humans. In the last few years, many nanosensors have been developed for the detection of internal and external conditions in food packaging [44]. It can also be applied to determine microbes, contaminants, pollutants, and finally the quality of the food [45]. By this technique, the amount of nanoparticles can be reduced for improving the quality of packaging. However, the quality of food depends on the composition of the nanomaterials used. Nanomaterials and nanoparticles could be of such nanoforms: nanotubes, fullerenes, nanofibres, nanoparticles, nanowhiskers, and nanosheets [46]. Although, nanoparticles in the food industry show various benefit, toxicity in the environment is a great concern by using nanoparticles in the industry, [47]. It needs time for regulations of nanomaterial before its implementation into food processing, packaging, and food contact [48]. In the beginning, the manufacturers should monitor any new product with the nanotechnology in the food system and its risk on living kingdom before commercialization [49].

5.3.2.7 Antimicrobial packaging

A suitable antimicrobial substance is necessarily be added for a better packaging system to prevent microbial growth [50]. Antimicrobial food packaging system has been achieved a remarkable milestone, as it controls the growth of pathogenic and spoilage microbes on food surfaces [51]. A multidisciplinary approach is required in the fields of food and material sciences for the development of antimicrobial packaging. So far, a thorough search dictates that various types of studies have been done on different nano-based materials [52]. The studies revealed that nanoparticles alone (i.e. nano Ag, Au, Fe, etc.) as well as their composites impart the antimicrobial properties to packaging materials and therefore improve the food's shelf-life [48–53]. The strong interconnection between the food products and the packaging materials results in potential applications of food packing for a number of perishable eatables [54].

5.3.2.8 Edible coating and film

It is composed of a thin layer of material that inhibits and functions as a barrier to moisture and O_2, followed by the easy movement for the food. The integral part of the food product contains edible films and coatings that unable to interfere with sensory characteristics [51]. Edible coatings and films function as a carrier for antimicrobial and antioxidant compounds to provide preservatives on the food surfaces. Due to the presence of edible coatings and films, the moisture remains unaffected during the storage of food. These also maintain lipid oxidation and brown coloration due to rancidity, reduce a load of spoilage and pathogen microorganisms on the surface of food, and useful in volatile flavor loss [55]. These are eco-friendly in nature, as biopolymers do not affect the food's surrounding environment [56]. By opting edible films and coatings, the recyclability of packaged food materials is enhanced, as compared to other traditional packaging materials, and in this way, these replace the synthetic polymer films [57]. Food graded antimicrobial packaging films show the potential of food packaging materials, due to biodegradability nature that gives sustainable improvement for the modern society [58].

5.3.2.9 Sustainable packaging

It is a well-known fact that sustainable and green protocols urge to use biodegradable and environmentally safe alternates for food packaging applications to reduce waste, which is being generated from other packaging materials. For example, an improved aim is created on sustainable packaging using materials based on polylactide acid (PLA), starch-based films, fiber composite sugar cane pulp, and many more [59]. Sustainable, biodegradable, and environmentally safe materials with significant biodegradability are being achieved from renewable sources such as biological products and microorganisms, obtained from naturally occurring biological probes such as starch, sugar, and natural fat or oil [60].

5.3.2.10 MAP

Advanced food packaging technologies, including MAP, maintain the quality and freshness of the food products and also enhance their shelf-life. Furthermore, the MAP is more appropriate packaging technology through which the shelf-life of the packed products increases, which are found fit for human consumption while maintaining their essential nutrients and bioactive compounds. The demand for MAP technology is increasing continuously to extend shelf-life, improve product image, and avoid wastage of a large range of fresh fruits and vegetables. The technology is found very suitable for products because it uses permeable films and at a specific temperature, the rate of respiration changes the concentration of carbon dioxide and oxygen around the products. By applying this technology, the respiration of fresh vegetables and fruits can be minimized. The gases most frequently used in MAP are N_2, CO_2, and O_2. The accepted limit of O_2 in MAP for fruits and vegetables is 1–5%, concerning the safety and quality [61–63].

5.4 Conclusion and Future Dimension

In the recent era, the first choice of the customers depends on new packaging technologies with upgraded innovations and developments. Continuous R&D activities are being running toward active and intelligent packaging materials to achieve dynamically and environment-friendly packaging pieces of stuff. Nevertheless, both the active and intelligent packaging technologies are very demanding and widely used for food products. Additionally, the resurgence in R&D will also improve the quality of the food packaging materials.

References

1. Yam, K.L., Takhistov, P.T., and Miltz, J. (2005). Intelligent packaging: Concepts and applications, *J. Food Sci.*, **70(1)**, pp.1–10.
2. Vanderroost, M., Ragaert, P., Devlieghere, F., and Meulenaer, B.D. (2014). Intelligent food packaging: The next generation, *Trends Food Sci. Technol.*, **39**, pp. 47–62.
3. Ishrat, M., Nayik, G.A., Dar, S.M., and Nanda, V. (2018). Novel food packaging technologies: Innovations and future prospective, *J. Saudi Soc. Agric. Sci.*, **17(4)**, pp. 454–462.

4. Cruz, A.G.D., Faria, J.D.A.F., and Dender, A.G.F.V. (2007). Packaging systems and probiotic dairy foods, *Food Res. Int.*, **40**, pp. 951–956.
5. Conte, A., Angiolillo, L., Mastromatteo, M., and Del Nobile, M.A. (2013). Food Industry, Chapter 13 "Technological options of packaging to control food quality", Muzzalupo I. (ed.), InTech, Rijeka, Croatia, pp. 354–379.
6. Kotsianis I.S., Giannou V., and Tzia C. (2002). Production and packaging of bakery products using MAP technology, *Trends Food Sci. Technol.*, **13**, pp. 319–324.
7. Sanchez-Garcia, M.D., Lopez-Rubio, A., and Lagaron, M.J. (2010). Natural micro and nanobiocomposites with enhanced barrier properties and novel functionalities for food biopackaging applications, *Trends Food Sci. Technol.*, **21**, pp. 528–536.
8. Greig, W. (2007). Evolutionary advances in conventional packaging technologies, Integrated Circuit Engineering Corporation, *Spri. Sci. Busi. Med.*, pp. 1–37.
9. Ankiel, M. and Grzybowska-Brzezińska, M. (2020). Informative value of packaging as a determinant of food purchase, *Market Scientific Res. Org.*, **36(2)**, pp. 31–44.
10. Labuza, T.P. and Breene, W.M. (1989). Applications of "active packaging" for improvement of shelf-life and nutritional quality of fresh and extended shelf-life foods 1, *J. Food Proc. Preserv.*, **13(1)**, pp. 1–69.
11. Brody, A.L., Budny, J.A., and Rooney, M.L. (1995). Active food packaging, Chapter 6 "Interactive packaging involving sachet technology", Rooney, M.L. (ed.), Blackie, Glasgow, the U.K., pp. 143–172.
12. Murat, O. and John, D.F. (2004). Active food packaging technologies, *Critical Rev. Food Sci. Nutri.*, **44(3)**, pp. 185–193.
13. Brody, A.L., Strupinsky, E.P., and Kline, L.R. (2001). Active packaging for food applications, CRC Press, pp. 263.
14. Popowicz, R. and Lesiów, T. (2014). Zasada Działania Innowacyjnych Opakowań Aktywnych w przemyśle żywnościowym, Artykuł przeglądowy [The principle of innovative active packaging operation in the food industry: A review paper], *NaukiInżynierskiei Technologie*, **1(12)**, pp. 82–101.
15. Wyrwa, J., Barska, A. (2017). Innovations in the food packaging market: Active packaging, *Eur. Food Res. Technol.*, **243**, pp. 1681–1692.
16. Dainelli, D., Gontard, N., Spyropoulos, D., Zondervan-van den Beuken, E., and Tobback, P. (2008). Active and intelligent food packaging: Legal aspects and safety concerns, *Trends Food Sci. Technol.*, **19**, pp. 103–112.

17. Han, J.W., Ruiz-Garcia, L., Qian, J.P., and Yang, X.T. (2018). Food packaging: A comprehensive review and future trends, *Food Sci. Food Safety*, **17(4)**, pp. 860–877.

18. Youssef, A.M. (2013). Polymer nanocomposites as a new trend for packaging applications, *Polymer-Plastics Tech. Eng.*, **52(7)**, pp. 635–660.

19. Dobrucka, R. and Cierpiszewski, R. (2014). Active and intelligent packaging food –Research and development – A review, *Polish J. Food Nutri. Sci.*, **64(1)**, pp. 7–15.

20. Schumann, B. and Schmid, M. (2018). Packaging concepts for fresh and processed meat –Recent progresses, *Innovative Food Science & Emerging Technologies*, **47**, pp. 88–100.

21. Barska, A. and Wyrwa, J. (2017). Innovations in the food packaging market – Intelligent packaging – A review, *Czech J. Food Sci.*, **35(1)**, pp. 1–6.

22. Borowy, T. and Kubiak, M.S. (2008). Active and intelligent packaging – A new look into the future, *Meat Econ.*, **3 (60)**, pp. 32–34.

23. Han, J.H., Ho, C.H., and Rodrigues, E.T. (2005). Intelligent packaging, Chapter 9 "Innovations in food packaging", Han, J.H. (ed.), Academic Press, pp. 138–155.

24. Vermeiren, L., Devlieghere, F., van Beest, M., de Kruijf, N., and Debevere, J. (1999). Developments in the active packaging of foods, *Trends Food Sci. Technol.*, **10(3)**, pp. 77–86.

25. Janicki, A. (2013). Opakowania Aktywnei Inteligentne [Active and intelligent packaging], *SystemyLogistyczne Wojsk*, **39**, pp. 81–93.

26. Nowacka, M. and Niemczuk, D. (2012). Nowoczesne Materiały I wyrobyprzeznaczone do kontaktu z żywnością orazichwpływnabez pieczeństwożywności, *Opakowanie*, **6**, pp. 64–69.

27. Malhotra, B., Keshwani, A., and Kharkwal, H. (2015). Antimicrobial food packaging: Potential and pitfalls, *Front. Microbiol.*, **611(6)**, pp. 1–9.

28. Rooney, M.L. (2005). Innovations in food packaging, Chapter 5 "Introduction to active food packaging technologies", Han, J.H. (ed.), Academic Press, pp. 63–79.

29. Komonwatthanapong, W., Manuspiya, H., Chandracha, A., and Pandejpong, T. (2019). Identification and validation of decision factors for selecting smart food packaging technology: A case of Thailand's food industry, *The Open Psy. J.*, **12**, pp. 25–34.

30. European Food Safety Authority (EFSA) (2009). Guidelines on submission of a dossier for safety evaluation by the EFSA of active or intelligent substances present in active and intelligent materials and articles intended to come into contact with food, *EFSA J.*, **7(8)** 1208, pp. 1–11.

31. Heising, J.K., Dekker, M., Bartels, P.V., and Van Boekel, M.A. (2014). Monitoring the quality of perishable foods: Opportunities for intelligent packaging, *Crit. Rev. Food Sci. Nutr.*, **54(5)**, pp. 645–654.

32. Vanderroost, M., Ragaert, P., Devlieghere, F., and De Meulenaer, B. (2014). Intelligent food packaging: The next generation, *Trend. food Sci. Technol.*, **39(1)**, pp. 47–62.

33. Kruss, S., Hilmer, A.J., Zhang, J., Reuel, N.F., Mu, B., and Strano, M.S. (2013). Carbon nanotubes as optical biomedical sensors, *Adv. drug deliv. Rev.*, **65(15)**, pp. 1933–1950.

34. da Silva, J.C.E. and Goncalves, H.M. (2011). Analytical and bioanalytical applications of carbon dots, *TrAC Trends Anal. Chem.*, **30(8)**, pp. 1327–1336.

35. Sikarwar, S., Yadav, S.B., Yadav, A.K., and Yadav, B.C. (2014). Nanocomposite material for packaging of electronic goods, *Int. J. Scient. Innov. Res.*, **1(2)**, pp. 93–108.

36. Atzori, L., Iera, A., and Morabito, G. (2010). The internet of things: A survey, *Comp. Net.*, **54(15)**, pp. 2787–2805.

37. Kuswandi, B., Wicaksono, Y., Abdullah, A., Heng, L.Y., and Ahmad, M. (2011). Smart packaging: Sensors for monitoring of food quality and safety, *Sens. & Instrumen. Food Qual.*, **5**, pp. 137–146.

38. Ortega-Rivas, E. (2012). Non-thermal food engineering operations, Chapter 15 "Protective and preserving food packaging", Springer Science & Business Media, pp. 325–336.

39. López-de-Dicastillo, C., Gómez-Estaca, J., Catalá, R., Gavara, R., and Hernández-Muñoz, P. (2012). Active antioxidant packaging films: Development and effect on lipid stability of brined sardines, *Food Chem.*, **131(4)**, pp. 1376–1384.

40. Hammond, S.T., Brown, J.H., Burger, J.R., Flanagan, T.P., Fristoe, T.S., Mercado-Silva, N., Nekola, J.C., and Okie, J.G. (2015). Food spoilage, storage, and transport: Implications for a sustainable future, *BioScience*, **65(8)**, pp. 758–768.

41. Riva, M., Piergiovanni, L., and Schiraldi, A. (2001). Performances of time–temperature indicators in the study of temperature exposure of packaged fresh foods, *Pack. Tech. Sci.: An Int. J.*, **14(1)**, pp. 1–9.

42. Nura, A. (2018). Advances in food packaging technology – A review, *J. Posthar. Technol.*, **6(4)**, pp. 55–64.

43. Brody, A.L. (2008). Packaging by the numbers, *Food Tech.*, **62(2)**, pp. 89–91.

44. Caon, T., Martelli, S.M., and Fakhouri, F.M. (2017). New trends in the food industry: Application of nanosensors in food packaging, Nanobio, Academic Press, pp. 773–804.

45. Neethirajan, S. and Jayas, D.S. (2011). Nanotechnology for the food and bioprocessing industries, *Food Bioprocess Technol.*, **4**, pp. 39–47.

46. Srivastava, A.K., Dev, A., and Karmakar, S. (2018). Nanosensors and nanobiosensors in food and agriculture, *Environ Chem. Lett.*, **16**, pp. 161–182.

47. Cushen, M., Kerry, J., Morris, M., Cruz-Romero, M., and Cummins, E. (2012). Nanotechnologies in the food industry – Recent developments, risks and regulation, *Trends Food Sci. Technol.*, **24(1)**, pp. 30–46.

48. Berekaa, M.M. (2015). Nanotechnology in food industry: Advances in food processing, packaging and food safety, *Int. J. Curr. Microbiol. App. Sci.*, **4(5)**, pp. 345–357.

49. Alfadul, S.M. and Elneshwy, A.A. (2010). Use of nanotechnology in food processing, packaging and safety: A review, *African J. Food, Agriculture, Nutri. Develop.*, **10(6)**, pp. 2719–2739.

50. Alhendi, A. and Choudhary, R. (2013). Current practices in bread packaging and possibility of improving bread shelf-life by nanotechnology, *Int. J. Food Sci. Nutr.*, **3**, pp. 55–60.

51. Raheem, D. (2013). Application of plastics and paper as food packaging materials – An overview, *Emirat. J. Food Agri.*, pp. 177–188.

52. Bratovčić, A., Odobašić, A., Ćatić, S., and Šestan, I. (2015). Application of polymer nanocomposite materials in food packaging, *Croat. J. Food Sci. Technol.*, **7(2)**, pp. 86–94.

53. Radusin, T.I., Ristić, I.S., Pilić, B.M., and Novaković, A.R. (2016). Antimicrobial nanomaterials for food packaging applications, *Food Feed Res.*, **43(2)**, pp. 119–126.

54. Garcia, C.V., Shin, G.H., and Kim, J.T. (2018). Metal oxide-based nanocomposites in food packaging: Applications, migration, and regulations, *Trends Food Sci. Technol.*, **82**, pp. 21–31.

55. Guilbert, S., Cuq, B., and Gontard, N. (1997). Recent innovations in edible and/or biodegradable packaging materials, *Food Additi. Contami.*, **14(6 & 7)**, pp. 741–751.

56. Perez-Perez, C., Regalado-González, C., Rodríguez-Rodríguez, C.A., Barbosa-Rodríguez, J.R., and Villaseñor-Ortega, F. (2006). Incorporation of antimicrobial agents in food packaging films and coatings, *Adv. Agri. Food Biotechnol.*, **11**, pp. 193–216.
57. Šuput, D.Z., Lazić, V.L., Popović, S.Z., and Hromiš, N.M. (2015). Edible films and coatings: Sources, properties and application, *Food Feed Res.*, **42(1)**, pp. 11–22.
58. Bourtoom, T. (2008). Edible films and coatings: Characteristics and properties, *Int. Food Res. J.*, **15(3)**, pp. 237–248.
59. Vodnar, D.C., Pop, O.L., Dulf, F.V., and Socaciu, C. (2015). Antimicrobial efficiency of edible films in food industry, *Notu. Bot. Hor. Agro. Cluj-Nap.*, **43(2)**, pp. 302–312.
60. Mahalik, N.P. (2014). Advances in packaging methods, processes and systems, *Challenges*, **5(2)**, pp. 374–389.
61. Pawar, P.A. and Purwar, A.H. (2013). Biodegradable polymers in food packaging, *Am. J. Eng. Res.*, **2(5)**, pp. 151–164.
62. Martínez-Romero, D., Guillén, F., Castillo, S., Valero, D., and Serrano, M. (2003). Modified atmosphere packaging maintains quality of table grapes, *J. Food Sci.*, **68(5)**, pp. 1838–1843.
63. Jain, S., Dhake, K., and Kohli. D. (2020). Research trends in horticulture sciences, 12 Ed., "Research trends in modified atmosphere packaging: An overview", AkiNik Publications.

Chapter 6

Edible Coatings: Recent Advancements on Protein- and Polysaccharide-Based Films

Anjali Gupta,[a] Divya Tripathy,[a] Meenu Aggarwal,[b] and Pooja Agarwal[a]

[a] *Division of Chemistry, School of Basic & Applied Sciences, Galgotias University, Greater Noida, Uttar Pradesh, India*
[b] *Department of Chemistry, Aggarwal College, Ballabgarh, Faridabad, Haryana, India*
anjaligupta@galgotiasuniversity.edu.in

Nowadays, the food industry has gained considerable attention by consumers and improved food quality has become priority for the ones involved in manufacturing, packaging, selling, and purchasing of food items. In this regard, edible films and coatings have drawn interest to improve the quality and safety of food products. The edible materials processed into thin layers and coated on the food products, in the form of coatings and films, which play a significant role in their storage, marketing, and selling. These coatings and films protect the eatables from physical or mechanical damage and also possess chemical and microbiological activities thereby increasing the shelf life of the food products.

Biomaterials in Food Packaging
Edited by Mohd Yusuf and Shafat Ahmad Khan

ISBN 978-981-4877-98-5 (Hardcover), 978-1-003-25678-6 (eBook)
www.jennystanford.com

6.1 Introduction

Edible coating is a thin layer used for the packaging of food items and can be safely consumed along with it. In recent years, biodegradable edible films or coatings have considerably grabbed attention to extend the shelf life of fresh fruits and minimally processed products. Polymer materials, which are obtained as by-products and products of agriculture, may be used as an alternative in the form of edible coatings to reduce postharvest loss. The major advantages with these biodegradable films and coatings are reduction of plastic packaging, retention of aroma and moisture, and restriction of lipid migration between food components. These properties differentiate them from other conventional packaging materials. It has been observed that most of the synthetic commercial packages hold average lifetime of around 200 years in marine environment; whereas, edible films decompose readily providing an environment-friendly solution.

Why do we require edible films? The food that directly comes from the nature is stored, transported, and then distributed to supermarkets and warehouses or to processing facilities. Hence, it takes appreciable time for products to finally reach the consumers, due to which the deterioration of their flavor, nutritional value, and other physiological reactions take place. This gives rise to the discovery of edible films and coatings, which cover the natural skin of fruits and remain as one of the most cost-effective ways to maintain the fresh food quality and safety. The demand for quality and processed food products, storage techniques, and increased awareness toward environmental protection has led researchers to focus on edible and biodegradable coatings [1]. Microbial attack on the upper layer of food is a major concern in fresh as well as processed items. In this regard, packaging with antimicrobial coatings minimizes or inhibits the growth of microbes on the food surface [2–4]. Edible coating is a semi-permeable layer over food surface that acts as a barrier to air, moisture, and microbes, increases the shelf life of the products [5–7], retains their nutrients and aroma, and caters to the needs of the food industry and consumer demand for fresh fruits and vegetables [8]. These days, varied edible coatings have been applied successfully for the preservation of strawberries, tomatoes, apples, oranges, grapefruit, capsicum, and many more. Hence, edible coatings could be one of the alternatives to control fresh quality, spoilage, and pathogenic bacteria. The different constituents and properties of coatings have been reviewed by varied researchers [9–12]. This article emphasizes the progress and development of protein- and polysaccharide-based coatings to preserve fresh vegetables and fruits, with an update of current research findings (Figure 6.1).

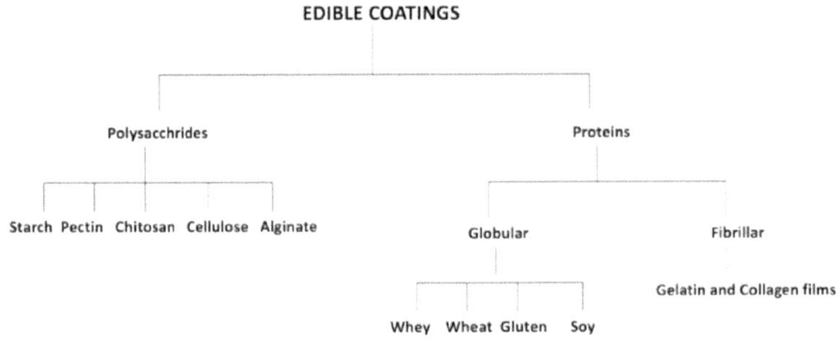

Figure 6.1 Types of edible coatings.

6.2 Protein-Based Edible Films and Coatings

Proteins function as promoters of varied biochemical processes and tissue building. They majorly exist in the form of fibrous or globular proteins as main classes having different physical and chemical behaviors. Due to their nutritional value, abundance availability, and other desirable mechanical and transparency properties, protein-based films have been widely explored [13].

In a research article, the effect of edible films from defatted sesame protein was discussed, and the consequences of variation in pH, concentration, plasticizer, and temperature were studied. It was found that all these variables drastically affected the properties of films like tensile strength, solubility, and water vapor permeability. Maximum tensile strength and low solubility was observed at pH 12, protein concentration 9%, and plasticizer 10% at temperature 90 °C; while at 50 °C, thermogravimetric analysis (TGA) scan showed good thermal stability for optimized film [14]. Potato proteins are considered to be of great nutritional quality as compared to other proteins due to their composition of balanced amino acids, and especially, high amount of lysine (around 8%). Moreover, they also possess superior functional properties like high foaming and emulsifying capacities. Recently, mechanical and barrier performance of potato protein isolate has been investigated to prove its appropriateness as a potential and new source for bio-based edible films. Potato protein isolate plasticized with glycerol possesses less cross-linking and less solubility in water as compared to whey protein at pH 7. The mechanical and barrier performance is also found to be 80%, as compared to whey protein [15]. Another source of edible films that is cheaper and has high nutritional value, but still unexplored, is legume protein. Montalvo-Paquini et al. developed a film from concentrated bean protein obtained from seven Mexican varieties (*Phaseolus vulgaris L.*) with glycerol used as a plasticizer. After evaluating various physical, mechanical, and chemical properties, it was concluded that this legume family can be considered as a potential alternative for biodegradable edible films [16].

6.2.1 Globular Proteins

6.2.1.1 Whey-based coatings and films

Whey protein constitutes around 20% soluble component of cow milk and its by-products obtained during manufacturing of casein and cheese. But, due to the high cost for its processing and transportation and large water content, whey is still underused. Although, researchers have started taking interest in whey protein, due to its highly nutritious constituents like α-lactalbumin and β-lactoglobulin and their integration with antipathogenic oils make whey-based coatings perfect for various food applications.

Sharma et al. found that edible films made up of whey protein and pectin integrated with cinnamon essential oil showed appreciable antibacterial activity with better physical parameters. The antibacterial activity for the two films were against *Listeria monocytogenes* (zones of inhibition 14.56 mm and 26.88 mm, respectively), *Escherichia coli* (zones of inhibition 11.66 mm and 15.55 mm, respectively), and *Staphylococcus aureus* (zones of inhibition 13.78 mm and 26.39 mm, respectively), which are food associated pathogens. Hence, as an alternative to sustainable packaging material, the whey-based film can be exclusively used in processing of food items like meat, vegetables, and bakery products [17]. In another research article, films made up of whey protein isolate were integrated with walnut oil and almond oil (with concentrations 1% and 0.5 %, respectively) to improve properties of the films. The combination increased the films' opacity, carbon dioxide and oxygen permeability, as well as improved their hydrophobic character. The addition of oil also decreased the water vapor permeability, swelling, and surface hydrophilicity of the films developed from whey protein [18]. Another whey-based coating used by researchers for the preservation of fresh cut apples, potatoes, and carrots was blend of whey and pectin protein film that was prepared with transglutaminase. A significant reduction in the weight loss was observed in the coated samples even after 10 days. The cross-linked blended coating also inhibited microbial growth in all the analyzed samples as well as preserving their phenolic components [19].

Nanocomposites were formed from gelatin and whey protein obtained from orange peel with different concentrations and Cloisite 30B (dry whey protein 5% w/w). The prepared edible films were biodegradable, as well as at 21% v/v concentration, they were highly antibacterial in nature with higher tensile strength and aqueous solubility, but with low transparency and moisture content. However, addition of Cloisite 30B and tripolyphosphate cross-linkers improved their transparency and other properties [20]. In another article, researchers prepared biodegradable films with whey protein isolate and glycerol, and the effects of variables, pH, whey protein isolate (WPI) to glycerol ratio, and sonication amplitude were analyzed. It was concluded that the packaging film was produced by WPI/glycerol in the ratio of 1.5 at pH 7.08 with sonication amplitude of 100 to obtain 0.275698 mm film thickness, with density 1.85064 gm/cm^3, transparency 2.32937, and moisture content 27.7102% [21].

6.2.1.2 Wheat gluten-based coatings and films

Wheat gluten, a by-product of the cereal processing industry, is another edible biopolymer that can be explored in the polymer market. But, its applications are limited due to its susceptibility to water and moisture that makes the polymer dimensionally unstable, prone to microbial attack with substandard mechanical properties. To increase its water and moisture resistance properties, chemical modification of wheat gluten by using cross-linkers was carried out, but still the majority of layers applied to wheat gluten lacked the appropriate adhesion.

Taking these problems into consideration, Das et al. tested a cross-linker Jeffamine diamine on wheat gluten for adhesion of polyethylene terephthalate film to make it more water resistant. The resultant layer also resolved the issue of dimensional instability to great extent [22]. It was also observed that gluten quality affects the properties of the film. Sharma et al. prepared gluten-based films from two Indian wheat cultivars like RAJ-3765 and C-306 with glycerol having different concentrations and their properties were compared. Water and acid solubility and water vapor transmission rate were found to be lower in case of C-306 [23].

In another research article, lipid phase was incorporated in wheat gluten-based edible films and its effect on their functional properties was studied [24]. It was very well established with the studies that the resultant films showed decreased hydrophilicity, water sorption and water transfer, decreased rigidity, and increased extensibility.

6.2.1.3 Soy protein-based coatings and films

Another potential isolate that can be explored as a renewable and biodegradable packaging material is soy protein isolate. But, its applications are limited due to high sensitivity to moisture and inferior mechanical properties. These properties are found to be enhanced by the incorporation of poly-vinyl alcohol, 1,2,3-propanetriol-diglycidyl-ether, and halloysite nanotubes by casting method without disturbing its transparency [25].

Carpine et al. developed and studied edible films based on emulsions whereby soy lecithin and virgin coconut oil were added to soy protein isolate and was plasticized with glycerol. Consequently, the moisture content decreased and elongation increased for the films containing these components. The peroxide value was also increased rapidly for olive oil that was stored in these film sachets, but did not attain the maximum value as recommended by Codex Alimentarius [26].

During slaughter process, meat can be contaminated with pathogenic bacteria that can cause serious illness and even death. In an article, antimicrobial films were studied using soy protein isolate, sodium lactate, Nisin, and Ethylenediamine tetraacetic acid (EDTA). These films showed bacteriostatic effect on meat against *Salmonella*, *E. coli,* and *Bacillus cereus* having zones of inhibition 33.33 mm, 32 mm, and 32.33 mm, respectively. The films extended the shelf life of meat by 3–6 days as compared to ordinary wraps [27].

In another article, inhibitory activities of soy-based edible coatings incorporated with thyme and oregano essential oils at different concentrations were evaluated. The studies were carried out on fresh beef against *E. coli* O157:H7, *S. aureus, and L. monocytogenes* during refrigeration. Maximum antibacterial activity was found against *S. aureus with addition of 3% essential oils to soy protein isolate films. These films* did not only control the pathogenic bacteria, but also found to enhance color stability along with satisfactory sensory characteristics [28].

Nowadays, attention has been drawn by bio-nanocomposite for coating eggshells, due to deterioration in egg quality. Yu et al. developed and studied the effect of integration of pine needle extracts and cellulose nanocrystals with soy protein on various properties of the coatings. The study clearly indicated that these coatings showed decreased moisture content and increased tensile strength with potential antioxidant activities [29]. Another edible coating that was found to enhance the apricots' shelf life is soybean protein isolate chitosan stored at 2 °C. This coating not only significantly prevented weight loss and decreased firmness in apricots, but also improved textural properties by inhibiting pectin degradation [30].

6.2.2 Fibrillar Proteins

6.2.2.1 Gelatin- and collagen-based coatings and films

Skin of goat is one of the by-products of slaughtering house that could be explored as a raw material for edible films. Said et al. combined collagen extracted from goatskin with glycerol and found improved properties of edible films like thickness and elongation at break and tensile strength [31].

Gelatin is also an important resource of protein obtained from collagen hydrolysis that can be explored to be used as edible films. Yang et al. studied the extraction of gelatin coating from duck feet, combined with cinnamon bark oil, and evaluated its physical properties and enhancement of microbial safety during packaging of cherry tomatoes. This coating was found to reduce the growth of *Salmonella typhimurium* on cherry tomatoes while storage, as well as inhibit the color change [32]. Aitboulahsen et al. integrated an essential oil *Mentha pulegium* (1%) with gelatin-based coating and studied its effect on different physicochemical parameters of strawberries in cold storage. The edible coating was found to inhibit various properties responsible for the deterioration of postharvest strawberries, such as weight loss, color, firmness, total phenolic content, total soluble solids, pH, and microbial activity [33]. In another article, citric acid was incorporated in gelatin-based coating and found that it inhibited the growth of *E. coli* [34]. It was concluded that films based on gelatin may emerge as potential biopolymers as compared to other protein-based films.

Wang et al. studied various properties of collagen-based films mixed with sodium alginate. It was concluded that the thermal stability and viscosity of the films were significantly improved with better tensile strength (26.49 MPa), water vapor permeability (1.79×10^{-10} g·cm^{-1}·s^{-1}·Pa^{-1}), and oxygen permeability (3.77×10^{-5} cm^{3}·m^{-2}·d^{-1}·Pa^{-1}) [35].

6.3 Polysaccharide-Based Edible Films and Coatings

Other commonly discussed natural polymers are polysaccharides that are extensively used for the preparation of edible coatings. They include starch, cellulose, pectin, chitosan, and alginate, and their chemical structures are given in Figure 6.2. They are considered to be potential oxygen blocker, but on the contrary being hydrophilic in nature, they possess reduced water vapor barrier properties. These coatings increase the shelf life of vegetables, fruits, and meat products by drastically inhibiting darkening of the surface, dehydration, and oxidative rancidity.

6.3.1 Starch-Based Coatings and Films

The most abundant natural carbohydrate polymer is starch that comprises amylose and amylopectin. The coatings made up of starch are found to be tasteless, odorless, transparent, and good O_2 and CO_2 barrier [36, 37]. However, they show poor water vapor barrier, due to their hydrophilicity, non-thermoplasticity, and poor mechanical properties [38–40].

Zuo et al. developed corn wheat starch based bilayer films incorporated with zein protein, and their physical and mechanical properties, antioxidant activity, and water vapor permeability were evaluated. It was found that the bilayer film exhibited excellent antioxidant activity, highly compact and smoother surface, with better cross-section that made it an excellent packaging material for seasoning, oil sauces, and dry vegetables [41]. In another research article, thermoplastic starch films were produced from different sources with plasticizer as glycerol and their physical and chemical properties were compared. It was found that films with high amylose content showed improved mechanical properties that might be attributed to highly crystalline domains [42]. Similar results were obtained when rice starch and corn starch were added; i.e. amylose content was increased to produce bioplastics. They were also found to possess better biodegradability and improved tensile properties [43]. Addition of nanoemulsion of nutmeg oil and glycerol further enhanced the functional and mechanical properties of the films. They were also found to show bactericidal activities against *E. coli* and *S. aureus* [44].

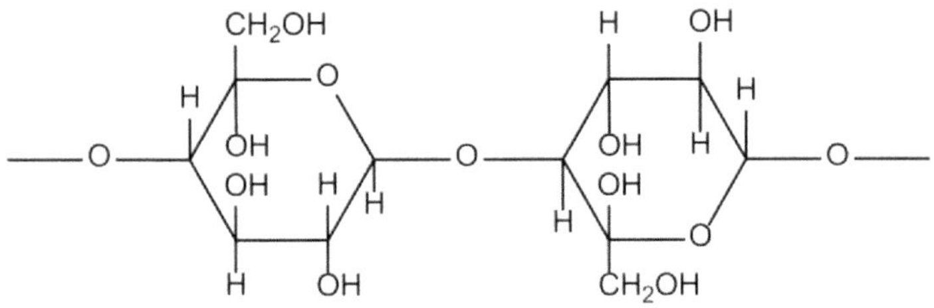

Cellulose

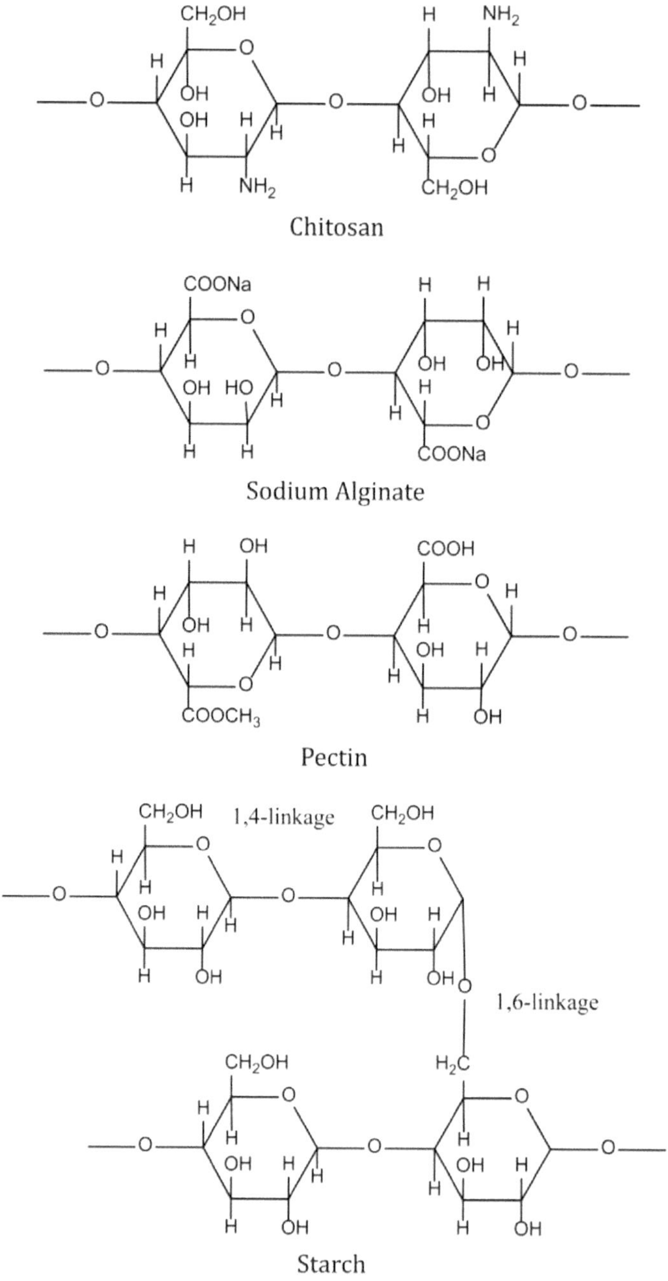

Figure 6.2 Chemical structures of some polysaccharides.

6.3.2 Cellulose-Based Coatings and Films

Da Silva et al. prepared different types of coatings incorporated with nanocrystals of cellulose, lemongrass essential oils, and glycerol. The coatings were applied on strawberries during refrigeration and their effects on acidity, weight loss, pH, total soluble content, and anthocyanin content were studied after different durations. It was found that as compared to uncoated fruits, the coatings not only minimized the weight loss and titratable acidity related to storage time, but at the same time controlled the increase of anthocyanin content and total soluble solids content, which were otherwise remarkably higher in uncoated fruits [45]. Similar results were obtained for the coatings based on cellulose nanofibers, hydroxypropyl methyl cellulose, glycerol, and *k*-carrageenan applied on *Vitis vinifera* L. grape surface for 41 days. It showed around 30–34% less water vapor permeability and weight loss, as compared to uncoated grapes, suggested an extension of shelf life of the product [46].

Moreover, hydroxypropyl methylcellulose and methylcellulose have been broadly used to decrease absorption of oil during deep frying foodstuffs. Methylcellulose has been applied on confectionery food items to block lipid passage [10].

6.3.3 Chitosan-Based Coatings and Films

Chitosan, derived from chitin, comprises exoskeleton of mollusks and crustaceans, and is produced by deacetylation of chitin in alkaline solution. It is a natural, safe, allergen-free, and sustainable polymer with improved medicinal properties [47]. Chitosan coatings have been successfully applied in the food industry, due to its properties [10] of forming continuous coating layer on food items. Kumar et al. recently reviewed its physicochemical properties and applications in the food industry [48].

Xinyu tangerine, a citrus fruit found in China with plentiful nutrients is prone to fungal infection by *Penicillium italicum* that reduces its commercial value. Recently, Chen et al. studied the effect of chitosan (CS)-based edible coating mixed with antimicrobial agent, i.e. fruit *Ficus hirta Vahl.* extract (FFE) on Xinyu tangerines at 5 °C. Antifungal activities were observed against *P. italicum*, with EC_{50}

value 12.543 mg/mL in vivo. It was also found that FFE–CS edible coating exhibited great reduction of titrable acid, total soluble solid, and ascorbic acid content, due to the reduction in weight loss, fruit decay rate, malondialdehyde content, and respiration rate at 5 °C. Furthermore, the activities of defensive enzymes like peroxidase, superoxide dismutase, and phenylalanine ammonia lyase were higher in the fruits coated with FFE–CS [49].

In another research article, the effect of chitosan-polyvinyl pyrrolidine coating incorporated with salicylic acid was studied on the shelf life of guava fruits at different color maturity stages. The edible coating was found to improve the antioxidant activities as well as browning parameters like total phenolic content, phenylalanine ammonia lyase, and polyphenol oxidase [50]. Similar effect was observed when CS/carboxymethyl/Moringa leaf extract coating was applied on avocado fruit. This coating was efficiently found to reduce the moisture loss, respiration rate, and firmness that significantly improved the shelf life and quality of the fruit [51]. Treviño-Garza et al. studied the extended shelf life and retained quality of cut fruit when CS/linseed or CS/Aloe mucilage edible coatings were applied on freshly cut pine apple [52]. Similar results were obtained when mango fruit was coated with CS/beeswax coating that improved its shelf life and maintained firmness up to three weeks, as compared to the uncoated fruit [53].

Another edible coating investigated for the increasing shelf life of blueberries was the blend of chitosan, quinoa protein, and sunflower oil at 4 °C. It was observed that this coating significantly controlled the growth of yeast and molds during the storage period of 32 days. Its elongation and tensile strength at break were also found to be 117.2% ± 7% and 0.45 ± 0.29 MPa, respectively, with water vapor permeability of $3.3 \times 10^{-12} \pm 4.0 \times 10^{-13}$ g s^{-1} m^{-1} Pa^{-1} [54].

6.3.4 Pectin-Based Coatings and Films

Pectin is a soluble component of plant fiber derived from cell walls of plants. It is also referred to as a detoxifying agent and prebiotics, as it exerts beneficial effects in the gastrointestinal system. Several researchers used pectin for layer-by-layer coating to increase the shelf life of freshly cut cantaloupe, pear, mango, peach, avocado, and cucumber [55–57]. Maftoonazad et al. studied the effect of pectin-based coating on the physical, chemical, and physiological properties of lime fruits. The moisture loss in the coated fruits was found to be lower as compared with the uncoated fruits along with better firmness, color retention, and other quality parameters [58]. In a research article, pectin-based coating was proved to improve the nutritional quality of dried red guava, as it acted as a barrier to carotenoid oxidation and also retained the total phenolic content of the fruit [59]. In another article, pectin gels were used as coatings to increase the lifetime of strawberries. Better results were obtained in terms of quality, color parameters, and humidity loss in coated strawberries [60]. In order to preserve black raspberry, pectin coating, enriched with thyme and lemon essential oils at 5 °C, was applied and its effect was studied for a period of 12 days. Due to its highly sensitive nature to fungal attack, high water content, and metabolic activity, the postharvest life of black raspberry is very short. The shelf life of the coated fruit was increased, due to delayed degradation of vitamin C, anthocyanin, and controlled loss of fruit moisture and weight, thereby preserved the quality and appearance of the fruit [61]. Other studied coatings that were designed to increase the shelf life of tomatoes were pectin, and beetroot- and corn flour-based coatings. After the application of these coatings, significant increase in the shelf life of tomatoes along with retention of quality was observed. This was attributed to the controlled loss of weight and other phenolic acids, respiration rate, ripening index, firmness, and antioxidant activity (2,2-diphenyl-1-picryl-hydrazyl-hydrate [DPPH]) [62].

6.3.5 Alginate-Based Coatings and Films

Alginate is a polysaccharide of great concern as an effective biopolymer film or coating constituent, due of its distinctive colloidal characteristics such as thickening, gel forming, film forming, and emulsion stabilizing agent. Vieira et al. observed the antipathogenic effect of alginate-based edible coating combined with essential oils against *Salmonella enterica* and *Listeria monocytogenes* on apples. It was shown that alginate coating enriched with citral and eugenol was found to be antipathogenic, thereby ensured the food safety and enhanced the shelf life in case of apples [63]. It was also observed that coatings made up of alginate (3%) and chitosan mixed with extract of olive leaves retarded the ripening process when applied over sweet cherries. This increased the shelf life of the fruit and retained the phytochemicals with better antioxidative property even after 20 days [64].

Similar results were obtained when strawberries and fresh cut cantaloupe were covered with sodium alginate-based layer. The coating significantly reduced the loss of water in cantaloupe pieces and, on the contrary, promoted water loss in case of strawberries [65]. Yin et al. also proved the significance of multilayer alginate-based coatings for the preservation of mangoes [66]. It was found that mangoes coated with five layers showed delayed decay, extended shelf life, nutrition preservation, and improved physical and chemical parameters. Alginate- and chitosan-based coatings incorporated with cinnamon oil microcapsules when applied over mangoes layer-by-layer helped in maintaining their commercial and nutritional value.

6.4 Conclusion

Since vegetables and fruits preservation is becoming a challenging task for the entire world, researchers have acknowledged edible films and coatings as a solution. This may be attributed to their characteristics, i.e. they are biodegradable, naturally occurring, renewable, and cheap, and also contain essential nutrients and natural antimicrobials. The researchers have also mixed various essential oils in coatings to minimize its shortcomings and maximize their health benefits (Table 6.1). These coatings and films not only possess varied nutraceutical features, but also retain the physical and chemical aspects of the fruits and vegetables to prevent postharvest losses. The coatings act as perfect barriers to moisture content, CO_2, O_2, and water vapor; thereby maintaining the aroma, flavor, texture, color, look, and nutritional value of the products, and protecting them from microbes. Keeping these benefits in mind, this article emphasizes on different polysaccharides and protein-based edible coatings to preserve various food items by increasing their shelf life.

Table 6.1 Effect of essential oils incorporation in various edible coatings

S. No.	Basic component of edible coating	Essential oil used	Food product	Activity	Reference
1	Whey protein	Cinnamon essential oil	Meat, bakery products, vegetables	Antibacterial activity against *Listeria monocytogenes*, *Escherichia coli,* and *Staphylococcus aureus*	[17]
2	Pectin	Cinnamon essential oil	Meat, bakery products, vegetables	Antibacterial activity against *L. monocytogenes*, *E. coli,* and *S. aureus*	[17]
3	Soy protein	Thyme and oregano essential oils	Fresh beef	Antibacterial activity against *E. coli O157:H7, S. aureus, and L. monocytogenes*	[28]
4	Gelatin	*Mentha pulegium* (1%)	Strawberries	Antimicrobial activity	[33]
5	Chitosan	*Ficus hirta Vahl.* extract	*Xinyu tangerine* (Chinese citrus fruit)	Antifungal activity against *Penicillium italicum* (EC_{50} = 12.543 mg/mL)	[49]
6	Pectin	Thyme and lemon essential oils	Black raspberry	Antifungal	[61]
7	Alginate	Citral and eugenol essential oils	Apples	Antibacterial activity against *Salmonella enterica* and *L. monocytogenes*	[63]

References

1. De Azeredo, H.M.C. (2012). Advances in fruit processing technologies, Rodrigues, S. and Fernandes, F.A.N. (eds.), Chapter 14 "Edible coatings", CRC Press, Boca Raton USA, pp. 345–361.
2. Frazão, G.G.S., Blank, A.F., and de Aquino Santana, L.C.L. (2017). Optimisation of edible chitosan coatings formulations incorporating Myrcia ovata Cambessedes essential oil with antimicrobial potential against foodborne bacteria and natural microflora of mangaba fruits, *LWT-Food Sci. Technol.*, **79**, pp.1–10.
3. Appendini, P. and Hotchkiss, J.H. (2002). Review of antimicrobial food packaging, *Innov. Food Sci. Emerg. Technol.*, **3(2)**, pp. 113–126.[4. Elsabee, M.Z. and Abdou, E.S. (2013). Chitosan based edible films and coatings: A review, *Mater. Sci. Eng.: C*, **33(4)**, pp. 1819–1841.
5. Trinetta, V. (2018). Reference module in food science, Chapter "Obsolete: Definition and function of food packaging", Elsevier, USA. https://doi.org/10.1016/B978-0-08-100596-5.03319-9.
6. Wyrwa, J. and Barska, A. (2017). Innovations in the food packaging market: Active packaging, *Eur. Food Res. Technol.*, **243(10)**, pp. 1681–1692.
7. Gonçalves, A.A. and Rocha, M.D.O.C. (2017). Safety and quality of antimicrobial packaging applied to seafood, *MOJ Food Process Technol.*, **4(1)**, 00079. DOI:10.15406/mojfpt.2017.04.00079.
8. Baldwin, E.A. (1994). Edible coatings and films to improve food quality, Baldwin, E.A., Hagenmaier, R., Bai, J., and Krochta, J.M. (eds.), Chapter 2 "Edible coatings for fresh fruits and vegetables: Past, present, and future", CRC Press, Boca Raton, USA, pp. 25–60.
9. Said, N.S. and Sarbon, N.M. (2019). Active antimicrobial food packaging, Var, I. and Uzunla, S. (eds.), Chapter 4 "Protein-based active film as antimicrobial food packaging: A review", InTech Open, London, pp. 53–70.
10. Hassan, B., Chatha, S.A.S., Hussain, A.I., Zia, K.M., and Akhtar, N. (2018). Recent advances on polysaccharides, lipids and protein based edible films and coatings: A review, *Int. J. Biolog. Macromolecules*, **109**, pp. 1095–1107.
11. Raghav, P.K., Agarwal, N., and Saini, M. (2016). Edible coating of fruits and vegetables: A review, *Int. J. Scientific Res. Modern Edu.*, **1(1)**, pp. 188–204.
12. Hammam, A.R. (2019). Technological, applications, and characteristics of edible films and coatings: A review, *SN Appl. Sci.*, **1(6)**, pp. 632. DOI: 10.1007/s42452-019-0660-8 .

13. Kaewprachu, P. and Rawdkuen, S. (2014). Mechanical and physico-chemical properties of biodegradable protein-based films: A comparative study, *Food Appl. Biosci. J.*, **2(1)**, pp. 15–30.

14. Sharma, L. and Singh, C. (2016). Sesame protein based edible films: Development and characterization, *Food Hydrocolloids*, **61**, pp. 139–147.

15. Schäfer, D., Reinelt, M., Stäbler, A., and Schmid, M. (2018). Mechanical and barrier properties of potato protein isolate-based films, *Coatings*, **8(2)**, pp. 58 (1–16).

16. Montalvo-Paquini, C., Avila-Sosa, R., López-Malo, A., and Palou, E. (2018). Preparation and characterization of proteinaceous films from seven Mexican common beans (Phaseolus vulgaris L.), *J. Food Quality*, 2018, 9782591, pp. 1–8. DOI: 10.1155/2018/9782591.

17. Sharma, D., Dhanjal, D.S., and Mittal, B. (2017). Development of edible biofilm containing cinnamon to control food-borne pathogen, *J. Appl. Pharmaceut. Sci.*, **7(01)**, pp. 160–164.

18. Galus, S. and Kadzińska, J. (2016). Whey protein edible films modified with almond and walnut oils, *Food Hydrocolloids*, **52**, pp. 78–86.

19. Marquez, G.R., Di Pierro, P., Mariniello, L., Esposito, M., Giosafatto, C.V., and Porta, R. (2017). Fresh-cut fruit and vegetable coatings by transglutaminase-crosslinked whey protein/pectin edible films, *LWT Food Sci. Technol.*, **75**, pp. 124–130.

20. Shams, B., Ebrahimi, N.G., and Khodaiyan, F. (2019). Development of antibacterial nanocomposite: Whey protein-gelatin-nanoclay films with orange peel extract and tripolyphosphate as potential food packaging, *Adv. Polym. Technol.*, 2019, 1973184, pp. 1–9. DOI: 10.1155/2019/1973184.

21. Abhishaben, M.S., Chandegara, V.K., Jithender, B., and Pankajkumar, M.S. (2019). Whey protein isolate based biodegradable food packaging film as affected by protein to glycerol ratio, pH and sonication amplitude, *Int. J. Curr. Microbiol. App. Sci.*, **8(3)**, pp. 895–909.

22. Das, O., Loho, T., Capezza, A., Lemrhari, I., and Hedenqvist, M. (2018). A novel way of adhering PET onto protein (wheat gluten) plastics to impart water resistance, *Coatings*, **8(11)**, pp. 388.

23. Sharma, N., Khatkar, B.S., Kaushik, R., Sharma, P., and Sharma, R. (2017). Isolation and development of wheat based gluten edible film and its physicochemical properties, *Int. Food Res. J.*, **24(1)**, pp. 94–101.

24. Rocca-Smith, J.R., Marcuzzo, E., Karbowiak, T., Centa, J., Giacometti, M., Scapin, F., and Debeaufort, F. (2016). Effect of lipid incorporation on functional properties of wheat gluten based edible films, *J. Cereal Sci.*, **69**, pp. 275–282.

25. Liu, X., Song, R., Zhang, W., Qi, C., Zhang, S., and Li, J. (2017). Development of eco-friendly soy protein isolate films with high mechanical properties through HNTS, PVA, and PTGE synergism effect, *Scientific Rep.*, **7**(1), pp. 1–9.

26. Carpiné, D., Dagostin, J.L.A., Bertan, L.C., and Mafra, M.R. (2015). Development and characterization of soy protein isolate emulsion-based edible films with added coconut oil for olive oil packaging: Barrier, mechanical, and thermal properties, *Food Bioprocess Technol.*, **8(8)**, pp. 1811–1823.

27. Liu, Y., Zhang, H., Chi, Y., Wu, Y., Cao, W., Li, T., and Xu, L. (2017). Properties of soy protein isolate antimicrobial films and its application in preservation of meat, *Emirates J. Food Agric.*, **8**, pp. 589–600.

28. Yemiş, G.P. and Candoğan, K. (2017). Antibacterial activity of soy edible coatings incorporated with thyme and oregano essential oils on beef against pathogenic bacteria, *Food Sci. Biotechnol.*, **26(4)**, pp. 1113–1121.

29. Yu, Z., Sun, L., Wang, W., Zeng, W., Mustapha, A., and Lin, M. (2018). Soy protein-based films incorporated with cellulose nanocrystals and pine needle extract for active packaging, *Ind. Crops Prod.*, **112**, pp. 412–419.

30. Zhang, L., Chen, F., Lai, S., Wang, H., and Yang, H. (2018). Impact of soybean protein isolate-chitosan edible coating on the softening of apricot fruit during storage, *LWT Food Sci. Technol.*, **96**, pp. 604–611.

31. Said, M.I., Erwanto, Y., and Abustam, E. (2016). Properties of edible film produced using combination of collagen extracts of bligon goatskin with glycerol, *Am. J. Anim. Vet. Sci.*, **11(4)**, pp. 151–159.

32. Yang, S.Y., Lee, K.Y., Beak, S.E., Kim, H., and Song, K.B. (2017). Antimicrobial activity of gelatin films based on duck feet containing cinnamon leaf oil and their applications in packaging of cherry tomatoes, *Food Sci. Biotechnol.*, **26(5)**, pp. 1429–1435.

33. Aitboulahsen, M., Zantar, S., Laglaoui, A., Chairi, H., Arakrak, A., Bakkali, M., and Hassani Zerrouk, M. (2018). Gelatin-based edible coating combined with mentha pulegium essential oil as bioactive packaging for strawberries, *J. Food Quality*, **2018**, 8408915, pp. 1–12. https://doi.org/10.1155/2018/8408915.

34. Uranga, J., Puertas, A.I., Etxabide, A., Dueñas, M.T., Guerrero, P., and de la Caba, K. (2019). Citric acid-incorporated fish gelatin/chitosan composite films, *Food Hydrocolloids*, **86**, pp. 95–103.

35. Wang, Z., Hu, S., and Wang, H. (2017). Scale-up preparation and characterization of collagen/sodium alginate blend films, *J. Food Quality*, **2017**, 4954259, pp. 1–10.

36. Jiang, Z., Neetoo, H., and Chen, H. (2011). Efficacy of freezing, frozen storage and edible antimicrobial coatings used in combination for control of Listeria monocytogenes on roasted turkey stored at chiller temperatures, *Food Microbiol.*, **28(7)**, pp. 1394–1401.

37. Neetoo, H., Ye, M., and Chen, H. (2010). Bioactive alginate coatings to control Listeria monocytogenes on cold-smoked salmon slices and fillets, *Int. J. Food Microbiol.*, **136(3)**, pp. 326–331.

38. Basiak, E., Lenart, A., and Debeaufort, F. (2017). Effect of starch type on the physico-chemical properties of edible films, *Int. J. Biolog. Macromolecules*, **98**, pp. 348–356.

39. Piñeros-Hernandez, D., Medina-Jaramillo, C., López-Córdoba, A., and Goyanes, S. (2017). Edible cassava starch films carrying rosemary antioxidant extracts for potential use as active food packaging, *Food Hydrocolloids*, **63**, pp. 488–495.

40. Das, M. and Chowdhury, T. (2016). Heat sealing property of starch based self-supporting edible films, *Food Packaging Shelf Life*, **9**, pp. 64–68.

41. Zuo, G., Song, X., Chen, F., and Shen, Z. (2017). Physical and structural characterization of edible bilayer films made with zein and corn-wheat starch, *J. Saudi Soc. Agric. Sci.*, **18**(3), pp. 324–331.

42. Domene-López, D., García-Quesada, J.C., Martin-Gullon, I., and Montalbán, M.G. (2019). Influence of starch composition and molecular weight on physicochemical properties of biodegradable films, *Polymer*, **11(7)**, pp. 1084.

43. Marichelvam, M.K., Jawaid, M., and Asim, M. (2019). Corn and rice starch-based bio-plastics as alternative packaging materials, *Fibers*, **7(4)**, pp. 32.

44. Aisyah, Y., Irwanda, L.P., Haryani, S., and Safriani, N. (2018). Characterization of corn starch-based edible film incorporated with nutmeg oil nanoemulsion, *IOP Conf. Ser.: Mater. Sci. Eng.*, **352**, pp. 012050.

45. Da Silva, I.S.V., Prado, N.S., De Melo, P.G., Arantes, D.C., Andrade, M.Z., Otaguro, H., and Pasquini, D. (2019). Edible coatings based on apple pectin, cellulose nanocrystals, and essential oil of lemongrass: Improving the quality and shelf life of strawberries (Fragaria Ananassa), *J. Renew. Mater.*, **7(1)**, pp. 73–87.

46. Silva-Vera, W., Zamorano-Riquelme, M., Rocco-Orellana, C., Vega-Viveros, R., Gimenez-Castillo, B., Silva-Weiss, A., and Osorio-Lira, F. (2018). Study of spray system applications of edible coating suspensions based on hydrocolloids containing cellulose nanofibers on grape surface (Vitis vinifera L.), *Food Bioprocess Technol.*, **11(8)**, pp. 1575–1585.

47. Bornet, A. and Teissedre, P.L. (2005). Applications and interest of chitin, chitosan and their derivatives in enology, *J. Int. Sci. Vigne Vin.*, **39**, pp.199–207.

48. Kumar, S., Ye, F., Dobretsov, S., and Dutta, J. (2019). Chitosan nanocomposite coatings for food, paints, and water treatment applications, *Appl. Sci.*, **9(12)**, pp. 2409.

49. Chen, C., Nie, Z., Wan, C., and Chen, J. (2019). Preservation of Xinyu tangerines with an edible coating using Ficus hirta Vahl fruits extract-incorporated chitosan, *Biomolecules*, **9(2)**, pp. 46.

50. Lo'ay, A.A. and Taher, M.A. (2018). Influence of edible coatings chitosan/PVP blending with salicylic acid on biochemical fruit skin browning incidence and shelf life of guava fruits cv.'Banati', *Scientia Horticult.*, **235**, pp. 424–436.

51. Tesfay, S.Z. and Magwaza, L.S. (2017). Evaluating the efficacy of moringa leaf extract, chitosan and carboxymethyl cellulose as edible coatings for enhancing quality and extending postharvest life of avocado (Persea americana Mill.) fruit, *Food Packaging Shelf Life*, **11**, pp. 40–48.

52. Treviño-Garza, M.Z., García, S., Heredia, N., Alanís-Guzmán, M.G., and Arévalo-Niño, K. (2017). Layer-by-layer edible coatings based on mucilages, pullulan and chitosan and its effect on quality and preservation of fresh-cut pineapple (Ananas comosus), *Postharvest Biol. Technol.*, **128**, pp. 63–75.

53. Eshetu, A., Ibrahim, A.M., Forsido, S.F., and Kuyu, C.G. (2019). Effect of beeswax and chitosan treatments on quality and shelf life of selected mango (Mangifera indica L.) cultivars, *Heliyon*, **5(1)**, pp. e01116.

54. Abugoch, L., Tapia, C., Plasencia, D., Pastor, A., Castro-Mandujano, O., López, L., and Escalona, V.H. (2016). Shelf-life of fresh blueberries coated with quinoa protein/chitosan/sunflower oil edible film, *J. Sci. Food Agric.*, **96(2)**, pp. 619–626.

55. Martiñon, M.E., Moreira, R.G., Castell-Perez, M.E., and Gomes, C. (2014). Development of a multilayered antimicrobial edible coating for shelf-life extension of fresh-cut cantaloupe (Cucumis melo L.) stored at 4 °C, *LWT-Food Sci. Technol.*, **56(2)**, pp. 341–350.

56. Moalemiyan, M., Ramaswamy, H.S., and Maftoonazad, N. (2012). Pectin-based edible coating for shelf-life extension of ataulfo mango, *J. Food Process Eng.*, **35(4)**, pp. 572–600.

57. Moalemiyan, M. and Ramaswamy, H.S. (2012). Quality retention and shelf-life extension in Mediterranean cucumbers coated with a pectin-based film, *J. Food Res.*, **1(3)**, pp. 159.

58. Maftoonazad, N. and Ramaswamy, H.S. (2019). Application and evaluation of a pectin-based edible coating process for quality change kinetics and shelf-life extension of lime fruit (Citrus aurantifolium), *Coatings*, **9(5)**, pp. 285.

59. Todisco, K.M., Janzantti, N.S., Santos, A.B., Galli, F.S., and Mauro, M.A. (2018). Effects of temperature and pectin edible coatings with guava by-products on the drying kinetics and quality of dried red guava, *J. Food Sci. Technol.*, **55(12)**, pp. 4735–4746.

60. Muñoz-Labrador, A., Moreno, R., Villamiel, M., and Montilla, A. (2018). Preparation of citrus pectin gels by power ultrasound and its application as an edible coating in strawberries, *J. Sci. Food Agric.*, **98(13)**, pp. 4866–4875.

61. Abdi, S. and Bakhshi, D. (2019). Pectin edible coatings enriched with essential oils of lemon and thyme to improve storage of black raspberry, *J. Postharvest Technol.*, **7(3)**, pp. 18–29.

62. Chaturvedi, K., Sharma, N., and Yadav, S.K. (2019). Composite edible coatings from commercial pectin, corn flour and beetroot powder minimize post-harvest decay, reduces ripening and improves sensory liking of tomatoes, *Int. J. Biolog. Macromolecules*, **133**, pp. 284–293.

63. Vieira, A.I., Guerreiro, A., Antunes, M.D., Miguel, M.D.G., and Faleiro, M.L. (2019). Edible coatings enriched with essential oils on apples impair the survival of bacterial pathogens through a simulated gastrointestinal system, *Foods*, **8(2)**, pp. 00057. DOI: 10.3390/foods8020057.

64. Zam, W. (2019). Effect of alginate and chitosan edible coating enriched with olive leaves extract on the shelf life of sweet cherries (Prunus avium L.), *J. Food Quality*, 2019, 8192964, pp. 1–7. DOI: 10.1155/2019/8192964.

65. Senturk Parreidt, T., Lindner, M., Rothkopf, I., Schmid, M., and Müller, K. (2019). The development of a uniform alginate-based coating for cantaloupe and strawberries and the characterization of water barrier properties, *Foods*, **8(6)**, pp. 00203. DOI: 10.3390/foods8060203.

66. Yin, C., Huang, C., Wang, J., Liu, Y., Lu, P., and Huang, L. (2019). Effect of chitosan-and alginate-based coatings enriched with cinnamon essential oil microcapsules to improve the postharvest quality of mangoes, *Material*, **12(13)**, pp. 02039. DOI: 10.3390/ma12132039.

Chapter 7

Characterization of Cassava Starch-Zinc Nanocomposite Film for Food Packaging

Adeshina Fadeyibi

Department of Food and Agricultural Engineering,
Faculty of Engineering and Technology,
Kwara State University Malete,
P.M.B. 1530, Ilorin, Kwara State 234/North Central, Nigeria
adeshina.fadeyibi@kwasu.edu.ng

Performance of a pure thermoplastic film can be enhanced by adding nanoparticles of the size ranging 1–100 nm for application in food packaging. This research was carried out to develop a nanocomposite by mechanically homogenizing different proportions of cassava starch (1000 g), glycerol (45–55%, w/w), and zinc nanocomposites (0–2%, w/w). A dispersed mixture of 24 g of these products was then mixed with distilled water (600 ml) and heated for 30 min at boiling temperature to form a thermoplastic solution. The film was formed in different sizes (15, 16, and 17 μm thickness) by casting the solution on a 350 mm ×180 mm plastic mold of different depths (8, 10, and 12 mm). The performance of the film, including barrier, thermal, and structural properties,

Biomaterials in Food Packaging
Edited by Mohd Yusuf and Shafat Ahmad Khan

ISBN 978-981-4877-98-5 (Hardcover), 978-1-003-25678-6 (eBook)
www.jennystanford.com

was determined using standard methods. The results showed that the oxygen and water vapor decreased with the thickness and increased with the concentration of the glycerol. Plastic d-spacing increased with an increase in thickness, and this might indicate the ability of the material to intercalate and exfoliate at some points during a prolonged packaging application. It might also indicate the short-range order of the material constituents in the film for a better service performance. A small degradation of the film was observed between 30 °C and 60 °C. This indicated that the film was thermally stable and might be suitable for packaging applications, especially in the tropical climes. The information of the characterized attributes and optimization of the cassava starch zinc-nanocomposite films justified their alternative application to pure thermoplastic and conventional films for food packaging.

7.1 Introduction

Nanocomposite films are prepared from the homogeneous mixture of renewable raw materials, such as starch, proteins, or any polysaccharide, with a nontoxic plasticizer like glycerol in boiling water at 95 °C [1–4]. Normally, the constituent materials are chosen based on their cost and availability, in order not to interfere with the primary need of the products for food packaging. All kinds of the nanocomposite films, regardless of their source, should have enhanced service performance, and those made from the starch are preferred for economic reason. Due to this, starch is often used in the production of nanocomposite film [3, 4– 6]. However, despite the huge potentials of the starch-based biodegradable materials, the poor service performance in terms of their physical, mechanical, thermal, and structural attributes have greatly limited their applications [7]. Hence, the nanoparticles are often added during the formulation process of the film to provide the attributes required for the packaging applications. The nanoparticles are organic or inorganic materials whose particle size does not exceed 100 nm [8, 9]. These materials can interact with the other material constituents in the biodegradable film because of their large reactive surface area to volume ratio [5–7]. The resulting blend will thus influence the structural build-up of the molecules, thereby inducing the desired improvement of the film's properties [7].

Many different formulations of nanocomposite films have been reported for application in food packaging [1–7, 9, 10]. For instance, Dang and Yoksan [1] developed a thermoplastic starch from a blend of a plasticized starch and chitosan nanoparticles. Fadeyibi et al. [2, 5, 6] developed a biodegradable film from a blend of the cassava starch and zinc nanoparticles. Chung at al. [3] prepared a biodegradable film from a blend of starch and clay nanocomposites. Dai et al. [7] developed a biodegradable film from a blend of corn starch and taro starch nanoparticles. Tall et al. [9] prepared a biodegradable film from a blend of starch and Ni–Ti nanoparticles. However, the issue with the above-mentioned films is how to solve the possibility of material migration into the packaged food. For this reason, many other types of nanocomposite films are being developed to address this challenge. One way out research has suggested to use a health-friendly nanocomposite material for inclusion in the film formulation. Typically, zinc nanoparticles have been reported to possess this quality and can be used in the formulation of the nanocomposite film, but the quantity added must not exceed 2% (w/w) [2, 5, 6, 10, 11]. Therefore, the study on the characterization of the nanocomposite film prepared from the blend of starch, plasticizer, and zinc nanoparticles cannot be over emphasized.

Application of this class of material in the food packaging has been widely reported. For instance, Fadeyibi et al. [5, 11] used the cassava starch-zinc nanocomposite film to preserve the nutrition qualities, hinder microbial growth, and extend the shelf-lives of tomatoes, cucumber, and garden eggs. Also, Zhang et al. [10] proposed the use of zinc oxide nanoparticles as an effective antimicrobial agent for the use in the preservation of fruits and vegetables. However, the mechanism of this striking behavior of the nanocomposite film has not been reported. Therefore, there is a need to study the characteristics of the nanocomposite film to fully understand its behavior. The objective of this research was to synthesize and characterize the cassava starch-zinc nanocomposite film based on its physical, mechanical, thermal, and structural attributes.

7.2 Materials and Methods

7.2.1 Synthesis and Particle Size Analysis of Nanoparticles

The nanoparticles were synthesized using the sol-gel method described by Fadeyibi et al. [2, 7, 12] and Surya [13] at the processing and storage laboratory, Federal University of Technology, Minna. In this method, a 30 ml of distilled water was added to a solution of 20 ml triethanolamine and 2 ml ethanol, and mixed thoroughly to form a homogeneous solution. This was transferred into a 500 ml beaker containing a 0.5 M solution of 5.39 g zinc acetate di-hydrate and 50 ml distilled water. Another solution of 2 g ammonium hydroxide and 10 ml distilled water was added to the homogeneous mixture in the beaker, which was placed on a Bunsen burner to heat for 20 min. A white bulky precipitate was formed, which was washed 10 times with distilled water before filtration to obtain a clean white residue and dried in an air-circulated oven dryer for 8 h. Moreover, the size distribution of the resulting nanoparticle was determined using a Zetasizer equipment (ver. 7.01) at the Center of Genetic Engineering and Biotechnology, Federal University of Technology Minna, Nigeria. The size of the nanoparticles was found in the range of 4–6 nm.

7.2.2 Development of Nanocomposite Film

The nanocomposite film was developed by mechanically homogenizing different proportions of cassava starch (1000 g), glycerol (45–55%, w/w), and zinc nanocomposites (0–2%, w/w). The choice of the glycerol instead of other types of plasticizers for binding the starch and the nanoparticles together was basically because it did not pose any health hazard when used to prepare materials for food packaging [7, 14]. A dispersed mixture of 24 g of the products was then mixed with distilled water (600 ml) and heated for 30 min at boiling temperature to form a thermoplastic solution. The film was formed in different sizes (15, 16 and 17 μm thickness) by casting the solution on a 350 mm ×180 mm plastic mold of different depths (8, 10, and 12 mm). The film was thereafter carefully peeled off from the mold and sealed in the polyethylene pack to avoid possible hydration. The process flow diagram for the preparation of cassava starch nanocomposite film is shown in Figure 7.1 [2, 15].

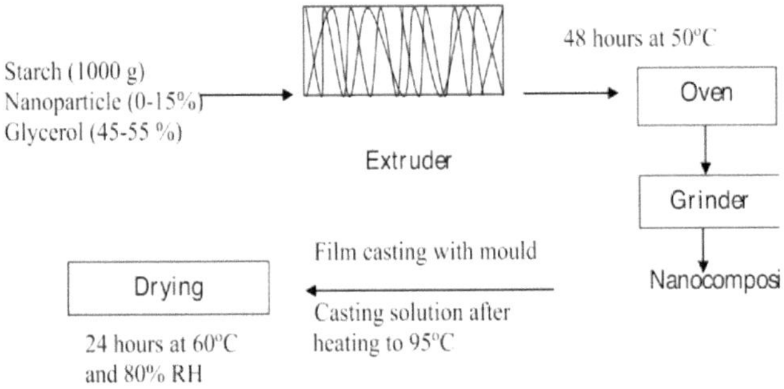

Figure 7.1 Process flow diagram for the development of the nanocomposite film.

7.2.3 Determination of Structural and Barrier Properties

A multipurpose X-ray diffractometer was used to study the structural properties, based on the full-width at half maximum and polymer d-spacing of the nanocomposite film, at the Sheda Science and Technology Complex (SHESTCO), Abuja, Nigeria. The range of orderliness of the molecular constituents in the film as well as the degree of intercalation and exfoliation processes in its structure was determined [2]. The ASTM Standard method was used to determine the permeability of the film due to water vapor [16]. This method involves placing an 8.75 cm diameter sample of the nanocomposite film inside a test dish filled with 50 ml deionized water at 100% relative humidity (RH). The setup was subjected to a temperature and RH of 25 °C and 65%, respectively. The test dish was arranged on a rotating turntable to expose it to the same atmospheric conditions for 2 h, before the initial weight of the sample was recorded. The final weight of the sample of the film was obtained after every 15 min, and average values and standard deviations were computed using the expression in Eq. (7.1) [16, 17].

$$W_{\text{vpt}} = \frac{\Delta m}{tA} \tag{7.1}$$

where, Δm is change in mass (g), t is the time (min), A is the area of the mouth of the test dish (m^2), and W_{vpt} = water vapor transmission rate (g^{-2}s^{-1}).

Thus, the water vapor permeability of each film sample was computed using Eq. (7.2) [16].

$$W_{vp} = \frac{W_{vpt}}{\Delta p} \times T \quad (7.2)$$

where T is the thickness of the film sample (mm), Δp the partial vapor pressure difference across the film, and W_{vp} the water vapor permeability (g/m.Pa.s).

Furthermore, the method described by Fadeyibi et al. [2] was used to determine the oxygen gas permeability of the nanocomposite film. In this method, 2 g of the film sample was placed inside an absorption column setup, which was stocked at one end, and stream of oxygen gas was passed into the column via the open end at the top [17, 18]. The concentration of the oxygen gas inside the column was measured using an electronic weighing balance for 1 h at an interval of 10 min. Thus, knowing the data of properties given in Table 7.1, the average value and standard deviation of the oxygen permeability of each sample of the nanocomposite film were computed using Eqs. (7.3–7.8).

Table 7.1 Data of properties used for the calculation of oxygen permeability

	Variable	Symbol	Value	SI unit
1	Volume of column	V_v	0.0174	cm^3
2	Density of oxygen	P_{02}	0.001331	g/m^3
3	Atmospheric pressure	P atm	101325	Pa
4	Saturated water vapor pressure	P^{sat}_{H20}	190.2× T °C	Pa
5	Area of film sample	A	0.000123	m^2
6	Atmospheric temperature	T	25	°C

$$[O_2] = \frac{mo_2}{1000M} \quad (7.3)$$

$$[O_2]_n = \frac{RH \times P^{sat}H_2O}{100RT} \quad (7.4)$$

$$[O_2]_t\% = \frac{O_2}{[O_2]_n} \times 100 \quad (7.5)$$

Therefore,

$$P_t = P_{atm} - P^{sat}H_2O \times T \quad (7.6)$$

$$\Delta PO_2 = ([O_2]_t - [O_2]_i)P_t \tag{7.7}$$

$$O_p = \frac{\Delta m \Delta y}{A\, \Delta t\, \Delta PO_2)} \tag{7.8}$$

where $[O_2]_t$ is the oxygen concentration inside the absorption column at time t, $[O_2]_i$ the initial column oxygen content, the mass of oxygen gas absorbed by the cassava starch nanocomposite film in the column, M the molar mass of oxygen gas (16 g/mol), R the molar gas constant, RH relative humidity (63%), T absolute temperature (25 + 273 K), O_p oxygen permeability (g/m.Pa.s), and Δy the thickness of the film (μm).

7.2.4 Determination of Thermal and Rheological Properties

A thermogravimetric analyzer was used to determine the thermal properties of the nanocomposite film. The weight degradation or loss of the film was determined by computing its difference in weight after exposure to a laboratory-controlled temperature 30–950 °C for 30 min. The rheological properties of the nanocomposite films were determined using a nano-mechanical testing machine (Model: Hysitron TI750) at the SHESTCO, Abuja, Nigeria. A record of the applied load on the film sample, depth of penetration of an indenter, and the corresponding indented area were determined by the machine's response. The values of the young modulus, hardness, activation volume, and tensile strength of the sample of the nanocomposite film were thereafter computed from a graph of the load versus the penetration depth or displacement (Figure 7.2.), and using the empirical relationships in Eqs. (7.9–7.14) [19].

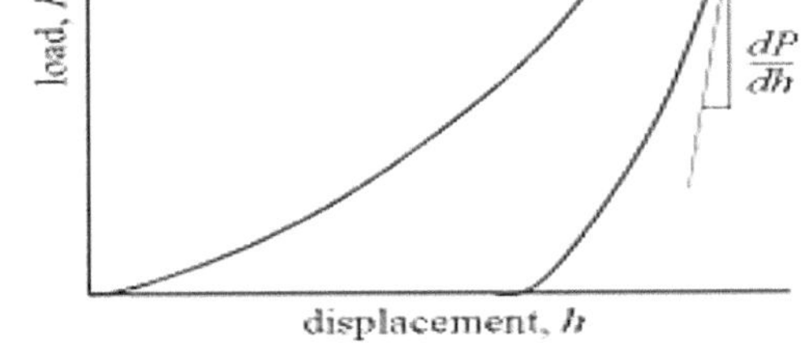

Figure 7.2 Load-displacement curve of a sample of nanocomposite film [19].

$$E = dP/dh \tag{7.9}$$

$$Y_r = \frac{S_2}{2\alpha} \times \sqrt{\frac{\pi}{A_i(h_a)}} \tag{7.10}$$

$$H = \frac{P_{max}}{A_p} \tag{7.11}$$

$$m = \frac{d \ln \sigma}{d \ln \dot{\delta}} \tag{7.12}$$

$$\dot{\upsilon} = \frac{9k_b \, T \ln \dot{\delta}}{dH} \tag{7.13}$$

$$T_s = \frac{P_{max}}{A_c} \tag{7.14}$$

where dP/dh is the slope of the load displacement curve (N/m), S stiffness of the contact of the indenter with film sample, α geometric constant of the material (= 1), h_a contact depth (μm), A_p projected area of the indenter (μm^2), Y_r the reduced modulus of the film sample (μN/mm^2), H hardness of the film sample (MPa), P_{max} the maximum applied load (μN), σ flow stress of the film due to indenter penetration (μN/mm^2), strain or drift rate of the film under indentation (nms^{-1}), m stress strain sensitivity of the flow (μN/mm^2), k_b Stephen Boltzmann's constant, T temperature of the film sample (°C), activation energy (kJ), A_c cross-sectional area of the film sample (mm^2), and T_s tensile strength of the film sample (N/mm^2).

7.3 Results and Discussion

7.3.1 Permeability of Nanocomposite Film

The study of water vapor permeability of food packaging materials is important for maintaining or extending the shelf life of the food product. The water vapor permeability coefficient indicates the amount of water vapor that permeates per unit of area and time in the nanocomposite film. The results of the permeability of the various categories of the nanocomposite film after 3 h exposure of water and oxygen gas are presented in Table 7.2. The variation of the water vapor and oxygen permeability with the concentration glycerol and the level of zinc nanoparticles are affected by the thickness of the film. The amount of water vapor, absorbed by the film, decreases with the concentration of the zinc nanoparticles and increases with the concentration of the glycerol. The molecular absorption behavior of the nanocomposite films may be influenced by the nanoparticles used. The water vapor permeability is low for the film without the nanoparticles and high for the film with 2% nanoparticles. The reason for this may be associated with the higher water absorption affinity of nanoparticles at higher concentrations. The addition of the nanoparticles to the film may have caused a modification of the polymer network of the film, thereby affecting its water vapor permeability. An increase in the glycerol concentration from 45% to 55% may even cause an increase in the mobility of the polymeric chains of the film [20]. The overall effect of the composition of the film on the water vapor permeability of the material can be enhanced with the addition of more nanoparticles and glycerol. The addition of more glycerol may decrease the intermolecular hydrogen bonds between the polymer chains, thereby increasing the flexibility of the nanocomposite film [20, 21]. The nanocomposite film is, therefore, an ideal material for packaging low-moisture food products because of its high-water vapor permeability coefficient.

The knowledge of oxygen gas permeability of the nanocomposite film is essential for understanding the rate of product respiration and shelf life. The amount of oxygen gas absorbed by the film decreases with the increase in concentrations of zinc nanoparticles and glycerol (Table 7.2). The oxygen permeability is generally higher for the film without the nanoparticles, and this decreases with the increase in level of nanoparticles and glycerol. The presence of the nanoparticles in the film is likely to alter the gas mobility, as they diffuse across the film. Although the reason for this behavior is not yet known, it is likely that the decrease in the porosity of the nanocomposite film as a result of an increase in amount of the nanoparticles may be responsible for this observation. It may also be related with the formation of a stable complex resulting from the filling up of the empty d-spaces of the zinc nanoparticles with the oxygen gas molecules [2]. Therefore, the rate of diffusion of the oxygen gas across the film layers may reduce due to the decrease in number of pore space or pinhole opening of the film. Thus, the nanocomposite film can help reduce the amount of atmospheric

Table 7.2 Permeability of cassava starch nanocomposite films for 3 h period

Permeability of cassava starch nanocomposite films (g/m.Pa.s)								
Film of 15 μm			Film of 16 μm			Film of 17 μm		
Treatments	water vapor permeability: WVP [× 10^{-10}]	OP = Oxygen permeability OP[× 10^{-11}]	Treatments	WVP [× 10^{-10}]	OP[× 10^{-11}]	Treatments	WVP [× 10^{-10}]	OP[× 10^{-11}]
45-0-15	2.240±1.076	1.568±0.084	45-0-16	1.723±0.101	1.521±0.018	45-0-17	0.613±0.023	1.240±0.002
45-1-15	2.360±0.049	1.530±0.016	45-11-6	1.730±0.119	1.500±0.017	45-1-17	0.797±0.006	1.109±0.082
45-2-15	2.450±1.027	1.437±0.030	45-2-16	1.837±0.112	1.287±0.081	45-2-17	0.813±0.053	1.192±0.008
50-0-15	3.170±0.001	1.347±0.041	50-0-16	1.840±0.008	1.169±0.009	50-0-17	1.221±0.003	1.151±0.043
50-1-15	3.380±0.001	1.320±0.087	50-1-16	1.873±0.001	0.886±0.004	50-1-17	1.323±0.002	0.633±0.058
50-2-15	3.420±0.000	1.280±0.011	50-2-16	1.874±0.002	0.789±0.005	50-2-17	1.340±0.001	0.614±0.063
55-0-15	3.450±0.000	1.184±0.020	55-0-16	2.032±0.001	0.657±0.004	55-0-17	1.424±0.003	0.502±0.004
55-1-15	3.640±0.001	0.917±0.050	55-1-16	2.040±0.001	0.589±0.017	55-1-17	1.428±0.004	0.486±0.014
55-2-15	3.680±0.003	0.807±0.075	55-2-16	2.059±0.002	0.481±0.006	55-2-17	1.513±0.001	0.456±0.023

Note: WVP, water vapor permeability; OP, Oxygen permeability.

oxygen gas influx into the system of the packaged food product. This may create a macro environment or modified atmosphere for the ideal storage or packaging of the food product.

Thickness is structurally related with the permeability of the nanocomposite film [22, 23]. The thickness of the film plays an important role in its absorption behavioral and characterization. Thus, the variations in barrier property of the nanocomposite film with an increase in its thickness are shown in Table 7.2. There is a clear decrease in the water vapor and the oxygen gas permeability from 2.240×10^{-10} to 0.613×10^{-10} g/m.Pa.s, with the increase in the thickness of the film. The permeability of the film is slightly higher as compared with the permeability of low- and high-density polyethene materials of similar thickness [23]. In a related investigation, Islam and Buschatz [24] observed an increase in the permeability of polyamide films with the increase in their thickness. Thus, the thinner the nanocomposite film, the higher will be the diffusion and water vapor coefficients, due to the less restriction of particle influx into the film layers.

7.3.2 Thermal Stability of Nanocomposite Film

A typical thermal degradation curve for different thicknesses of the cassava starch nanocomposite film, which were obtained for the film with 45% glycerol and 0% zinc nanoparticles, is shown in Figure 7.3. Similar trends were observed for the remaining 26 film formulations. Less than 2% of the weight of the film was degraded between 30 °C and 85 °C. The residual weights of the film with 45% glycerol and 15–17 μm thickness were 14%, 12%, and 9% at 30 °C.

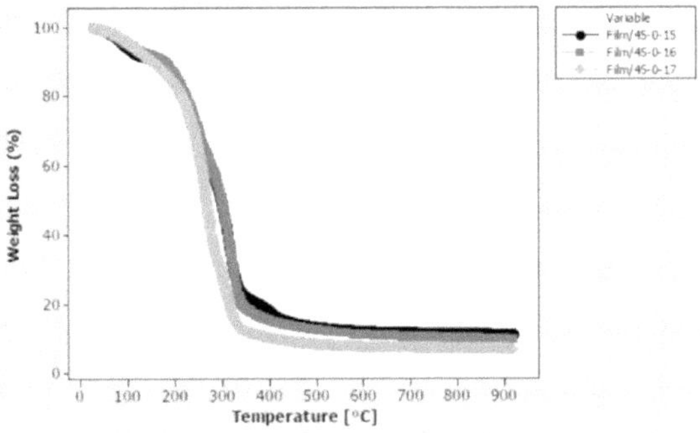

Figure 7.3 Thermal degradation curves of nanocomposite film.

The residual weights of the films with 45% glycerol and 1% zinc nanoparticles were 13% for 15 μm, 10% for 16 μm, and 12% for 17 μm. The films containing 45% glycerol and 2% zinc nanoparticles offered 17%, 16%, and 10% residual weight loss with the increase in thickness. Also, less than 2% of the weights of the films were degraded between 30 °C and 70 °C. Furthermore, only 2% of the weights of the cassava starch nanocomposite films were degraded between 30 °C and 50 °C, in the case of the films with 50% glycerol and 1% zinc nanoparticles. In general, since more than 80% of the films were degraded when subjected to temperature above their glass transition range, the residual weight represented almost complete decompositions of the films.

Beyond this temperature range, the cassava starch nanocomposite film may become unstable and the weight loss of the film may increase with the increase in temperature. The differences in the values of the glass transition temperatures of the films and their residual weights can be partly associated with their varied thicknesses, which affect the physical states, volumes, and shapes of the samples at higher temperature. It can also be argued that the increased concentrations of the glycerol and the zinc nanoparticles are partly responsible the difference in the glass transition temperature.This behavior agrees with the low-density polyethylene film and other synthetic polymers, which show a single mass loss step with significant decomposition [25]. Therefore, the knowledge of thermal properties of the nanocomposite film is necessary for understanding its decomposition pattern for packaging application, especially in a high temperature environment.

7.3.3 Mechanical Properties of Nanocomposite Film

The mechanical properties of the nanocomposite film may be necessary for studying the influence of defects such as pinholes that may affect other properties, such as the resistance to breakage and abrasion. The effects of nanoparticles and glycerol concentrations on the mechanical properties of the nanocomposite film are shown in Table 7.3. The reduced modulus and the strain energy decrease with the increase in the nanoparticles and glycerol concentrations. Also, the hardness, drift rate, tensile strength, and deformation increase with the increase in the concentrations of glycerol and zinc nanoparticles. Although the reason for this is not well known, from previous research and preliminary investigation [26–30], the morphology of the film and its displacement when load is applied are evaluated. This may create a large interface between the film constituents', thereby affecting its molecular mobility and relaxation behavior. This corroborates the findings of Park et al. [31], who reported that the polymer structure, plasticizer selection and concentration, molecular weight of the film forming materials, type of solvents, and film thickness affect the mechanical properties.

Furthermore, the results of the effects of thickness on the loading and unloading behaviors of cassava starch nanocomposite films are typically shown in Figure 7.4. The variations of the applied load on the cassava starch nanocomposite films with the displacement, observed in the nano-indentation analyses, are also shown in Figure 7.4. The strain energy absorbed by the cassava starch nanocomposite film is higher for the 15 μm film than for the 16 μm and 17 μm, thus a magnitude of 190 pNm is released during the loading part of the hysteresis loop. But, only 5 pNm of the strain energy is dissipated as heat during the unloading part of the hysteresis loop, as most of the energy is recovered in the process. It is possible that the energy dissipated as heat during the unloading part of the loop by the cassava starch nanocomposite films is lower than those of the pure thermoplastic films, irrespective of the concentration of the glycerol [32]. The energy dissipated also decreases with the thickness of the films.

Table 7.3 Mechanical properties of cassava starch nanocomposite films

Variables			**Mechanical properties of the films**						
g	np	t	E	H	φ	d	σ_t	δ	F
45	0	15	1.39	941.1	101.0	11.34	0.400	4040.2	50.00
45	1	15	2.47	765.6	80.80	3.862	0.000	3231.6	50.00
45	2	15	3.19	515.3	74.28	−7.63	0.000	2977.0	49.90
45	0	16	5.39	1.010	55.91	−1.57	0.000	2236.5	50.00
45	1	16	43.7	2.690	24.95	12.64	0.200	998.10	50.00
45	2	16	7.01	701.6	55.02	8.478	0.100	2200.7	50.00
45	0	17	2.76	163.7	103.3	−0.76	0.100	4132.8	50.00
45	1	17	6.34	14.11	155.4	−2.31	0.100	6214.7	50.00

(Continued)

Table 7.3 (*Continued*)

Variables			Mechanical properties of the films						
g	np	t	E	H	φ	d	σ_t	δ	F
45	2	17	10.2	1.480	42.01	−3.60	0.100	1680.5	50.00
50	0	15	4.73	433.9	68.75	5.46	0.100	2749.9	50.00
50	1	15	5.09	311.3	33.11	−3.05	0.200	1324.4	50.00
50	2	15	5.62	286.5	75.94	−0.17	0.100	3037.6	50.00
50	0	16	5.52	1.110	54.82	11.72	0.000	2192.8	50.00
50	1	16	92.4	73.50	14.54	−1.54	0.100	581.50	50.00
50	2	16	11.3	1.090	43.49	6.90	0.100	1739.5	50.00
50	0	17	17.2	1.410	36.78	24.3	0.100	1471.2	50.00
50	1	17	12.2	391.3	61.28	−3.84	0.200	2451.0	50.00
50	2	17	13.5	4.040	34.22	25.4	0.100	1368.8	50.00
55	0	15	168	67.86	9.556	−0.28	0.200	383.00	49.90
55	1	15	60.7	5.600	18.67	−1.41	0.200	746.80	50.00
55	2	15	45.2	4.980	20.8180	−10.1	0.100	832.70	50.00
55	0	16	41.3	33.87	22.0480	−0.70	0.100	881.90	50.00
55	1	16	167	67.86	9.55590	−0.28	0.200	383.00	49.90
55	2	16	12.9	7.060	542.909	−7.47	0.1	3620.6	299.9
55	0	17	43.40	35.51	21.5030	−2.56	0.1	860.10	50.00
55	1	17	185.2	11.43	11878	4.88	0.4	475.10	50.00
55	2	17	84.32	9.070	15170	−1.55	0.2	606.80	50.00

Note: = Drift rate (nm/s), = Maximum force (μN), = Maximum depth (nm), = Tensile strength (μN/mm^2), = Hardness (MPa), = Reduced modulus (μN/mm^2), = Strain energy (pNm), g = Glycerol concentration (%, w/w), np = Concentration of nanoparticles (%, w/w), t = Thickness of the film (μm).

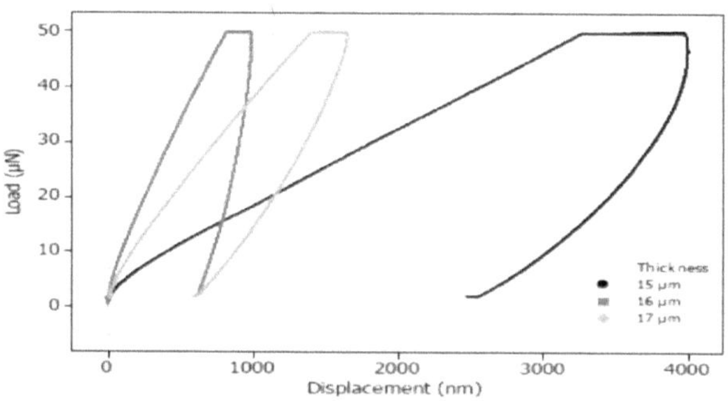

Figure 7.4 Load-displacement profiles of nanocomposite film [29].

7.3.4 Structural Stability of Nanocomposite Film

The results of some structural properties, showing the X-ray diffraction patterns of the cassava starch nanocomposite films of different thicknesses and concentrations of glycerol and nanoparticles, are typically shown in Figure 7.5. The peak values of the structural properties at 15 μm, 16 μm, and 17 μm film thicknesses, which were deduced from the X-ray diffraction patterns, are presented in Table 7.4.

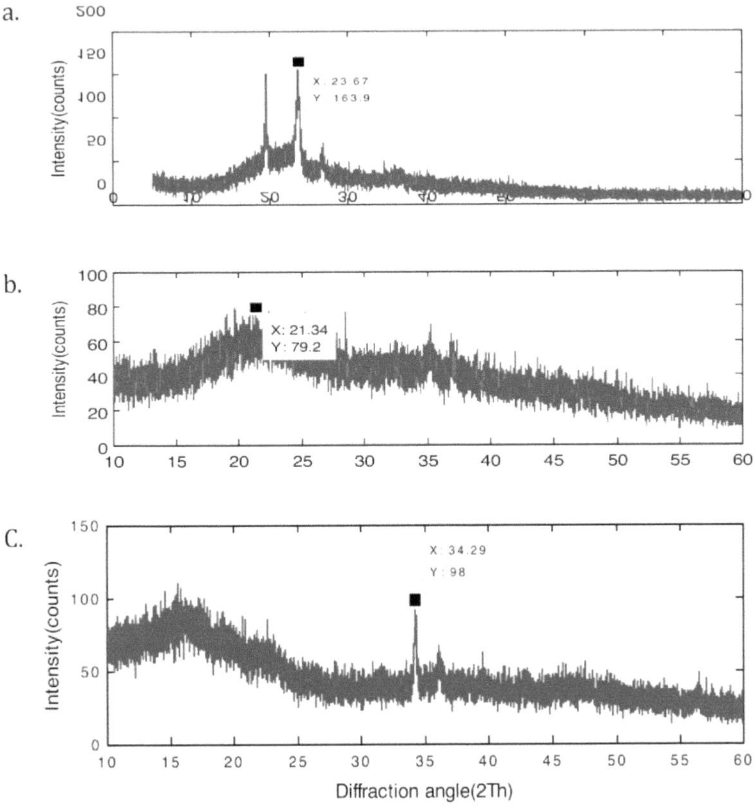

Figure 7.5 Profile of nanocomposite film with 2% nanoparticles: (a) 45% glycerol, (b) 50% glycerol, and (c) 55% glycerol.

Table 7.4 Peak values of cassava starch nanocomposite films

Variables			Structural properties of cassava starch Nanocomposite films				
			Position	**Height**	**FWHM**	**d-spacing**	**Relative intensity**
g	*np*	*t*	[°2Th]	[cts]	[°2Th]	[Å]	[%]
45	0	15	10.12	10.79	0.19	8.74	65.22
45	1	15	5.27	6.71	0.5	16.77	7.7
45	2	15	35.06	12.54	0.76	2.56	65.74
45	0	16	10.12	10.79	0.19	8.74	65.22
45	1	16	10.12	10.79	0.19	8.74	65.22
45	2	16	23.19	34.56	0.25	3.84	48.35
45	0	17	5.27	6.71	0.5	16.8	7.7
45	1	17	34.39	21.02	0.25	2.61	100
45	2	17	23.29	15.96	0.76	3.82	75.92
50	0	15	35.12	11.62	0.19	2.56	100
50	1	15	23.33	27.86	0.5	3.81	67.68
50	2	15	22.87	20.09	0.63	3.89	100
50	0	16	22.87	20.09	0.63	3.89	100
50	1	16	5.27	6.71	0.5	16.8	7.7
50	2	16	24.63	8.3	0.92	3.61	100
50	0	17	34.54	17.44	0.19	2.6	100
50	1	17	23.88	13.64	0.61	3.72	100
50	2	17	17.38	9.05	0.76	5.1	15.34
55	0	15	34.54	17.44	0.19	2.6	100
55	1	15	17.23	10.39	0.76	5.15	27.55
55	2	15	10.48	10.82	0.5	8.44	65.4
55	0	16	19.56	104.3	0.16	4.54	100
55	1	16	35.24	15.62	0.5	2.55	100
55	2	16	18.62	41.17	0.88	4.76	100
55	0	17	34.26	43.48	0.16	2.62	100
55	1	17	32.16	17.64	0.5	2.78	71.76
55	2	17	34.42	7.21	0.38	2.61	68.25

Note: FWHM, full width at half maximum; *g*, glycerol concentration (%, w/w); *np*, concentration of nanoparticles (%, w/w); *t*, thickness of the film (μm).

The X-ray diffraction is a commonly used technique for the evaluation of crystal structure because it relies on coherent scattering from many unit cells in a spatially time averaged fashion [33]. The effects of the intensity of incident radiation on the pure thermoplastic films of different thicknesses, containing 45% glycerol, are shown in Table 7.4. The diffraction patterns depicted show a defect in the 15 μm thermoplastic film, which occurs at a diffraction angle of 19.68 °2Th, unlike the other thermoplastic films that show no observable defect with diffraction. The effect of the intensity of the incident radiation on the nanocomposite film, which contains 45% glycerol, 1% zinc nanoparticle with thickness in the range of 15 to 17 μm, was shown in Table 7.4. The diffraction patterns depicted show defects in the 15 μm and 16 μm cassava starch nanocomposite films, which occur at diffraction angles of 19.62 °2Th and 19.68 °2Th, respectively; unlike the 17 μm film that shows no observable defect with diffraction. The effects of the intensity of incident radiation on the cassava starch nanocomposite films of different thicknesses, containing 45% glycerol and 2% zinc nanoparticles, are also shown in Table 7.4. The diffraction patterns depicted show defects in the 15 μm and 17 μm cassava starch nanocomposite films, which occur at diffraction angles of 23.2 °2Th and 19.44 °2Th, respectively; unlike the 16 μm film that shows no observable defect with diffraction. It can be seen, from the foregoing, that thickness is important in determining the crystalline structures of cassava starch nanocomposite films, in terms of the diffraction position along the films' geometry. The observed patterns for the variations of the intensity with the diffraction angles are relatively homogeneous. The patterns show relatively sharp peaks, which indicate that the nanocomposite film may be crystalline in its structure. This resembles the structure of a face-centered cubic zinc particle with a lattice constant of approximately 0.4085 nm and is consistent with the available literature value (0.4086 nm) [2, 27]. The film-forming materials and especially their structural cohesion have important effects on the structural properties of edible films and coatings [27]. In addition, an area of almost 100% structural defect intensity in the molecular structure of the nanocomposite film is found, as shown in Table 7.4. The intensity of the structure of the film depends on the material composition, as evident in the different patterns exhibited by the various film categories and treatments. The defect is more pronounced in the film with 55% glycerol than the film with either 50% or 45% glycerol concentration. Also, the values of the full width at half maximum can help further explain the size of the observed crystals in the film. An understanding of this phenomenon may be vital for the food packaging engineer in the designing of systems for the food packaging application. It may also aid in the understanding of barrier mechanism of nanocomposite film, which is yet to be reported.

7.4 Conclusion

The service performance analysis of the cassava starch-based zinc nanocomposite film is carried out to ascertain its suitability for packaging application. This is done by characterizing the film based on its mechanical, thermal, structural, and barrier properties. The results show that the amount of water vapor absorbed by the film decreases with the increase in concentrations of the zinc nanoparticles and the glycerol. The molecular absorption behavior of the nanocomposite films may be influenced by the nanoparticles used. The water vapor permeability is low for the film without the nanoparticles and high for the film with 2% nanoparticles. The residual weights of the films with 45% glycerol and 1% zinc nanoparticles are 13% for 15 μm, 10% for 16 μm, and 12% for 17 μm. The films containing 45% glycerol and 2% zinc nanoparticles offer 17%, 16%, and 10% residual weight loss with the increase in their thickness. The reduced modulus and the strain energy decrease with the increase in the nanoparticles and glycerol concentrations. The hardness, drift rate, tensile strength, and deformation increase with the rise in concentrations of glycerol and zinc nanoparticles. The diffraction patterns depicted show a defect in the 15 μm thermoplastic film, which occurs at a diffraction angle of 19.68 °2Th, unlike the other thermoplastic films, which show no observable defect with diffraction. Thus, the nanocomposite film has a potential for use as an active packaging material in the food processing.

References

1. Dang, K.M. and Yoksan, R. (2015). Development of thermoplastic starch blown by incorporating plasticized chitosan, *Carbohydr. Polym.*, **115**, pp. 575–581. doi: 10.1016/j.carbpol.2014.09.005.
2. Fadeyibi, A., Osunde, Z.D., Agidi, G., Idah, P.A., and Egwim, E.C. (2016). Development and optimisation of cassava starch-zinc-nanocomposite film for potential application in food packaging, *J. Food Process Technol.*, **7**, pp. 591–599. doi:10.4172/2157-7110.1000591
3. Chung, Y., Ansari, S., Estevez, L., Hayrapetyan, S., Giannelis, E.P., and Lai, H. (2010). Preparation and properties of biodegradable starch–clay nanocomposites, *Carbohydr. Polym.*, **79(2)**, pp. 391–396.
4. Huang, M.F. and Yu, J.G. (2006). Structure and properties of nanocomposites and their characteristics, *J. Appl. Polym. Sci.*, **99(2)**, pp. 170–176. doi: 1021/am506260j
5. Fadeyibi, A., Osunde Z.D, Egwim, E.C, and Idah P.A. (2017). Performance evaluation of cassava starch zinc nanocomposite film for tomatoes packaging, *J. Agric. Eng.*, **47(3)**, pp. 137–146.
6. Fadeyibi, A., Osunde, Z.D., and Yisa, M.G. (2017). Effects of glycerol and diameter of holes in breaker plate on performance of a screw mixer for nanocomposites, *Agric. Eng.*, **21(4)**, pp. 15–26.
7. Dai, L., Qui, C., Xiong, L., and Sun, Q. (2015). Characterisation of corn starch-based films reinforced with taro starch nanoparticles, *Food Chem.*, **174(1)**, pp. 82–88. doi: 10.1016/j.foodchem.2014.11.005
8. Mehyar, G.F. and Han, J.H. (2014). Physical and mechanical properties of high-amylose rice and pea starch films as affected by relative humidity and plasticiser, *J. Food Sci.*, **69(1)**, pp. 449–459. doi: 10.1111/j.1365- 2621.2004.tb09929.x
9. Tall, P.D., Ndiaye, S., Beye, A C., Zong, Z., Soboyejo, W.O., Lee, H.J., Ramirez, A.G., and Rajan, K. (2007). Nanoindentation of Ni–Ti thin films, *Mater. Manuf. Proc.*, **22(1)**, pp. 175–179.
10. Zhang, L., Jiang, Y., Ding, Y., et al. (2008). ZnOnanofluids – A potential antibacteria agent, *Prog. Nat. Sci.*, **18**, pp. 939–944.
11. Fadeyibi, A., Osunde, Z.D., and Yisa, M.G. (2020). Effects of period and temperature on quality and shelf-life of cucumber and garden-eggs packaged using cassava starch-zinc nanocomposite film, *J. Appl. Pack. Res.*, **12 (1)**, pp. 36–55.

12. Fadeyibi, A.,Yisa, M.G., Adeniji, F.A.,Katibi, K.K., Alabi, K.P., and Adebayo, K.R. (2018). Potentials of zinc and magnetite nanoparticles for contaminated water treatment, *Agric. Res.,* **39(2)**, pp. 175–180.

13. Surya, P.G. (2012). Synthesis and characterisation of zinc nanoparticles by sol-gel process, A Master Degree Dissertation, Department of Physics, National Institute of Technology, Rourkela-769008, Orissa, India.

14. Wang, S., Sue, H.J., and Jane, J. (1996). Effects of polyhydric alcohols on the mechanical properties of soy protein plastics, *J. Macromol. Sci. Part A*, **33(5)**, pp. 557–569.

15. Kumar, P., Sandeep, K.P., Alavi, S., Truong, V.D., and Gorga, R.E. (2010). Preparation and characterisation of bio-nanocomposite films based on soy protein isolate and montmorillonite using melt extrusion, *J. Food Eng.*, **100**, pp. 480–489.

16. Standards, ASTM. (2005). E96-05. Standard test methods for water vapor transmission of materials, Philadelphia, PA, *Carbohydr. Polym.*, **79**, pp. 391–399. doi: 10.1016/j.carbopol.2009.08.01

17. Fadeyibi, A.,Osunde, Z.D., and Yisa, M.G. (2019). Optimisation of processing parameters of nanocomposite film for fresh-sliced okra packaging, *J. Appl. Pack. Res.*, **11 (2)**, pp. 1–20.

18. Castelló, R., Ferreira, A.R., Costa, N., Fonseca, I.M., Alves, V.D., and Coelhoso, I.M. (2010). Nanocomposite films obtained from carrageenan/pectin biodegradable polymers, *International Conference on Food Innovation*, October 25–29, pp. 1–4.

19. Fadeyibi, A.,Osunde, Z.D., Yisa, M.G., and Okunola, E.C. (2017). Investigation into properties of starch-based nanocomposite films for fruits and vegetables packaging, *FUTA J. Eng. Eng. Technol.,* **11 (1)**, pp. 12–17.

20. Gontard, N., Guilbert, S. and Cuq, J.L. (1993). Water and glycerol as plasticizers affect mechanical and water vapor barrier properties of an edible wheat gluten film, *J. Food Sci.*, **58(1)**, pp. 206–211.

21. Rojas-Graü, M., Tapia, M., and Martín-Belloso, O. (2008). Using polysaccharide-based edible coatings to maintain quality of fresh-cut Fuji apples, *Lebensm.-Wiss. U-Technol.*, **41(1)**, pp. 139–147.

22. McHugh, T.H., Avena-Bustillos, R., and Krochta, J.M. (1993). Hydrophylic edible films: Modified procedure for water vapour permeability and explanation of thickness effects, *J. Food Sci.*, **58(4)**, 899–903.

23. Zeman, S. and Kubik, L. (2007). Permeability of polymeric packaging materials, *Technical Sci.*, **10**, pp. 26–34. doi: 10.2478/v10022-007-0004-6

24. Islam, M.A. and Buschatz, H. (2005). Assessment of thickness-dependent gas permeability of polymer membranes, *Ind. J. Chem. Technol.*, **12**, pp. 88–92.

25. Chatloff, R.P. and Sircar, A.K. (2005). Encyclopedia of polymer science and technology, Chapter 1 "Thermal analysis of polymers", John Wiley & Sons, Hoboken NJ, pp. 1–86.

26. Lagaron, J.M. and Sanchez-Garcia, M. (2008). Environmentally friendly food packaging, Chiellini, E. (ed.), Chapter 3 "Thermoplastic nanobiocomposites for rigid and flexible food packaging applications", Woodhead Publishing, Boca Raton, FL, pp. 62–89.

27. Chen, H. (1995). Functional properties and applications of edible films made of milk proteins, *J. Dairy Sci.*, **78**, pp. 2563–2583.

28. Guilbert, S., Gontard, N., and Gorris, G.M. (1996). Prolongation of the shelf life of perishable food products using biodegradable films and coatings, *Lebensm-Wiss. U. Technol.*, **29(1)**, pp. 10–17.

29. Gennadios, A., Rhim, J.W., Handa, A., Weller, C.L., and Hanna, M.A. (1998). Ultraviolet radiation affects physical and molecular properties of soy protein films, *J. Food Sci.*, **63(2)**, pp. 225–228.

30. Fadeyibi, A., Osunde, Z.D., Agidi, G., and Evans, E.C. (2014). Flow and strength properties of cassava and yam starch-glycerol composites essential in the design of handling equipment for granular solids, *J. Food Eng.*, **129**, pp. 38–46. DOI: 10.1016/j.jfoodeng.2014.01.006.

31. Park, H.J., Weller, C.L., Vergano, P.J., and Testin, R.F. (1993). Permeability and mechanical properties of cellulose based edible films, *J. FoodSci.*, **58(1)**, pp. 1361–1365.

32. Fadeyibi, A., Osunde, Z.D., Agidi, G., and Egwim, E.C. (2016). Nano-rheological behaviour of cassava starch-zinc-nanocomposite film under dynamic loading for high speed transportation of packaged food, Matheus Poletto (ed.), Composite from renewable and sustainable materials, In Tech Publication, Rijeka, Croatia.

33. Pawar, S.G., Patil, S.L., Chougule, M.A., Raut, B.T., Jundale, D.M., and Patil, V.B. (2010). Polyaniline: TiO_2 nanocomposites: Synthesis and characterisation, *Arch. Appl. Sci. Res.*, **2(2)**, pp. 194–201.

Chapter 8

Review on Advanced Food Packaging Materials Based on Functional Biopolymer Matrix

Md. Aftab Alam, Rizwana Khatoon, Shamsul Huda, and Pramod Kumar Sharma

Department of Pharmacy, School of Medical and Allied Sciences, Galgotias University, Greater Noida, Uttar Pradesh, India

aftab.alam@galgotiasuniversity.edu.in

The chapter describes the functional food packaging materials based on biodegradable polymeric moiety, with biological activeness that can be suitable to use to their synthetic counterparts. Functional-polymer-based have many characteristics to the subsequent materials of packaging. In the present scenario of eco-preservation, the polymeric systems are under great interest such as polysaccharides (cellulose, starch, pullulan, alginate); proteins (gelatin, casein); polyesters and polyhydroxy esters; and many more. Such bio-functional polymeric alternatives to synthetic ones have significant applications in the packaging industry, as safe materials to make cleaner as well as greener world.

Biomaterials in Food Packaging

Edited by Mohd Yusuf and Shafat Ahmad Khan

ISBN 978-981-4877-98-5 (Hardcover), 978-1-003-25678-6 (eBook)

www.jennystanford.com

8.1 Introduction

The main functions of food packaging are to protect food products from deterioration, retaining and extending valuable impacts of processing, and keeping food quality and security with extended shelf life [1] and extensive external impacts on the growth of the global food market, accessibility of raw materials, customer demands, etc. [2].

Non-renewable, non-biodegradable packaging constituents have severe environmental disadvantages. Consumers and environmental activists regarded them as a significant source of strong waste and environmental pollution [3–5]. To address this issue, businesses and scientists worked on ways to create fresh packaging approaches with eco-friendly, plentiful biodegradable packaging products produced from natural renewable polymers [4, 6].

These polymers have elevated molecular weight, and are non-metallic compounds [7]. They consist of repeating macromolecules and, depending on their structure, they have different features. A range of plastics is used as feedstock sources for contemporary plastic products (both renewable and non-renewable). However, plastics have created serious environmental concern [8]. Feedstock for renewable resources includes microbial-grown and starch-extracted polymers. Such materials can be reinforced by natural fibers from crops like hemp, flax, jute, and other sources of cellulose [9].

The perfect biodegradable packaging constituents are acquired from renewable bio-based resources, commonly known as biopolymers, with outstanding mechanical and barrier characteristics and end-of-life options [10–13]. The biopolymers are a group of natural resource molecules that display distinctive polymer characteristics. These are molecules consisting of repeating units (aka monomers) in a specific order of the same or distinct sort. Their molecular weight relies on the number of units that are repeated [14, 15]. The 'bio' prefix indicates living organisms as the sources of biopolymer, which is often biodegradable. The biopolymers are considered as alternatives to petroleum-based plastics because they are biodegradable, renewable, and available in large quantity [16–19]. Figure 8.1 illustrates the categories and sources of biopolymers.

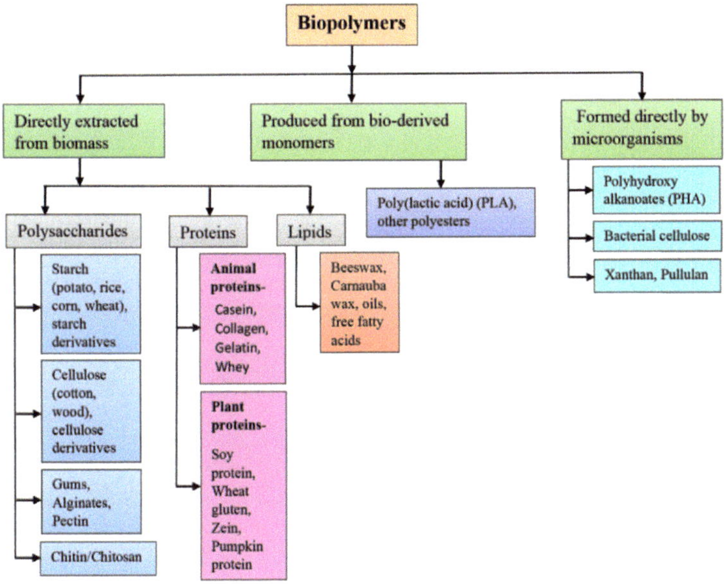

Figure 8.1 Illustration of categories and sources of biopolymers.

The natural biopolymers can also be altered or functionalized mildly to serve a particular purpose. Food hydrocolloids are biopolymers with elevated molecular weight. They are usually used in many food formulations as functional food additives. They are also used in several applications in the food industry, including thickening, stabilizing, encapsulating, gelling, and film-forming, the drug sector depending on their molecular forms and weights [14, 20, 21].

Biopolymers have been used as food, apparel, and furnishings by humankind since ancient times, based on biomaterials such as silk, wool, leather, and cellulose. The increasing dependence on synthetic polymers has raised several environmental and human health problems. An emerging issue is the advancement of materials from natural polymers, due to growing environmental concerns [22].

The biopolymers generated from multiple natural resources, like starch, cellulose, chitosan, and various plant and animal proteins, have been regarded as appealing options for non-biodegradable plastic packaging products, as they are plentiful, renewable, cheap, eco-friendly, biodegradable, and biocompatible [23–25]. Thus, they will play a major part in the bio-economy and certainly shape the packaging industry's future [26].

8.2 Origin and Description of Bio-Based Polymers

Renewable resource-based polymers are gaining increasing attention as environmentally friendly materials. The bio-based polymers, depending on their sources and construction, can be split into three primary classifications:

(i) Polymers obtained straight from biomass like polysaccharides (such as starch, cellulose, and chitosan) and proteins (such as caseinates, soy, zein, whey, and gluten) [27];

(ii) Synthetic polymers mainly derived from alcohols like polyvinyl alcohol (PVA), polycaprolactone (PCL), copolymers of ethylene-vinyl alcohol (EVA), and polylactic acid (PLA) [28–30]; and

(iii) Polymers generated by natural or genetically-modified microbes like polyhydroxyalkanoates (PHAs) or bacterial cellulose [31–33].

These bio-based materials are particularly helpful in food-related applications in three primary fields: food packaging, food coating, and food and encapsulation edible films.

8.3 Polysaccharides

Polysaccharides are polymeric structures in biological systems, also known as carbohydrates. These are made up of numerous monosaccharides, consisting mainly carbon, oxygen, and hydrogen. There are five or six carbon atoms in the most prevalent kinds of monosaccharide units [34].

They are used an attractive food packaging platform. Most polysaccharide biopolymers are used to safeguard food from pathogenic and spoilage microbes in active packaging. In the food industry, many polysaccharides are discovered for food packaging applications because they are non-toxic, biodegradable, environmentally friendly, biocompatible, and eatable particles. These biopolymers depend on the bonds linking the sugar residues at different positions [35].

8.3.1 Plant-Based Polysaccharides

8.3.1.1 Starch

Due to its commercial accessibility, intriguing property balance, and industrial productivity, the starch is the most important material for food packaging among all biopolymers (Figure 8.2) [36]. It is the most widely used renewable raw material and a natural resource that is easily biodegradable. It is acquired from seeds, corn, wheat, rice, potato, cassava, and sweet potato [37, 38]. Starch is a biocompatible and biodegradable low-priced renewable source material [39, 40] whose films can be manufactured either by extrusion [41] or casting [42].

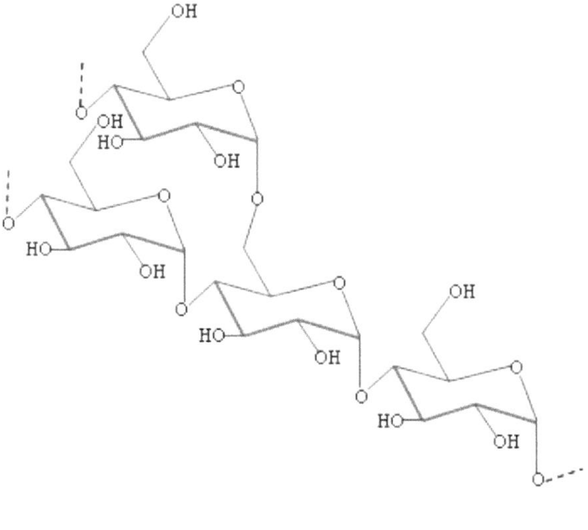

Figure 8.2 Chemical structure of starch.

Starch is composed of a combination of two α-glucose-linear amylose (containing 200–20,000 glucose units) polymers and amylopectin (heavily branched polymer with brief side chains of 30 glucose units connected along the chain to every 20–30 glucose units) [43].

It can be used to make films that are eatable or biodegradable. Dissimilar starch source, high amylose starch, and modified starch [44–46] are used by casting from aqueous solution to create self-supporting films. These films seem to have mild oxygen obstacle characteristic, but bad humidity obstacle and mechanical characteristics that restrict their broad uses. The chemical starch alteration was proved more than 60 years ago by adding butan-1-ol [47].

A fascinating theme regarding starch is the incorporation of various guest molecules into hydrocarbons to create complicated constructions for applied purposes [48]. In the food industry, the distinctive starch structure has many applications, such as water retention, gelling, thickening, stabilizing, and bulking agent [49]. In the industry, a partly degraded starch with particular enzymes is used as low-calorie bulking agent [50]. The intake of low-calorie sugars is a major problem in diabetes management [51].

8.3.1.2 Cellulose

In nature, cellulose is the richest occurring biopolymer and is a major component of fibrous plants, and certain bacterial species including the genus Agrobacterium, Gluconacetobacter, Sarcina, etc. (Figure 8.3) [52, 53]. It is biocompatible, low-cost, and intelligent biopolymer with improved biological and chemical properties [54]. Cellulose molecules are amphiphilic and have a large density of hydroxyl groups and consist of a chain of residues of β-(1→4)-linked glucose [55]. The cellulose is very hard to use in packaging because it has bad mechanical characteristics in its raw form and is hydrophilic and crystalline. Therefore, cellophane with excellent mechanical characteristics must be treated with chemicals, such as NaOH, H_2SO_4, CS_2, etc. [56].

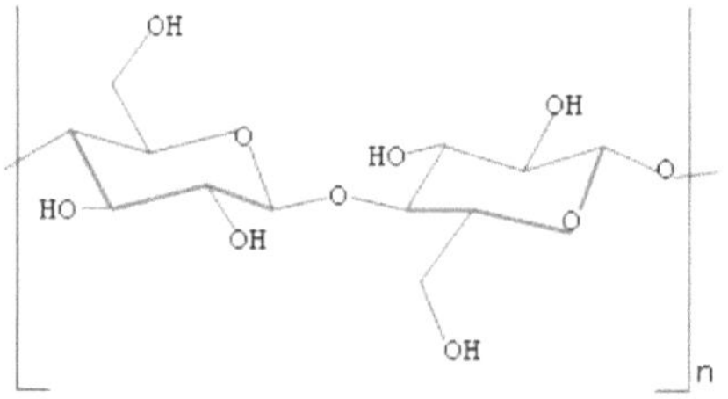

Figure 8.3 Chemical structure of cellulose.

There are several commercially available cellulose derivatives. The most frequently used cellulose derivatives are carboxymethyl cellulose, methylcellulose, ethylcellulose, hydroxyethylcellulose, hydroxypropyl cellulose, and cellulose acetate (CA). However, CA and its derivatives are chiefly used in food packaging applications (boiled products and new products). The CA has comparatively small moisture and gas barrier characteristics and its film production must be plasticized [57, 58].

The cellulose derivatives can be generated through the derivation of cellulose from the solved state by esterification or hydroxyl group etherification. The incorporation of hydrophobic compounds is one way to increase the humidity barrier property of the cellulose derivatives, e.g. fatty acids into the matrix of cellulose to create a composite film [59]. If this scenario is to alter, research is needed to create effective handling techniques for the manufacturing of cellulose derivatives.

8.3.2 Animal-Based Polysaccharides

8.3.2.1 Chitosan

Chitosan is a chitin de-N-acetylated derivative of amino polysaccharide (Figure 8.4) [60–64]. Chitosan is, next to cellulose, the second most abundant natural product, structurally made up of glucosamine and N-acetylglucosamine units connected via the glucosidic bond β-1-4. As chitosan is biodegradable, non-toxic, and easily biocompatible, it has been widely researched for multiple applications in the food and packaging industries [65, 66].

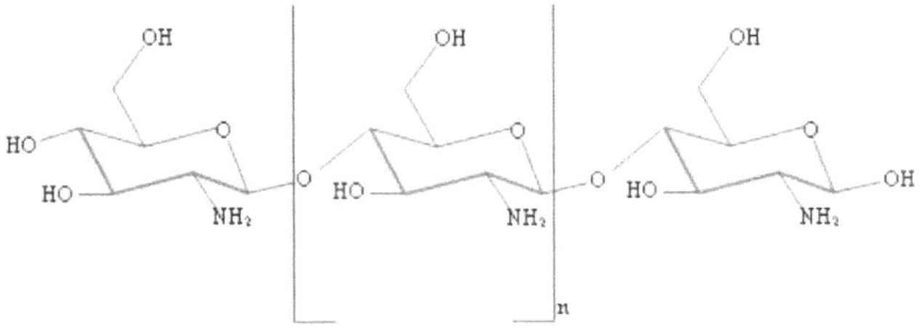

Figure 8.4 Chemical structure of chitosan.

It can also be acquired from the chitin element of fungal cell walls (*Aspergillus fumigates, Histoplasma farciminosum*) and the *Saccharomyces cerevisiae* fungus [64, 67]. *Mucor rouxii* forms (filamentous and yeast-like) contain big amounts of chitosan [60]. Chitin is an outer skeleton structural polysaccharide of crustacean creatures, such as crabs, shrimps, lobsters, insects, and some fungi and algae cell walls [68, 69]. Chitosan forms films without adding additives display good carbon dioxide and oxygen permeability, along with outstanding mechanical and antimicrobial characteristics that reduce the oxidation process and increase the food product's shelf life and quality [70–72]. Hydrophobic elements (vital oil, beeswax) can be added to decrease the water vapor permeability of chitosan films [73–76].

8.3.2.2 Pullulans

Pullulan, comprises a sequence of units of the sort of maltotry, i.e. α-(1→6)-linked (1→4)-α-D-triglucosides, is a non-ionic polysaccharide generated from linear, water-soluble, and exopolysaccharide substrates by the fungus *Aureobasidium pullulans* (Figure 8.5). Pullulan is easily soluble in water, but it is very sparingly soluble in inorganic and non-water-miscible solvents. The pullulan is currently used for numerous applications in food, medicine, and cosmetic products [77–80]. It can be used efficiently in edible films because of its applications [81–83].

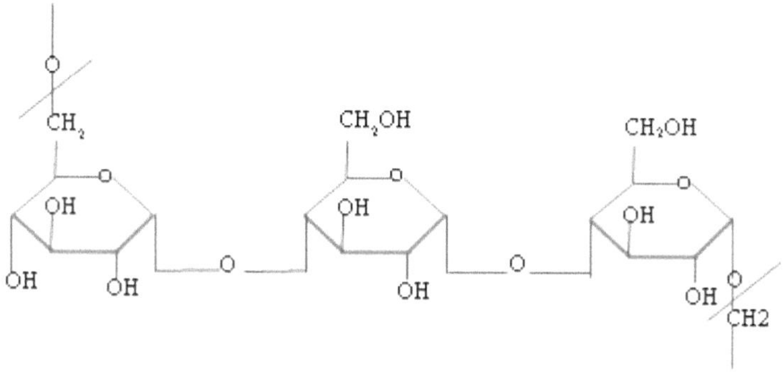

Figure 8.5 Chemical structure of pullulan.

8.3.2.3 Carrageenan

Carrageenan is found in red seaweed (*Kappaphycusalvarezii*, formerly *Eucheuma cottonii*) [84] and has excellent gas barrier characteristic (Figure 8.6) [85]. It is a linear anionic sulfate polysaccharide that consists of alternate residue of d-galactopyranose connected via α-(1→3) and β-(1→4) glycosidic linkages [86, 87]. Coatings based on carrageenan have been revealed to extend shelf life of poultry products [88, 89]. In addition, carrageenan-based products are being utilized as thickeners in the food industry. Edible carrageenan-based films are not widely stated in the literature. More data on applying carrageenan as a layer is available elsewhere [90].

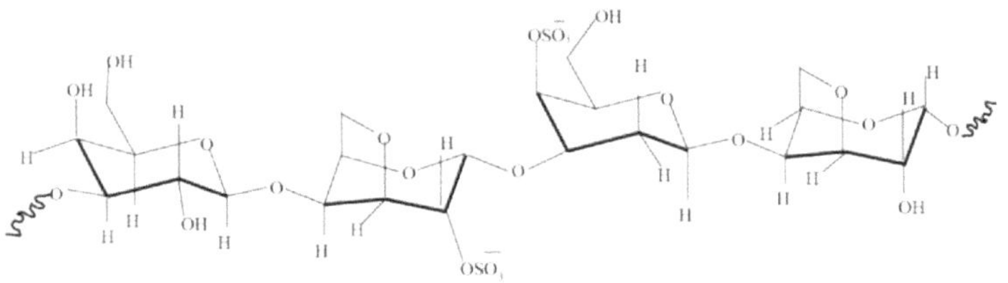

Figure 8.6 Chemical structure of carrageenan.

Carrageenan is generally categorized into three industrially appropriate kinds: λ-carrageenan, which is a heavily sulfated galactan with a viscosity enhancing characteristics; ι-carrageenan, which forms smooth gels that are thermo reversible; and κ-carrageenan, which provides water syneresis to make powerful and fragile gels. There is also a hybrid type composed of κ- carrageenan and ι-carrageenan. Its existence in seaweed relies on the basis of algal, harvest season, and process of extraction used [91, 92].

However, carrageenan biopolymers have a complicated hybrid chemical composition consisting of monomers of λ-, ι-, or κ-carrageenan together with monomers of non-gelling biological precursors like monomers of ν- or μ-carrageenan [91]. Due to powerful contact, the anionic character of the ι-carrageenan is higher in water [93, 94]. More than 100 carrageen products are accessible today, worldwide, which exceeds the manufacturing of the agar [95, 96].

8.3.2.4 Xanthan

Xanthan gum is a polysaccharide bacterium formed by the bacterium *Xanthomonas campestris* aerobic fermentation (Figure 8.7) [97, 98]. The structure contains β-d-mannose (1→4) β-d-acetyl glucose (1-2)-α-d-mannose backbone of β (1→4) C3 substituted d-glucose.

The fragile gel is prepared in elevated viscosity of low-shear rate. In essence, xanthan specimens enable process control like spreading, pumping, pouring, and sparing. Xanthan thixotropy has resulted in the growth of multiple dry-mix preparations like desserts, gravies, and sauces, with their thermally stable textual properties toward cooling and heating. Xanthan is the highest gluten substitute. Galactomannan mixtures based on xanthan are used to make thickened or gelled food [99].

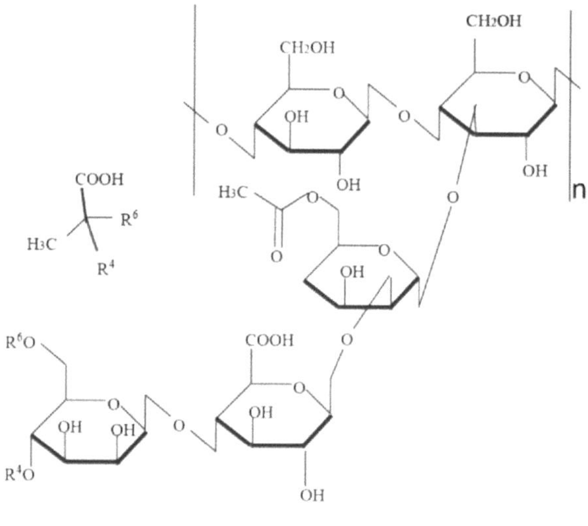

Figure 8.7 Chemical structure of xanthan.

It is a colorful, free-flowing white-to-cream powder that is soluble in warm and cold water, but insoluble in most organic solvents [100, 101]. Xanthan interacts with other components of food like whey protein. These interactions result in a synergistic impact on emulsifying capacity and an increase in emulsion droplet stability against flocculation and coalescence [102].

8.4 Proteins

Proteins are complicated structures consisting of amino acids connected via peptide bonds and organized in a three-dimensional structure. It can be derived from crop sources (gluten, wheat, maize, zein, and soy) and animal sources (whey, casein, keratin, and gelatin). Because of the distinctive side chain in their structure, they are extremely desirable to alter the necessary features of packaging products. Moreover, due to their renewable nature, proteins and protein-based materials are used in many industrial applications for biodegradability and their outstanding gas barrier characteristics. However, they are adversely impacted as starch-based polymers by their hydrophilic nature. They, therefore, need to be mixed with other polymers or must be changed chemically or microbiologically [77, 103].

Most protein biopolymers are used as hydrostatic agents, tissue engineering scaffolds, suture biomaterials, and drug delivery cars in many applications. Biopolymers based on proteins undergo natural procedures of degradation [35].

8.4.1 Plant-Based Proteins

8.4.1.1 Soy protein

Since the 19th century, soy protein has been used in a various kind of products. It occurs as soy flour and is used in adhesives, plastics, and packaging materials manufacturing. It is a nice solution to the pro polymers as well [104]. There has also been an extensive study of soy proteins for their film-forming characteristics and use in the production of biodegradable materials [105]. Soy proteins are now accessible commercially from soybean processing plants in three distinct forms: soy protein concentrate, soy flour, and soy protein isolate, which include more than 90% protein and are acquired through isoelectric rainfall of alkali solubilized protein dispersions. Soy proteins consist of a combination of globular proteins, with the two primary ones being beta-conglycinin (7S globulin) and glycinin (11S globulin), representing 37% and 31%, respectively, of the protein mix [106–108]. Additionally, glycerol and water significantly enhance soy protein polymer's flexibility, but significantly reduce tensile strength [109].

8.4.1.2 Zein protein

Zein is a maize protein (*Zea mays* Linn.). It consists of an elevated non-polar amino acid content that makes zein more hydrophobic. This characteristic makes it an outstanding barrier to oxygen, but has bad mechanical properties [110]. Different kinds of thermophilic products have been developed [111]. Technically, films produced from a protein like zein that is soluble in alcohol have comparatively elevated barrier characteristics as compared to other proteins. Zein has great characteristics of film-forming and can be used to make biodegradable films. Zein-film is created by lipophilic, hydrogen, and restricted disulfide bonds between zein chains [112]. Table 8.1 depicts some example of biopolymer matrix-enabled antimicrobial agents.

The formation of maize zein films involves the growth in film matrix of lipophilic, hydrogen, and restricted disulfide bonds between zein chains [113]. Corn zein is commercially generated by aqueous alcohol extraction and dried into a granular powder. However, due to the weak interaction between protein and glycerol molecules, glycerol tends to migrate through the film matrix. Glycerol migration results in the film's loss of flexibility. It was shown that a combination of glycerol and polyethylene glycol slowed the rate of migration in zein films [114, 115].

Table 8.1 Some examples of biopolymer matrix-enabled antimicrobial agents

S. no.	Biopolymeric matrix	Antimicrobial agent	Microorganism tested	Reference
1.	Gelatin	Olive leaf extract	*L. monocytogenes*	[136]
2.	Calcium alginate	Acetic acid	*L. monocytogenes*, *S. Typhimurium*, and *E. coli* O157:H7	[137,138]
3.	Cellulose	Pediocin	*L. monocytogenes*	[139]
4.	Borley bran protein/ gelatin	Grapefruit seed extract	*E. coli* O157:H7 and *L. monocytogenes*	[140]

(*Continued*)

Table 8.1 (*Continued*)

S. no.	Biopolymeric matrix	Antimicrobial agent	Microorganism tested	Reference
5.	Whey protein	Lactoperoxidase	*L. monocytogenes*	[141]
6.	Wheat gluten	Sorbic acid, Nisin	*P. notatum* and *L. monocytogenes*	[142–144]
7.	Chitosan	Acetic acid	*S. liquefaciens, L. Sakei*, and *Enterobacteriaceae*	[145, 146]
8.	Pullulan			[147]
9.	Corn zein	Nisin	*L. monocytogenes* and *S. enteriditis*	[148, 149]
10.	Starch film	Chitosan	*B. subtilis, E. coli*, and *S. aureus*	[150, 151]
11.	Corn zein	Lauric acid	*L. monocytogenes* and *S. enteriditis*	[149]
12.	Carrageenan	Grapefruit seed extract	*Micrococcus luteus, L. innocua, S. enteritides, E. coli*, and *S. aureus*	[152]
13.	Soy protein Corn zein	EDTA	*L. plantarum* and *E. coli*	[153, 154]

8.4.2 Animal-Based Proteins

8.4.2.1 Keratin

The keratin protein has a significant structure that gives exterior coverage to hair, wool, feathers, horns, and nails of most mammals, birds, and reptiles. It is an organic, biodegradable polymer in which biodegradation occurs due to powerful covalent bonds within its composition and its continuous crosslinking [116–118].

Naturally, keratin is non-burning, lipophobic, biocompatible, and biodegradable protein that can be used for chemical processing in various respects. The keratin protein obtained from chicken feathers can be used for biotechnological applications in various forms, such as sponges, films, and fibers, alone or mixed with other natural or synthetic polymers [119–124]. Keratin does not have excellent mechanical characteristics, so it should be used in combination with a synthetic polymer that is not biodegradable [125, 126].

8.4.2.2 Whey protein

Whey protein forms elastic films that are used as biodegradable packaging materials [127–130]. It consists of several individual proteins, the primary components being α-lactalbumin, β-lactoglobulin, bovine-serum albumin, and immunoglobin [131, 132].

It contains a combination of enzymes like β-lactoglobulin (18kDa molecular weight), α-lactalbumin (14kDa molecular weight), bovine serum albumin, and immunoglobulins. β-lactoglobulin includes about 57% of complete whey protein and two disulfide groups such as free sulfhydryl group and lipophilic group. Approximately 20% of complete whey protein is α-lactalbumin and it includes four disulfide bonds. In aqueous solutions, whey protein film formation includes heat denaturation. The denaturation breaks current disulfide bonds and forms fresh hydrophobic and intermolecular disulfide bonds. Relative humidity, type, and plasticizer concentration affect the water vapor permeability of whey protein films [133–155].

8.4.2.3 Collagen

Collagen is the richest biopolymer and is an important element of bone, cartilage, ligament, skin, and tendon [155]. It consists of three parallel α-chains combined to create a triple-stranded super helical structure. A repeated series of Gly-X-Y residues (where X and Y are often proline and hydroxyproline, respectively) dominates the amino acid sequence of collagen [156]. It is a flexible polymer, but it hardly degrades and requires some therapy due to the complicated and helical-fiber framework. Different techniques of crosslinking using chemical agents are feasible to enhance mechanical characteristics, decrease solubility, and improve thermal stability of collagen-based films [157]. To date, 22 distinct collagens have been recognized in the human. However, types I–IV were widely studied collagens. Collagen is the best-selling edible protein film. The films based on elevated hydrolyzed collagen have more homogeneous surfaces [158–160].

8.4.3 Microbial-Based Proteins

8.4.3.1 PHA

Under unbalanced growth circumstances, PHA is a family of bacterial polyesters with biodegradable and biocompatible nature. The bacterial fermentation of sugar and lipids produces these polymers in nature. The PHAs consist structurally of easy macromolecules consisting of monomers of 3-hydroxy fatty acids. The thermo-mechanical characteristics of PHAs are comparable to synthetic polymers like polypropylene [161]. These polymers offer great packaging films, either alone or in conjunction with starch or synthetic plastic [162].

Hydroxybutyrate (PHB) and poly(hydroxy valerate) (PHV) are the most prevalent forms of PHA. The PHAs with thermoplastic characteristics and medium-chain length (PHAmcl) with elastomeric characteristics are categorized as brief chain length (PHAscl) [163]. The thermo-mechanical characteristics of PHAs are comparable to synthetic polymers like polypropylene [164, 165]. Many microorganisms produce PHB, but *Ralstoniaeutropha* is the most commonly researched bacterium because of its capacity to produce large amount of PHB [166]. Other PHB-accumulating microorganisms are *Haloferax mediterranei*, *Halomonas boliviensis, Bacillus megaterium,* and others [167–169]. PHB-based polymers or composites are currently used in the agriculture, medicine, and food packaging industries [170].

8.4.3.2 Bacterial cellulose (BC)

Bacterial cellulose (BC) is a naturally occurring biomaterial of C- and N-rich medium generated by certain bacterial species such as exopolysaccharides including *Komagataeibacter* genus (formerly *Gluconacetobacter*) [171]. So far, BC is unexploited extensively, but it possesses great potential. Its chemical and physical composition is the same as plant-forming cellulose. However, to remove lignin and hemicellulose, plant cellulose must undergo severe chemical therapy. The BC was generated by fermenting waste from the fermentation of beer culture. Static cultivation on the grounds of manufacturing was discovered to be superior to shaking cultivation [172]. In addition, it can be used for multiple applications, including gelling, thickening, and binding to water. It has been used to boost tofu gel power, to avoid precipitation of cocoa in a chocolate drink and to maintain the viscosity of the drink after heat treatment [173]. It is an outstanding food packaging option, but sadly it does not have any antibacterial and antioxidant characteristics to avoid contamination of the food [170–172]. Its composites are therefore used to acquire these characteristics [174].

8.5 Edible Films and Coatings

The thin layers of edible films and coatings usually <0.3 mm in thickness are used to coat food products for increasing their shelf life and these can be safely consumed along with the edible portion [6, 175, 176]. These also follow the general legislations and regulations on food [106]. Unlike synthetic polymers, this form of the material originates from natural sources and is biodegradable as well as consumable safely [177–180]. The benefits of biopolymer-based films to their synthetic ones are [181–186]:
manufactured from renewable resources;

- consumed in conjunction with the edible packed materials and prevent solid waste;
- even if they do not consume with the packed products, create no pollution/hazard due to biodegradable in nature;
- encourage the sensory features of packaged products by incorporating multiple substances, like colors, flavors, and sweetness;
- useable for individual packaging of tiny parts of food, especially berries, beans, peas, and nuts;
- may complementing the product's dietary significance;
- can provide active finishing to the food (antimicrobial and antioxidant);
- useable in combination with inedible as part of multi-layer packaging materials, where the edible film is in direct contact;
- can be used for heterogeneous food products to avoid moisture and other components from migrating between them, i.e. pies, sweets, and pizza; and

- can be used successfully for substance microencapsulation (mainly aroma) to effectively and accurately administer the same dosage in the food.

8.6 Properties of Edible Films

8.6.1 Antimicrobial Activity

Packaging products with antimicrobial function have been acknowledged as one of the most emerging active packaging technologies, helping to enhance food security and shelf life by destroying or inhibiting food-contaminating spoilage and pathogenic microorganisms [187, 188]. In recent trend, nanomaterials are smart alternates with high surface-to-volume ratio and enhanced antimicrobial features, which are being the main reasons for their effectiveness [189]. The addition of antimicrobial as well as antioxidant agents to edible films is recognized very useful because it provides the chance to release agents gradually and maintain a critical concentration for an extended period [190–194].

8.6.2 Antioxidant Activity

The use of antioxidants in packaged food products is another promising area to improve the oxidative stability of food toward residual O_2. The presence of residual O_2 may results in texture change, discoloration, microbial growth, formation of aldehydes, and dietary losses with severe toxicity [195, 196], which are the main causes of decreasing shelf life of meat, poultry, fish, and seafood products [197]. To solve this problem, the use of reactive-free radicals (hydroxyl, peroxide, superoxide, alkoxy, etc.), non-radicals (hypochlorous, H_2O_2), and various kinds of antioxidants have been employed [198, 199].

In the development of active packaging, both natural and synthetic antioxidants are used that offer promising results. The most frequently used synthetic antioxidant to form packaging films is butylatedhydroxytoluene (BHT; E-321). Huang and Weng (1998), when it integrated into polyethylene films, noted the efficacy of BHT in stopping the oxidation of fish muscle and oil [200].

8.7 Alginate-Based Edible Films

Naturally occurring alginates are indigestible polysaccharides produced either from algae or bacterial species [201, 202]. Alginates-based films are currently being found appropriate in formation of films due to their eco-safe, biodegradability, and non-toxicity characteristics [201–204]. Edible films made from alginates have excellent mechanical characteristics, but due to their hydrophilic nature, they have bad water resistance [178].

Sodium alginate form water-soluble glossy gel, tasteless, and odorless films with low permeability for oxygen and moisture [205]. Amongst other alginate types, it is now commonly used as a stabilizing, emulsifying, swelling, thickening, and chelating agent [206, 207] and scavenger for various ions like Hg(II) [208], Cr(VI) [209], Pb(II), Cu(II), Cd(II) [210]. Stem cell differentiation [211] is efficient in alginate hydrogels in flame retardation, and barium alginate films have shown promise results in this [212].

8.8 Protein-Based Edible Films

In general, proteins constitute structural components of various body tissues composed of several amino acids joined by peptide bonding. Protein-based biopolymers are used in several sectors, including drug delivery and tissue engineering. Biopolymers based on proteins undergo natural procedures of degradation [35].

Protein-directed films have been obtained from either animal-sourced proteins, including egg white protein, keratin, collagen, casein, gelatine, and fish myofibrillar protein, or from plant sourced-proteins such as wheat gluten, soy, zein, and other proteins [213]. Because of the powerful cohesive energy density of the polymer, protein films are fragile [211]. Addition of compatible plasticizers enhances the films' extensibility and viscoelasticity [214]. In recent decades, protein-based edible films have gained interests, due to their excellent use as safe edible packaging materials, over synthetic films. Due to its outstanding biodegradable and biocompatible nature, pectin is regarded as a wonder biopolymer of natural origin [211], which is manufactured at commercial scale from citrus peels [215]. Additionally, wheat gluten, stored in wheat and maize, is another permissible protein with brilliant effectiveness, and it is a natural origin edible film and coating material [216]. After successful assessment, wheat gluten is recognized as "green plastic for future", although more challenging applications are required in its strengthening because wheat gluten-based films are fragile without using plasticizers [217].

Gluten films have excellent barrier characteristics for oxygen and carbon dioxide, but they have comparatively elevated WVP. Wheat gluten is also used to maintain fruit and vegetable quality [217]. Consequently, Lee et al. (2005) investigated the impact of gamma-irradiation on gluten film's physicochemical properties [218]. The absence of fundamental knowledge of the tertiary and quaternary structures of gluten is one of the main challenges connected with gluten processing to create packaging products [219].

8.9 Fossil-Based Biopolymeric Blends

The objective of mixing two or more bioactive polymers is to obtain a mixture of each polymer's favorable characteristics within the blend [215]. The benefits of polymer blends are that these can be combined into fresh compositions, thus obtaining products having specific characteristics. This approach triggers the research and development of new materials [220].

8.10 Future Perspective

Benefit of the biopolymers/bioplastics is that the disposal of used products has a decreased environmental impact. The bioplastics are generally biodegradable and thus enable the materials to be cycled. These are the least environmentally toxic polymers among all the plastics, and can decrease reliance on plastics based on petroleum. The depletion of fossil fuels and increasing petroleum costs are the major problems for the plastics industry's long-term sustainability. On the other hand, agricultural products must be regarded as alternatives for the production of bioplastics, with the dwindling supply of petroleum products.

Research involves the development of bacteria strains that require distinct plant sources, as well as the genetic engineering of a food crop with the capacity to create plastics or a biointermediate from the plant's stem for bioplastic manufacturing. Recently, studies have been extended on film-forming and coating food products with edible biopolymers. It is possible to use edible or biodegradable films and coatings to maintain the quality during the product's shelf life. Furthermore, due to human responsibility toward the production, research, and use of biodegradable polymers, their demand is increasing. This is the primary driving force implementing biopolymers that shows their enormous potential in the future.

Issues other than the effect on the environment and reliance on fossil fuel influence the feasibility of replacing standard plastics with bioplastics. The development of the bioplastics is presently limited by factors such as economical and mechanical characteristics. There are a few types of bioplastics being commercially produced, which means that the production of bioplastics can be economical, but to what extent it is unknown. The bioplastics are in their technological infancy, while standard plastics have reached a degree of maturity, since they were produced in the 1920s [221].

8.11 Conclusion

The growing awareness of the environment has resulted in rising demand for biodegradable packaging, composed of raw materials obtained from agricultural feedstock, marine sources, or agro-food sector by-products. Biopolymers have a wide variety and their applications in food packaging are therefore varied, as these have less environmental impacts. Because biodegradable films are produced from renewable feedstock and agricultural waste, these are economical and widely used in packaging applications. Edible films can be used for packaging products that require solubilization or rehydration before use in the food industry and other related sectors.

The descriptions of the biopolymers; applications of the most frequently researched natural polysaccharides such as starch, cellulose, chitosan; and proteins such as collagen, keratin, and whey protein, are given in this chapter. Recent developments in biopolymeric products with antioxidant and antimicrobial characteristics, and extensive discussion of alginate-based, protein-based edible films and coatings, are also included in the text.

References

1. Marsh, K. and Bugusu, B. (2007). Food packaging—roles, materials, and environmental issues, *J. Food Sci.*, **72**, pp. R39–R55.
2. Coles, R. (2003). Food packaging technology, 5th Ed., Coles, R., McDowell, D., and Kirwan, M. J. (eds.), Chapter 1 "Introduction", CRC Press, USA, pp. 1–31.
3. Parreidt, T., Müller, K., and Schmid, M. (2018). Alginate-based edible films and coatings for food packaging applications, *Foods*, **7**, pp. 170. doi:10.3390/foods7100170
4. Risch, S. J. (2000). Food packaging testing methods and applications, Risch, S. J. (ed.), ACS Symposium Series 753, Chapter 1 "New developments in packaging materials", American Chemical Society, Washington, DC, USA, pp. 1–7.
5. Ramos, Ó. L., Reinas, I., Silva, S. I., Fernandes, J. C., Cerqueira, M. A., Pereira, R. N., Vicente, A. A., Poças, M. F., Pintado, M. E., and Malcata, F. X. (2013). Effect of whey protein purity and glycerol content upon physical properties of edible films manufactured therefrom, *Food Hydrocoll.*, **30**, pp. 110–122.
6. Gontard, N. and Guilbert, S. (1994). Food packaging and preservation, Mathlouthi, M. (ed.), Chapter 9 "Bio-packaging: technology and properties of edible and/or biodegradable material of agricultural origin", Springer, Boston, pp. 159–181.
7. Callister, W. D. (1999). Materials science and engineering: An introduction, 5th Ed., John Wiley & Sons, New York.
8. Williams, G. I. and Wool, R. P. (2000). Composites from natural fibers and soy oil resins, *Appl. Comp. Mater.*, **7**, pp. 421–432.
9. Bismarck, A., Aranberri-Askargorta, I., Springer, J., Lampke, T., Wielage, B., Stamboulis, A., Shenderovich, I., and Limbach, H. H. (2002). Surface characterization of flax, hemp and cellulose fibers, surface properties and the water uptake behaviour, *Polym. composites*, **23(5)**, pp. 872–894.
10. Han, J. H. (2003). Novel food packaging techniques, Ahvenainen, R. (ed.), Chapter 4 "Antimicrobial food packaging", Woodhead Publishing, Cambridge, pp. 50–70.
11. Imran, M., Revol-Junelles, A. M., Martyn, A., Tehrany, E. A., Jacquot, M., Linder, M., and Desobry, S. (2010). Active food packaging evolution: Transformation from micro-to nanotechnology, *Critical Rev. Food Sci. Nutr.*, **50**, pp. 799–821.

12. Clarinval, A. M. and Halleux, J. (2005). Biodegradable polymers for industrial applications, Smith, R. (ed.), Chapter 1 "Classification of biodegradable polymers", Woodhead Publishing, Cambridge, pp. 3–31.

13. Floros, J. D., Dock, L. L., and Han, J. H. (1997). Active packaging technologies and applications, *Food Cosm. Drug Pack.*, **20(1)**, pp. 10–17.

14. Ozilgen, S. and Bucak, S. (2018). Biopolymers for food design, Grumezescu, A. M. and Holban, A. M. (eds.), Chapter 6 "Functional Biopolymers in Food Manufacturing", Academic Press, London, pp. 157–189.

15. Pattanashetti, N. A., Heggannavar, G. B., and Kariduraganavar, M. Y. (2017). Smart biopolymers and their biomedical applications, *Procedia Manufac.*, **12**, pp. 263–279.

16. Rhim, J. W., Park, H. M., and Ha, C. S. (2013). Bio-nanocomposites for food packaging applications, *Prog. Polym. Sci.*, **38**, pp.1629–1652.

17. Muratore, G., Nobile, D., Buonocore, G. G., Lanza, C. M., and Asmundo, N. (2005). The influence of using biodegradable packaging films on the quality decay kinetic of plum tomato, *J. Food Eng.*, **67**, pp. 393–399.

18. Philip, J. C., Bartsev, A., Ritchie, R. J., Baucher, M. A., and Guy, K. (2013). Bioplastics science from a policy vantage point, *New Biotechnol.*, **30**, pp. 635–646.

19. Liu, W., Misra, M., Askeland, P., Drzal, L. T., and Mohanty, A. K. (2005). Green composites from soy-based plastic and pineapple leaf fiber: Fabrication and properties evaluation, *Polymer*, **46**, pp. 2710–2721.

20. Mensitieri, G., Di Maio, E., Buonocore, G. G., Nedi, I., Oliviero, M., Sansone, L., and Iannace, S. (2011). Processing and shelf life issues of selected food packaging materials and structures from renewable resources, *Trends Food Sci. Technol.*, **22**, pp. 72–80.

21. Sabiha-Hanim, S. and Siti-Norsafurah, A. M. (2012). Physical properties of hemicellulose films from sugarcane bagasse, *Proc. Eng.*, **42**, pp. 1390–1395.

22. Laine, C., Harlin, A., Hartman, J., Hyvärinen, S., Kammiovirta, K., Krogerus, B., Pajari, H., Rautkoski, H., Setälä, H., Sievänen, J., and Uotila, J. (2013). Hydroxyalkylated xylans – Their synthesis and application in coatings for packaging and paper, *Ind. Crop. Prod.*, **44**, pp. 692–704.

23. Luckachan, G. E. and Pillai, C. K. S. (2011). Biodegradable polymers: A review on recent trends and emerging perspectives, *J. Polym. Environ.*, **19**, pp. 637–676.

24. Sorrentino, A., Gorrasi, G., and Vittoria, V. (2007). Potential perspectives of bio-nanocomposites for food packaging applications, *Trends Food Sci. Technol.*, **18**, pp. 84–95.

25. Tang, X. Z., Kumar, P., Alavi, S., and Sandeep, K. P. (2012). Recent advances in biopolymers and biopolymer-based nanocomposites for food packaging materials, *Crit. Rev. Food Sci. Nutr.*, **52**, pp. 426–442.

26. Montava-Jordà, S., Torres-Giner, S., Ferrandiz-Bou, S., Quiles-Carrillo, L., and Montanes, N. (2019). Development of sustainable and cost-competitive injection-molded pieces of partially bio-based polyethylene terephthalate through the valorization of cotton textile waste, *Inter. J. Mol. Sci.*, **20**, pp. 1378.

27. Popović, S. Z., Lazić, V. L., Hromiš, N. M., Šuput, D. Z., and Bulut, S. N. (2018). Popović, S. Z., Lazić, V. L., Hromiš, N. M., Šuput, D. Z., and Bulut, S. N. (2018). : *Biopolymers for Food Design*, vol. 20, eds. Grumezescu, A.M., and Holban A.M., Chapter 8 "Biopolymer packaging materials for food shelf-life prolongation" Elsevier US, pp. 223–277.

28. Arvanitoyannis, I., Psomiadou, E., Biliaderis, C. G., Ogawa, H., Kawasaki, N., and Nakayama, A. (1997). Biodegradable films made from low density polyethylene (LDPE), ethylene acrylic acid (EAA), polycaprolactone (PCL) and wheat starch for food packaging applications: Part 3, *Starch-Stärke*, **49**, pp. 306–322.

29. Haugaard, V. K., Udsen, A. M., Mortensen, G., Høegh, L., Petersen, K., and Monahan, F. (2001). Potential food applications of biobased materials: An EU-concerted action project, *Starch-Stärke*, **53**, pp. 189–200.

30. Petersen, K., Nielsen, P. V., Bertelsen, G., Lawther, M., Olsen, M. B., Nilsson, N. H., and Mortensen, G. (1999). Potential of biobased materials for food packaging, *Trends Food Sci. Technol.*, **10**, pp. 52–68.

31. Martínez-Sanz, M., Olsson, R. T., Lopez-Rubio, A., and Lagaron, J. M. (2011). Development of electrospun EVOH fibres reinforced with bacterial cellulose nanowhiskers, Part I: Characterization and method optimization, *Cellulose*, **18**, pp. 335–347.

32. Martínez-Sanz, M., Olsson, R. T., Lopez-Rubio, A., and Lagaron, J. M. (2012). Development of bacterial cellulose nanowhiskers reinforced EVOH composites by electrospinning, J. *Appl. Polym. Sci.*, **124**, pp. 1398–1408.

33. Plackett, D. and Siró, I. (2011). Multifunctional and nanoreinforced polymers for food packaging, Lagarón, J. (ed.), Chapter 18 "Polyhydroxyalkanoates (PHAs) for food packaging", Woodhead Publishing, Cambridge, pp. 498–526.

34. Asimov, I. (12013). The elements of polymer science & engineering, 3rd Ed., Rudin, A. & Choi, P. (eds.), Chapter 13 "Biopolymers", Academic Press, Waltham MA, pp. 521–535.

35. Prameela, K., Mohan, C. M., and Ramakrishna, C. (2018). *Biopolymers for Food Design*, vol. 20, eds. Grumezescu, A.M., and Holban A.M., Chapter 1 "Biopolymers for food design: Consumer-friendly natural ingredients" Elsevier US, pp. 1–32.

36. Valdés, A., Mellinas, A. C., Ramos, M., Garrigós, M. C., and Jiménez, A. (2014). Natural additives and agricultural wastes in biopolymer formulations for food packaging, *Front. Chem.*, **2**, pp. 6.

37. BeMiller, J. N. (2007). Starches, modified food starches, and other products from starches, Carbohydrate chemistry for food scientists, 2nd Ed, AACC International, Minnesota, pp. 173–223.

38. Weber, C. J., Haugaard, V., Festersen, R., and Bertelsen, G. (2002). Production and applications of biobased packaging materials for the food industry, *Food Addit. Contam.*, **19**, pp. 172–177.

39. Ning, W., Xingxiang, Z., Na, H., and Jianming, F. (2010). Effects of water on the properties of thermoplastic starch poly (lactic acid) blend containing citric acid, *J. Thermoplastic Comp. Mater.*, **23(1)**, pp. 19–34.

40. Ismail, H. and Zaaba, N. F. (2014). The mechanical properties, water resistance and degradation behaviour of silica-filled sago starch/PVA plastic films, *J. Elastic Plastics*, **46**, pp. 96–109.

41. Forssell, P. M., Hulleman, S. H., Myllärinen, P. J., Moates, G. K., and Parker, R. (1999). Ageing of rubbery thermoplastic barley and oat starches, *Carbohydr. Polym.*, **39**, pp. 43–51.

42. Shamekh, S., Myllärinen, P., Poutanen, K., and Forssell, P. (2002). Film formation properties of potato starch hydrolysates, *Starke/Stärch*, **54**, pp. 20–24.

43. Ray, S. S. and Bousmina, M. (2005). Biodegradable polymers and their layered silicate nanocomposites: In greening the 21st century materials world, *Prot. Mater. Sci.*, **50**, pp. 962–1079.

44. Mali, S. and Grossmann, M. V. E. (2003). Effects of yam starch films on storability and quality of fresh strawberries (Fragaria ananassa), *J. Agric. Food Chem.*, **51**, pp. 7005–7011.

45. Roth, W. B. and Mehltretter. C. L (1967). Some properties of hydroxypropylated amylomaize starch films, *Food Technol.*, **21**, pp. 72.

46. Wolff, I. A., Davis, H. A., Cluskey, J. E., Gundrum, L. J., and Rist, C. E. (1951). Preparation of films from amylose, *Ind. Eng. Chem.*, **43**, pp. 915–919.

47. Schoch, T. J. (1942). Fractionation of starch by selective precipitation with butanol, *J. Am. Chem. Soc.*, **64**, pp. 2957–2961.

48. Tomasik, P. and Schilling, C.H. (1998). Complexes of starch with organic guests, *Adv. Carbohydr. Chem. Biochem.*, **53**, pp. 345–426.

49. Singh, J., Kaur, L., and McCarthy, O. J. (2007). Factors influencing the physico-chemical, morphological, thermal and rheological properties of some chemically modified starches for food applications: A review, *Food Hydrocoll.*, **21**, pp. 1–22.

50. Escarpa, A. and González, M. C. (1997). Tecnología del almidón resistente/Technology of resistant starch, *Food Sci. Technol. Int.*, **3**, pp. 149–161.

51. Kelley, D. E. (2003). Sugars and starch in the nutritional management of diabetes mellitus, *Am. J. Clin. Nutr.*, **78**, pp. 858S–864S.

52. Carreira, P., Mendes, J. A., Trovatti, E., Serafim, L. S., Freire, C. S., Silvestre, A. J., and Neto, C. P. (2011). Utilization of residues from agro-forest industries in the production of high value bacterial cellulose, *Bioresour. Technol.*, **102**, pp. 7354–7360.

53. Pandey, J. K., Takagi, H., Nakagaito, A. N., Saini, D. R., and Ahn, S. H. (2012). An overview on the cellulose based conducting composites, *Comp. Part B: Eng.*, **43**, pp. 2822–2826.

54. Mitra, B. C. (2014). Environment friendly composite materials: Biocomposites and green composites, *Defe. Sci. J.*, **64**, pp. 244–261.

55. Rein, D. M., Khalfin, R., and Cohen, Y. (2012). Cellulose as a novel amphiphilic coating for oil-in-water and water-in-oil dispersions, *J. Coll. Interface Sci.*, **386**, pp. 456–463.

56. Majid, I., Thakur, M., and Nanda, V. (2018). Majid, I., Thakur, M., and Nanda, V. (2018). *Encyclopedia of Renewable and Sustainable Materials, Reference Module in Materials Science and Materials Engineering*, vol 2, eds. Hashmi, S., and Choudhury, I.A., Chapter 62 "Biodegradable packaging materials". Elsevier USA, pp. 688–697.

57. Weber, C. J. (2000). Biobased packaging materials for the food industry: Status and perspectives, a European concerted action, KVL Publishing, the U.S.

58. Pawar, P. A. and Purwar, A. H. (2013). Biodegradable polymers in food packaging, *Am. J. Eng. Res.*, **2**, pp. 151–164.

59. Morillon, V., Debeaufort, F., Blond, G., Capelle, M., and Voilley, A. (2002). Factors affecting the moisture permeability of lipid-based edible films: A review, *Crit. Rev. Food Sci. Nutr.*, **42**, pp. 67–89.

60. Muzzarelli, R. A., Boudrant, J., Meyer, D., Manno, N., DeMarchis, M., and Paoletti, M. G. (2012). Current views on fungal chitin/chitosan, human chitinases, food preservation, glucans, pectins and inulin: A tribute to Henri Braconnot, precursor of the carbohydrate polymers science, on the chitin bicentennial, *Carbohydr. Polym.*, **87**, pp. 995–1012.

61. Janssen, L. and Moscicki, L. (eds.) (2009). Thermoplastic starch: A green material for various industries, John Wiley & Sons, London.

62. Pillai, C. K. S., Paul, W., and Sharma, C. P. (2009). Chitin and chitosan polymers: Chemistry, solubility and fiber formation, *Prog. Polym. Sci.*, **34**, pp. 641–678.

63. Sergeev, A. A., Voznesenskiy, S. S., Bratskaya, S. Y., Mironenko, A. Y., and Lagutkin, R. V. (2012). Investigation of humidity influence upon waveguide features of chitosan thin films, *Phys. Processes*, **23**, pp. 115–118.

64. Yusuf, M. (2018). Food packaging and preservation, Grumezescu, M.A. and Holban, M.A. (eds.), Chapter 12 "Natural Antimicrobial Agents for Food Biopreservation", Academic Press, London USA, pp. 409–438.

65. Rhim, J. W. and Ng, P. K. (2007). Natural biopolymer-based nanocomposite films for packaging applications, *Crit. Rev. Food Sci. Nutr.*, **47**, pp. 411–433.

66. Wallenberger, F. T. and Weston, N. (2004). Natural fibers, plastics and composites natural, Materials Source Book, CHIPS, Texas.

67. Merzendorfer, H. (2011). The cellular basis of chitin synthesis in fungi and insects: Common principles and differences, *Eur. J. Cell Biol.*, **90**, pp. 759–769.

68. Rinaudo, M. (2006). Chitin and chitosan: Properties and applications, *Prog. Polym. Sci.*, **31**, pp. 603–632.

69. Kandra, P. and Kalangi, H. P. J. (2015). Current understanding of synergistic interplay of chitosan nanoparticles and anticancer drugs: Merits and challenges, *Appl. Microbiol. Biotechnol.*, **99**, pp. 2055–2064.

70. Gemili, S., Yemenicioğlu, A., and Altınkaya, S. A. (2009). Development of cellulose acetate based antimicrobial food packaging materials for controlled release of lysozyme, *J. Food Eng.*, **90**, pp. 453–462.

71. Zhang, Y., Rempel, C., and Mclaren, D. (2014). *Innovations in Food Packaging*, vol 2, ed. Han J.H., Chapter 12 "Edible coating and film materials: Carbohydrates" Elsevier USA, pp. 305–323.

72. Zhang, Y. B., Wang, J. W., Jiang, P. P., Li, Y. X., and Liu, X. Y. (2014). Physical and mechanical properties of chitosan films incorporated with Florida mandarin oil, *Adv. Mater. Res.*, **881**, pp. 1153–1156.

73. Bonilla, J., Atarés, L., Vargas, M., and Chiralt, A. (2012). Edible films and coatings to prevent the detrimental effect of oxygen on food quality: Possibilities and limitations, *J. Food Eng.*, **110**, pp. 208–213.

74. Bonilla, J., Talón, E., Atarés, L., Vargas, M., and Chiralt, A. (2013). Effect of the incorporation of antioxidants on physicochemical and

antioxidant properties of wheat starch–chitosan films, *J. Food Eng.*, **118**, pp. 271–278.

75. Hromiš, N. M., Lazić, V. L., Markov, S. L., Vaštag, Ž. G., Popović, S. Z., Šuput, D. Z., Džinić, N. R., Velićanski, A. S., and Popović, L. M. (2015). Optimization of chitosan biofilm properties by addition of caraway essential oil and beeswax, *J. Food Eng.*, **158**, pp. 86–93.

76. Krkić, N., Lazić, V., and Gvozdenović, J. (2011). Chitosan biofilm properties as affected by the addition of oregano essential oil, *J. Proc. Energy Agric.*, **15**, pp. 165–168.

77. Rahman, R., Sood, M., Gupta, N., Julie, D., Bandral, Hameed, F., and Ashraf, S. (2019). Bioplastics for food packaging: A review, *Int. J. Curr. Microbiol. Appl. Sci.*, **8**, pp. 2311–2321.

78. Dias, A. M. G. C., Hussain, A., Marcos, A. S., and Roque, A. C. A. (2011). A biotechnological perspective on the application of iron oxide magnetic colloids modified with polysaccharides, *Biotechnol. Adv.*, **29**, pp. 142–155.

79. Cheng, K. C., Demirci, A., Catchmark, J. M., and Puri, V. M. (2011). Effects of initial ammonium ion concentration on pullulan production by Aureobasidium pullulans and its modelling, *J. Food Eng.*, **103**, pp. 115–122.

80. Wu, J., Zhong, F., Li, Y., Shoemaker, C. F., and Xia, W. (2013). Preparation and characterization of pullulan–chitosan and pullulan–carboxymethyl chitosan blended films, *Food Hydrocoll.*, **30**, pp. 82–91.

81. Yatmaz, E. and Turhan, I. (2012). Pullulan production by fermentation and usage in food industry, *GIDA. J. Food Sci.*, **37**, pp. 95–102.

82. Krochta, J.M. and Mulder-Johston, C. (1997). Food technology: Edible and biodegradable polymer films challenges and opportunities, *Food Technol.*, **51**, pp. 61–74.

83. Singh, R. S., Saini, G. K., and Kennedy, J. F. (2008). Pullulan: Microbial sources, production and applications, *Carbohydr. Polym.*, **73**, pp. 515–531.

84. Chan, S. W., Mirhosseini, H., Taip, F. S., Ling, T. C., and Tan, C. P. (2013). Comparative study on the physicochemical properties of κ-carrageenan extracted from Kappaphycus alvarezii (doty) doty ex Silva in Tawau, Sabah, Malaysia and commercial κ-carrageenans, *Food Hydrocoll.*, **30**, pp. 581–588.

85. Hamzah, H. M., Osman, A., Tan, C. P., and Ghazali, F. M. (2013). Carrageenan as an alternative coating for papaya, *Postharvest Biol. Technol.*, **75**, pp. 142–146.

86. Karbowiak, T., Gougeon, R. D., Rigolet, S., Delmotte, L., Debeaufort, F., and Voilley, A. (2008). Diffusion of small molecules in edible films: Effect of water and interactions between diffusant and biopolymer, *Food Chem.*, **106**, pp. 1340–1349.

87. Martins, J. T., Cerqueira, M. A., Bourbon, A. I., Pinheiro, A. C., Souza, B. W., and Vicente, A. A. (2012). Synergistic effects between κ-carrageenan and locust bean gum on physicochemical properties of edible films made thereof, *Food Hydrocoll.*, **29**, pp. 280–289.

88. Pearce, J. A. and Layers, C. G. (1949). Frozen storage of poultry: V. effects of some processing factors on quality, *Canad. J. Res.*, **27**, pp. 253–265.

89. Meyer, R. C., Winter, A. R., and Weiser, H. H. (1959). Edible protective coatings for extending the shelf life of poultry, *Food Technol.*, **13**, pp. 146–148.

90. Campos, C. A., Gerschenson, L. N., and Flores, S. K. (2011). Development of edible films and coatings with antimicrobial activity, *Food Bioproc. Technol.*, **4**, pp. 849–875.

91. Lahaye, M. (2001). Developments on gelling algal galactans, their structure and physico-chemistry, *J. Appl. Phycol.*, **13**, pp. 173–184.

92. Rhein-Knudsen, N., Ale, M. T., and Meyer, A. S. (2015). Seaweed hydrocolloid production: An update on enzyme assisted extraction and modification technologies, *Marine Drugs*, **13(6)**, pp. 3340–3359.

93. Glück, U. and Thier, H. P. (1980). Quantitative determination of some thickeners in dairy products, *Z. Lebens. Unter. Forsc.*, **170**, pp. 272–279.

94. Yabe, Y., Ninomiya, T., Tatsuno, T., and OKAD, A. (1991). Simple colorimetric determination of carrageenan in jellies and salad dressings, *J. Associ. Office. Anal. Chem.*, **74**, pp. 1019–1022.

95. De Ruiter, G. A. and Rudolph, B. (1997). Carrageenan biotechnology, *Trends Food Sci. Technol.*, **8**, pp. 389–395.

96. Indergaard, M. and Ostgaard, K. (1991). Seaweed resources in Europe: Uses and potential, Guiry, M.D. and Blunden, G. (eds.), Chapter 6 "Polysaccharides for food and pharmaceutical uses", John Wiley & Sons, Chichester, pp. 169–183.

97. Laneuville, S. I., Turgeon, S. L., and Paquin, P. (2013). Changes in the physical properties of xanthan gum induced by a dynamic high-pressure treatment, *Carbohydr. Polym.*, **92**, pp. 2327–2336.

98. Morris, E. R., Rees, D. A., Young, G., Walkinshaw, M. D., and Darke, A. (1977). Order-disorder transition for a bacterial polysaccharide in solution: A role for polysaccharide conformation in recognition between Xanthomonas pathogen and its plant host, *J. Molecular Biol.*, **110**, pp. 1–16.

99. Flores, S. K., Costa, D., Yamashita, F., Gerschenson, L. N., and Grossmann, M. V. (2010). Mixture design for evaluation of potassium sorbate and xanthan gum effect on properties of tapioca starch films obtained by extrusion, *Mater. Sci. Eng.*, **30**, pp. 196–202.

100. Yang, S. T., Lo, Y. M., and Chattopadhyay, D. (1998). Production of cell-free xanthan fermentation broth by cell adsorption on fibers, *Biotechnol. Prog.*, **14**, pp. 259–264.

101. Born, K., Langendorff, V., and Boulenguer, P. (2005). Polysaccharides and polyamides in the food industry: Polysaccharides, Vol. 1: Properties, production, and patents, Steinbuchel, A. and Rhee, S. K. (eds.), Chapter 23 "Xanthan in polysaccharides and polyamides in the food industry", Wiley-VCH, New York, pp. 481–518.

102. Bertrand, M. E. and Turgeon, S. L. (2007). Improved gelling properties of whey protein isolate by addition of xanthan gum, *Food. Hydrocoll.*, **21**, pp. 159–166.

103. Bourtoom, T. (2008). Edible films and coatings: Characteristics and properties, *Int. Food Res. J.*, **15**, pp. 237–248.

104. Tian, H., Xu, G., Yang, B., and Guo, G. (2011). Microstructure and mechanical properties of soy protein/agar blend films: Effect of composition and processing methods, *J. Food Eng.*, **107**, pp. 21–26.

105. Song, F., Tang, D. L., Wang, X. L., and Wang, Y. Z. (2011). Biodegradable soy protein isolate-based materials: A review, *Biomacromolecules*, **12**, pp. 3369–3380.

106. Olivas, G.I. and Barbosa-Canovas, G. (2009). Edible films and coatings for food applications, Embuscado, M. E. and Huber, K. C. (eds.), Chapter 7 "Edible films and coatings for Fruits and Vegetables", Springer, New York, pp. 211–244.

107. Cho, S.Y. and Rhee, C. (2004). Mechanical properties and water vapor permeability of edible films made from fractionated soy proteins with ultrafiltration, *Food Sci. Technol.*, **37**, pp. 833–839.

108. Zhang, J., Mungara, P., and Jane, J. L. (2001). Mechanical and thermal properties of extruded soy protein sheets, *Polymer*, **42**, pp. 2569–2578.

109. Mo, X., Sun, X. S., and Wang, Y. (1999). Effects of molding temperature and pressure on properties of soy protein polymers, *J. Appl. Polym. Sci.*, **73**, pp. 2595–2602.

110. Ozcalik, O. and Tihminlioglu, F. (2013). Barrier properties of corn zein nanocomposite coated polypropylene films for food packaging applications, *J. Food Eng.*, **114**, pp. 505–513.

111. Reddy, N. and Yang, Y. (2011). Novel green composites using zein as matrix and jute fibers as reinforcement, *Biomacromolecules Bioenergy*, **35**, pp. 3496–3503.

112. Guilbert, S. (1986). Food packaging and preservation: Theory and practice, Mathlouthi, M. (ed.), Chapter 18 "Technology and application of edible protective films", Elsevier Applied Science Publishers, London, the U.K., pp. 371–399.

113. Gennadios, A., McHugh, T. H., Weller, C. L., and Krochta, J. M. (1994). Edible and coating and films to improved food quality, Krochta, J. M., Baldwin, E. A., and Nisperos-Carriedo, M. O. (eds.), Chapter 9 "Edible coating and film based on proteins", CRC Press, Boca Raton, pp. 201–278.

114. Padua, G. W. and Wang, Q. (2002). Protein-based films and coatings, Gennadios, A. (ed.), Chapter 2 "Formation and properties of corn zein films and coatings", CRC Press, Boca Raton, pp. 43–67.

115. Hernandez-Izquierdo, V. M. and Krochta, J. M. (2008). Thermoplastic processing of proteins for film formation: A review, *J. Food Sci.*, **73**, pp. R30–R39.

116. Kumawat, T. K., Sharma, A., Sharma, V., and Chandra, S. (2018). Keratin, Blumenberg, M. (ed.), Chapter "Keratin waste: The biodegradable polymers", IntechOpen, London. doi: 10.5772/intechopen.79502

117. Vasconcelos, A. and Cavaco-Paulo, A. (2013). The use of keratin in biomedical applications, *Cur. Drug Targets*, **14**, pp. 612–619.

118. Sharma, S., Gupta, A., Chik, S. M. S. B. T., Kee, C. Y. G., and Poddar, P. K. (2017). Dissolution and characterization of biofunctional keratin particles extracted from chicken feathers, In IOP conference series: Materials science and engineering, *IOP Publishing*, **191**, pp. 012013.

119. Khot, S. N., Lascala, J. J., Can, E., Morye, S. S., Williams, G. I., Palmese, G. R., Kusefoglu, S. H., and Wool, R. P. (2001). Development and application of triglyceride-based polymers and composites, *J. Appl. Polym. Sci.*, **82**, pp. 703–723.

120. Sharma, S. and Gupta, A. (2016). Sustainable management of keratin waste biomass: Applications and future perspectives, Braz. Arch. Biol. Technol.: An Int. J., **59**, 1-14. https://doi.org/10.1590/1678-4324-2016150684

121. Boakye, M., Rijal, N., Adhikari, U., and Bhattarai, N. (2015). Fabrication and characterization of electrospun PCL-MgO-keratin-based composite nanofibers for biomedical applications, *Material*, **8**, pp. 4080–4095.

122. Edwards, A., Jarvis, D., Hopkins, T., Pixley, S., and Bhattarai, N. (2015). Poly (ε-caprolactone)/keratin-based composite nanofibers

for biomedical applications, *J. Biomacromolecules Mater. Res. Part B: Appl. Biomater.*, **103**, pp. 21–30.

123. Flores-Hernández, C., Colín-Cruz, A., Velasco-Santos, C., Castaño, V., Rivera-Armenta, J., Almendarez-Camarillo, A., García-Casillas, P., and Martínez-Hernández, A. (2014). All green composites from fully renewable biopolymers: Chitosan-starch reinforced with keratin from feathers, *Polymer*, **6(3)**, pp. 686–705.

124. Khosa, M. A., Wu, J., and Ullah, A. (2013). Chemical modification, characterization, and application of chicken feathers as novel biosorbents, *RSC Adv.*, **3**, pp. 20800–20810.

125. Aluigi, A., Tonetti, C., Vineis, C., Tonin, C., and Mazzuchetti, G. (2011). Adsorption of copper (II) ions by keratin/PA6 blend nanofibers, *Eur. Polym. J.*, **47**, pp. 1756–1764.

126. Aluigi, A., Vineis, C., Tonin, C., Tonetti, C., Varesano, A., and Mazzuchetti, G. (2009). Wool keratin-based nanofibres for active filtration of air and water, *J. Biobased Mater. Bioenergy*, **3**, pp. 311–319.

127. Morr, C. V. and Ha, E. Y. W. (1993). Whey protein concentrates and isolates: Processing and functional properties, *Crit. Rev. Food Sci. Nutr.*, **33**, pp. 431–476.

128. Popović, S., Peričin, D., Vaštag, Ž., Lazić, V., and Popović, L. (2012). Pumpkin oil cake protein isolate films as potential gas barrier coating, *J. Food Eng.*, **110**, pp. 374–379.

129. Wang, Y., Xiong, Y. L., Rentfrow, G. K., and Newman, M. C. (2013). Oxidation promotes cross-linking but impairs film-forming properties of whey proteins, *J. Food Eng.*, **115**, pp. 11–19.

130. Wang, L., Xue, Z., Zhao, B., Yu, B., Xu, P., and Ma, Y. (2013). Jerusalem artichoke powder: A useful material in producing high-optical-purity L-lactate using an efficient sugar-utilizing thermophilic Bacillus coagulans strain, *Bioresour. Technol.*, **130**, pp. 174–180.

131. deWit, J. N. and Klarenbeek, G. (1984). Effects of various heat treatments on structure and solubility of whey proteins, *J Dairy Sci.*, **67**, pp. 2701–2710.

132. Kinsella, J. E. and Morr, C. V. (1984). Milk proteins: Physicochemical and functional properties, *Crit. Rev. Food Sci. Nutr.*, **21**, pp. 197–262.

133. Krochta, J. M. (1997). PhD Dissertation, "Food proteins and their applications", pp. 529–550.

134. Perez-Gago, M. B. and Krochta, J. M. (2002). Protein-based films and coatings, Gennadios, A. (ed.), Chapter 6 "Formation and properties of whey protein films and coatings", CRC Press, Boca Raton, pp. 159–180.

135. Pérez-Gago, M. B., Nadaud, P., and Krochta, J. M. (1999). Water vapor permeability, solubility, and tensile properties of heat-denatured versus native whey protein films, *J. Food Sci.*, **64**, pp. 1034–1037.

136. Albertos, I., Avena-Bustillos, R. J., Martín-Diana, A. B., Du, W. X., Rico, D., and McHugh, T. H. (2017). Antimicrobial olive leaf gelatin films for enhancing the quality of cold-smoked salmon, *Food Pack. Shelf Life*, **13**, pp. 49–55.

137. Siragusa, G. R. and Dickson, J. S. (1992). Inhibition of Listeria monocytogenes on beef tissue by application of organic acids immobilized in a calcium alginate gel, *J. Food Sci.*, **57**, pp. 293–296.

138. Siragusa G. R. and Dickson J. S. (1993). Inhibition of Listeria monocytogenes, Salmonella Typhimurium and Escherichia coli 0157:H7 on beef muscle tissue by lactic or acetic acid contained in calcium alginate gels, *J. Food Safety*, **13**, pp. 147–158.

139. Ming, X., Weber, G. H., Ayres, J. W., and Sandine, W. E. (1997). Bacteriocins applied to food packaging materials to inhibit Listeria monocytogenes on meats, *J. Food Sci.*, **62**, pp. 413–415.

140. Song, H. Y., Shin, Y. J., and Song, K. B. (2012). Preparation of a barley bran protein–gelatin composite film containing grapefruit seed extract and its application in salmon packaging, *J. Food Eng.*, **113**, pp. 541–547.

141. Min, S., Harris, L. J., Han, J. H., and Krochta, J. M. (2005). Listeria monocytogenes inhibition by whey protein films and coatings incorporating lysozyme, *J. Food Prot.*, **68**, pp. 2317–2325.

142. Carlin, F., Gontard, N., Reich, M., and Nguyen-The, C. (2001). Utilization of zein coating and sorbic acid to reduce Listeria monocytogenes growth on cooked sweet corn, *J. Food Sci.*, **66**, pp. 1385–1389.

143. Guilbert, S. and Biquet, B. (1996). Edible films and coatings, *Food Pack. Technol.*, **1**, pp. 315–353.

144. Ko, S., Janes, M. E., Hettiarachchy, N. S., and Johnson, M. G. (2001). Physical and chemical properties of edible films containing nisin and their action against Listeria monocytogenes, *J. Food Sci.*, **66**, pp. 1006–1011.

145. Ouattara, B., Simard, R. E., Piette, G., Begin, A., and Holley, R. A. (2000). Diffusion of acetic and propionic acids from chitosan-based antimicrobial packaging films, *J. Food Sci.*, **65**, pp. 768–773.

146. Ouattara, B., Simard, R. E., Piette, G., Bégin, A., and Holley, R. A. (2000). Inhibition of surface spoilage bacteria in processed meats by application of antimicrobial films prepared with chitosan, *Interna. J. Food Microbiol.*, **62**, pp. 139–148.

147. Trinetta, V., Floros, J. D., and Cutter, C. N. (2010). Sakacin A-containing pullulan film: An active packaging system to control epidemic clones of Listeria monocytogenes in ready-to-eat foods, *J. Food Safety*, **30**, pp. 366–381.

148. Janes, M. E., Kooshesh, S., and Johnson, M. G. (2002). Control of Listeria monocytogenes on the surface of refrigerated, ready-to-eat chicken coated with edible zein film coatings containing nisin and/or calcium propionate, *J. Food Sci.*, **67**, pp. 2754–2757.

149. Hoffman, K. L., Han, I. Y., and Dawson, P. L. (2001). Antimicrobial effects of corn zein films impregnated with nisin, lauric acid, and EDTA, *J. Food Prot.*, **64**, pp. 885–889.

150. Salleh, E., Muhamad, I. I., and Khairuddin, N. (2007). Inhibition of *Bacillus subtilis* and *Escherichia coli* by antimicrobial starch-based film incorporated with lauric acid and chitosan. In Proceedings of the 3rd CIGR section VI international symposium on food and agricultural products: Processing and innovation, Naples, Italy.

151. Shen, X. L., Wu, J. M., Chen, Y., and Zhao, G. (2010). Antimicrobial and physical properties of sweet potato starch films incorporated with potassium sorbate or chitosan, *Food Hydrocoll.*, **24**, pp. 285–290.

152. Kanmani, P. and Rhim, J. W. (2014). Development and characterization of carrageenan/grapefruit seed extract composite films for active packaging, *Intern. J. Biol. Macromolecules*, **68**, pp. 258–266.

153. Padgett, T., Han, I. Y., and Dawson, P. L. (1998). Incorporation of food-grade antimicrobial compounds into biodegradable packaging films, *J. Food Prot.*, **61**, pp. 1330–1335.

154. Padgett, T., Han, Y., and Dawson, P. L. (2000). Effect of lauric acid addition on the antimicrobial efficacy and water permeability of corn zein films containing nisin, *J. Food Proc. Preserv.*, **24**, pp. 423–432.

155. Liu, D., Liang, L., Regenstein, J. M., and Zhou, P. (2012). Extraction and characterisation of pepsin-solubilised collagen from fins, scales, skins, bones and swim bladders of bighead carp, *Food Chem.*, **133**, pp. 1441–1448.

156. Haug, I. J., Draget, K. I., and Smidsrød, O. (2004). Physical and rheological properties of fish gelatin compared to mammalian gelatin, *Food Hydrocoll.*, **18**, pp. 203–213.

157. Lacroix, M. and Vu, K. D. (2014). Innovations in food packaging, 2nd Ed., Han, J.H. (ed.), Chapter 11 "Edible coating and film materials: Proteins", Academic Press, London, pp. 277–304.

158. Bubnis, W. A. and Ofner III, C. M. (1992). The determination of ϵ-amino groups in soluble and poorly soluble proteinaceous materials by a spectrophotometrie method using trinitrobenzenesulfonic acid, *Anal. Biochem.*, **207**, pp. 129–133.

159. Matmaroh, K., Benjakul, S., Prodpran, T., Encarnacion, A. B., and Kishimura, H. (2011). Characteristics of acid soluble collagen and pepsin soluble collagen from scale of spotted golden goatfish, *Food Chem.*, **129**, pp. 1179–1186.

160. Fadini, A. L., Rocha, F. S., Alvim, I. D., Sadahira, M. S., Queiroz, M. B., Alves, R. M. V., and Silva, L. B. (2013). Mechanical properties and water vapour permeability of hydrolysed collagen–cocoa butter edible films plasticised with sucrose, *Food Hydrocoll.*, **30**, pp. 625–631.

161. Galego, N., Rozsa, C., Sánchez, R., Fung, J., Vázquez, A., and Santo Tomas, J. (2000). Characterization and application of poly (β-hydroxyalkanoates) family as composite biomaterials, *Polym. Test.*, **19**, pp. 485–492.

162. Tharanathan, R. N. (2003). Biodegradable films and composite coatings: Past, present and future, *Trends Food Sci. Technol.*, **14**, pp. 71–78.

163. Rincones, J., Zeidler, A. F., Grassi, M. C. B., Carazzolle, M. F., and Pereira, G. A. (2009). The golden bridge for nature: The new biology applied to bioplastics, *J. Macromolecular Sci. Part C: Polym. Rev.*, **49**, pp. 85–106.

164. Steinbüchel, A. and Füchtenbusch, B. (1998). Bacterial and other biological systems for polyester production, *Trends Biotechnol.*, **16**, pp. 419–427.

165. Tripathi, A. D., Srivastava, S. K., and Yadav, A. J. A. Y. (2014). Polymers for packaging applications, Alvi, S., Thomas, S., Sandeep, K. P., Kalarikkal, N., Verghese, J., and Yaragall, S. (eds.), Chapter 7 "Biopolymers: Potential biodegradable packaging material for food industry", CRC Press, Boca Raton, pp. 153–172.

166. Khanna, S. and Srivastava, A. K. (2005). Statistical media optimization studies for growth and PHB production by Ralstonia eutropha, *Proc. Biochem.*, **40**, pp. 2173–2182.

167. Quillaguaman, J., Delgado, O., Mattiasson, B., and Hatti-Kaul, R. (2006). Poly (β-hydroxybutyrate) production by a moderate halophile, Halomonas boliviensis LC1, *Enz. Microbiol. Technol.*, **38**, pp. 148–154.

168. Tan, D., Xue, Y. S., Aibaidula, G., and Chen, G. Q. (2011). Unsterile and continuous production of polyhydroxybutyrate by Halomonas TD01, *Bioresour. Technol.*, **102**, pp. 8130–8136.

169. Liu, Z., Wang, Y., He, N., Huang, J., Zhu, K., Shao, W., Wang, H., Yuan, W., and Li, Q. (2011). Optimization of polyhydroxybutyrate (PHB) production by excess activated sludge and microbial community analysis, *J. Hazardous Mater.*, **185**, pp. 8–16.

170. Valappil, S. P., Misra, S. K., Boccaccini, A. R., Keshavarz, T., Bucke, C., and Roy, I. (2007). Large-scale production and efficient recovery of PHB with desirable material properties, from the newly characterised Bacillus cereus SPV, *J. Biotechnol.*, **132**, pp. 251–258.

171. Rajwade, J. M., Paknikar, K. M., and Kumbhar, J. V. (2015). Applications of bacterial cellulose and its composites in biomedicine, *Appl. Microbiol. Biotechnol.*, **99**, pp. 2491–2511.

172. Ha, J. H., Shehzad, O., Khan, S., Lee, S. Y., Park, J. W., Khan, T., and Park, J. K. (2008). Production of bacterial cellulose by a static cultivation using the waste from beer culture broth, *Kor. J. Chem. Eng.*, **25**, pp. 812–820.

173. Okiyama, A., Motoki, M., and Yamanaka, S. (1993). Bacterial cellulose IV: Application to processed foods, *Food Hydrocoll.*, **6**, pp. 503–511.

174. Gao, C., Yan, T., Du, J., He, F., Luo, H., and Wan, Y. (2014). Introduction of broad-spectrum antibacterial properties to bacterial cellulose nanofibers via immobilising ε-polylysine nanocoatings, *Food Hydrocoll.*, **36**, pp. 204–211.

175. Guilbert, S., Gontard, N., and Cuq, B. (1995). Technology and applications of edible protective films, *Pack. Technol. Sci.*, **8**, pp. 339–346.

176. Guilbert, S., Gontard, N., and Gorris, L. G. (1996). Prolongation of the shelf-life of perishable food products using biodegradable films and coatings, *LWT-Food Sci. Technol.*, 29, pp. 10–17.

177. Yusuf, M. (2020). Natural materials and products from insects: Chemistry, Kumar, D. and Shahid, M. (eds.), Chapter 7 "Present and future prospects on nutritious feeding using insects", Springer, Cham, pp. 111–119.

178. Kester, J. J. and Fennema, O. R. (1986). Edible films and coatings: A review, *Food Technol.*, **40**, pp. 47–59.

179. Yusuf, M. (2017). Food packaging and preservation, Grumezescu, A.M. and Holban, A.M. (eds.), Chapter 12 "Natural antimicrobial agents for food biopreservation with future prospects", Academic Press, London, pp. 409–438.

180. Gennadios, A. and Weller, C. L. (1994). Moisture adsorption by grain protein films, *Trans. ASAE*, **37**, pp. 535–539.

181. Baldwin, E. A. (2007). Handbook of food preservation, 2nd Ed., Rahman, M.S. (ed.), Chapter 21 "Surface treatments and edible coatings in food preservation", CRC Press, Boca Raton, pp. 477–508.

182. Cutter, C. N. and Sumner, S. S. (2002). Protein-based films and coatings, Gennadios, A. (ed.), Chapter 18 "Application of edible coatings on muscle foods", CRC Press, Boca Raton, pp. 467–484.

183. Ustunol, Z. (2009). Edible films and coatings for food applications, Embuscado, M. E. and Huber, K. C. (eds.), Chapter 8 "Edible films and coatings for meat and poultry", Springer, New York, pp. 245–268.

184. Véronique, C. O. M. A. (2008). Bioactive packaging technologies for extended shelf life of meat-based products, *Meat Sci.*, **78**, pp. 90–103.

185. Kilincceker, O., Dogan, I. S., and Kucukoner, E. (2009). Effect of edible coatings on the quality of frozen fish fillets, *LWT Food Sci. Technol.*, **42**, pp. 868–873.

186. Robertson, G. L. (2016). Food packaging: Principles and practice, 3rd Ed., CRC press, Boca Raton.

187. Cha, D. S. and Chinnan, M. S. (2004). Biopolymer-based antimicrobial packaging: A review, *Crit. Rev. Food Sci. Nutr.*, **44**, pp. 223–237.

188. Falguera, V., Quintero, J. P., Jiménez, A., Muñoz, J. A., and Ibarz, A. (2011). Edible films and coatings: Structures, active functions and trends in their use, *Trends Food Sci. Technol.*, **22**, pp. 292–303.

189. Radusin, T. I., Ristić, I. S., Pilić, B. M., and Novaković, A. R. (2016). Antimicrobial nanomaterials for food packaging applications, *Food Feed. Res.*, **43**, pp. 119–126.

190. Nisperos-Carriedo, M. O. (1994). Edible coatings and films to improve food quality, 1st Ed., Baldwin, E. A., Hagenmaier, R., Bai, J., and Krochta, J. M. (eds.), Chapter 11 "Edible coatings and films based on polysaccharides", CRC Press, Boca Raton, pp. 305–336.

191. Zactiti, E. M. and Kieckbusch, T. G. (2009). Release of potassium sorbate from active films of sodium alginate crosslinked with calcium chloride, *Pack. Technol. Sci. Int. J.*, **22**, pp. 349–358.

192. Vojdani, F. and Torres, J. A. (1990). Potassium sorbate permeability of methylcellulose and hydroxypropyl methylcellulose coatings: Effect of fatty acids, *J. Food Sci.*, **55**, pp. 841–846.

193. Appendini, P. and Hotchkiss, J. H. (2002). Review of antimicrobial food packaging, *Innov. Food Sci. Emeg. Technol.*, **3**, pp. 113–126.

194. Quintavalla, S. and Vicini, L. (2002). Antimicrobial food packaging in meat industry, *Meat Sci.*, **62**, pp. 373–380.

195. Liu, L., Hu, J., Zhang, J., Fu, Z., and Zhang, J. (2010). Development of time-temperature data collection program for frozen fish in the cold chain, *Sensor Lett.*, **8**, pp. 47–51.

196. Vital, A. C. P., Guerrero, A., de Oliveira Monteschio, J., Valero, M. V., Carvalho, C. B., de Abreu Filho, B. A., and do Prado, I. N. (2016). Effect of edible and active coating (with rosemary and oregano essential oils) on beef characteristics and consumer acceptability, *PloS one*, **11(8)**, pp. 1–15. doi:10.1371/journal.pone.0160535

197. Shankar, S. and Rhim, J. W. (2016), Tocopherol-mediated synthesis of silver nanoparticles and preparation of antimicrobial PBAT/silver nanoparticles composite films, *LWT – Food Sci. Technol.*, **72**, pp. 149–156.

198. Gómez-Estaca, J., Montero, P., and Gómez-Guillén, M. C. (2014). Shrimp (Litopenaeus vannamei) muscle proteins as source to develop edible films, *Food Hydrocoll.*, **41**, pp. 86–94.

199. Mitsumoto, M., O'Grady, M. N., Kerry, J. P., and Buckley, D. J. (2005). Addition of tea catechins and vitamin C on sensory evaluation, colour and lipid stability during chilled storage in cooked or raw beef and chicken patties, *Meat Sci.*, **69**, pp. 773–779.

200. Huang, C. H. and Weng, Y. M. (1998). Inhibition of lipid oxidation in fish muscle by antioxidant incorporated polyethylene film, *J. Food Proce. Preserv.*, **22**, pp. 199–209.

201. Goh, C. H., Heng, P. W. S., and Chan, L. W. (2012). Alginates as a useful natural polymer for microencapsulation and therapeutic applications, *Carbohydr. Polym.*, **88**, pp. 1–12.

202. De Lima D. O., Aimoli C. G., and Beppu M. M. (2009). Investigation on the biomimetic influence of biopolymers on calcium phosphate precipitation-Part 1: Alginate, *Mater. Sci. Eng.*, **29**, pp. 1109–1113.

203. Vu, C. H. T. and Won, K. (2013). Novel water-resistant UV-activated oxygen indicator for intelligent food packaging, *Food Chem.*, **140**, pp. 52–56.

204. Blanco-Pascual, N., Montero, M. P., and Gómez-Guillén, M. C. (2014). Antioxidant film development from unrefined extracts of brown seaweeds Laminaria digitata and Ascophyllum nodosum, *Food Hydrocoll.*, **37**, pp. 100–110.

205. Yoo, S. and Krochta, J. M. (2011). Whey protein–polysaccharide blended edible film formation and barrier, tensile, thermal and transparency properties, *J. Sci. Food Agric.*, **91**, pp. 2628–2636.

206. Sriamornsak, P. and Kennedy, R. A. (2008). Swelling and diffusion studies of calcium polysaccharide gels intended for film coating, *Int. J. Pharmacol.*, **358**, pp. 205–213.

207. Dekamin, M. G., Peyman, S. Z., Karimi, Z., Javanshir, S., Naimi-Jamal, M. R., and Barikani, M. (2016). Sodium alginate: An efficient biopolymeric catalyst for green synthesis of 2-amino-4H-pyran derivatives, *Int. J. Biol. Macromolecules*, **87**, pp. 172–179.

208. Sarkar, K., Ansari, Z., and Sen, K. (2016). Detoxification of Hg (II) from aqueous and enzyme media: Pristine vs. tailored calcium alginate hydrogels, *Int. J. Biol. Macromolecules*, **91**, pp. 165–173.

209. Huang, J. F., Li, Y. T., Wu, J. H., Cao, P. Y., Liu, Y. L., and Jiang, G. B. (2016). Floatable, macroporous structured alginate sphere supporting iron nanoparticles used for emergent Cr (VI) spill treatment, *Carbohydr. Polym.*, **146**, pp. 115–122.

210. Wang, F., Lu, X., and Li, X. Y. (2016). Selective removals of heavy metals (Pb2+, Cu2+, and Cd2+) from wastewater by gelation with alginate for effective metal recovery, *J. Hazardous Mater.*, **308**, pp. 75–83.

211. Lee, J. W., Kim, H., and Lee, K. Y. (2016). Effect of spacer arm length between adhesion ligand and alginate hydrogel on stem cell differentiation, *Carbohydr. Polym.*, **139**, pp. 82–89.

212. Liu, Y., Zhang, C. J., Zhao, J. C., Guo, Y., Zhu, P., and Wang, D. Y. (2016). Bio-based barium alginate film: Preparation, flame retardancy and thermal degradation behaviour, *Carbohydr. Polym.*, **139**, pp. 106–114.

213. Gennadios, A. (ed.) (2002). Protein-based films and coatings, CRC Press, London.

214. Brault, D., D'Aprano, G., and Lacroix, M. (1997). Formation of free-standing sterilized edible films from irradiated caseinates, *J. Agric. Food. Chem.*, **45**, pp. 2964–2969.

215. Makhijani, K., Kumar, R., and Sharma, S. K. (2015). Biodegradability of blended polymers: A comparison of various properties, *Crit. Rev. Environ. Sci. Technol.*, **45**, pp. 1801–1825.

216. Mishra, R. K., Banthia, A. K., and Majeed, A. B. A. (2012). Pectin based formulations for biomedical applications: A review, *Asiain J. Pharmacol. Clin. Res.*, **5**, pp. 1–7.

217. Gennadios, A. and Weller, C. L. (1990). Edible films and coatings from wheat and corn proteins, *Food Technol.*, **44**, pp. 63–69.

218. Lee, S. L., Lee, M. S., and Song, K. B. (2005). Effect of gamma-irradiation on the physicochemical properties of gluten films, *Food Chem.*, **92**, pp. 621–625.

219. Petersen, K., Nielsen, P. V., Bertelsen, G., Lawther, M., Olsen, M. B., Nilsson, N. H., and Mortensen (1999). Potential of biobased materials for food packaging, *Trends Food Sci. Technol.*, **10**, pp. 52–68.

220. Silva, W. S. (2011). Modeling and optimization of ternary of polypropylene (PP), ethylenepropylene-diene monomer (EPDM) and scrap rubber tire (SRT), Master's Thesis, Polytechnic Institute of Rio de Janeiro State University.

221. Queiroz, A. U. and Collares-Queiroz, F. P. (2009). Innovation and industrial trends in bioplastics, *J. Macromolecular Sci., Part C: Polym. Rev.*, **49**, pp. 65–78.

Chapter 9

Prospects of Nanomaterials in Active Finishing of Food and Food Packaging: A Review

Divya Bajpai Tripathy,[a] Anjali Gupta,[b] and Pooja Agarwal[b]

[a] *Department of Forensic Science,*
Galgotias University, Greater Noida, Uttar Pradesh, India
[b] *Department of Chemistry, Galgotias University,*
Greater Noida, Uttar Pradesh, India
divyabaj@gmail.com

Continuous advancements in the area of nanotechnology enable the use of nanomaterials in vast range of sectors from electronizcs to environment science, and from detergents to agriculture. In almost each and every sector, research and development is exploiting nanomaterials for different applications. Along with the development of the smart and active packaging for agri-food, nanomaterials have been used as nanosensors, nanofertilizers, and nanopesticides. Various novel nanomaterials have proved promising excellence and safety of food, crop growth, as well as in monitoring environmental settings. A vast variety of nanomaterials, from inorganic to nanoorganic materials, and from metal to metal oxides, have been exploited in the food sector. In the present review, the usage of nanomaterials in active finishing of food and food packaging

Biomaterials in Food Packaging
Edited by Mohd Yusuf and Shafat Ahmad Khan

ISBN 978-981-4877-98-5 (Hardcover), 978-1-003-25678-6 (eBook)
www.jennystanford.com

are discussed that also involve the toxicological rudiments and risk calculation of nanomaterials in food and food materials. More emphasis has been taken to underline the sustainable development of biologically-synthesized nanomaterials for their prospective applications.

9.1 Introduction

Approximately all of the foodstuffs that are eaten and all types of drinks that are consumed are packaged in some way. Maximum of the food packaging materials formed by using non-degradable synthetic polymers signify not only a serious global environmental problem, but also have adverse effects on nutritional values of food products. In addition, the dependence on fossil resources also bounds the sustainability of food packaging and production. In the packaging of food materials, the main purpose is the protection of the products without disturbing their nutritional values and their own shelf life as well as environmental concerns. The commerce of the food packaging industry has to follow some protocol in order to accomplish this target. The industry is continuously in quest of new technologies for further improvement in critical parameters, such as quality, safety, shelf life, and traceability of food material [1].

Recently, the introduction of nanotechnology has generated more prospects for the improvement and exploitation of new materials, such as nanomaterials, on food packaging. It is less than two decades since nanomaterials have been found to have applications in food packaging. The innovative technology for food packaging claimed to have various value-added characteristics like antimicrobial activity and active packaging [2]. Nanomaterials are the particles that have size of less than 100 nm. These materials are known to have various properties that make them suitable to be exploited in varied food science applications. As the analysis of nanomaterials in food matrices is on its initial stages, there is still an increasing demand to emphasize on the modern analytical techniques that make nanomaterials to be exploited more conveniently in this field [3]. Although, nanomaterials possess a number of advantages over conventional food-packaging system, there is a great disagreement about their production cost, and safety and fitness of these materials when in contact with foodstuffs [4]. In the current review, properties, limitations, and current status of different aspects of nanomaterials exploitation in food packaging have been discussed.

9.2 Food Packaging Applications of Nanomaterials

Due to their nanosize and versatility, nanomaterials have widely been exploited in food packaging applications (Figure 9.1) which can be summarized as below:

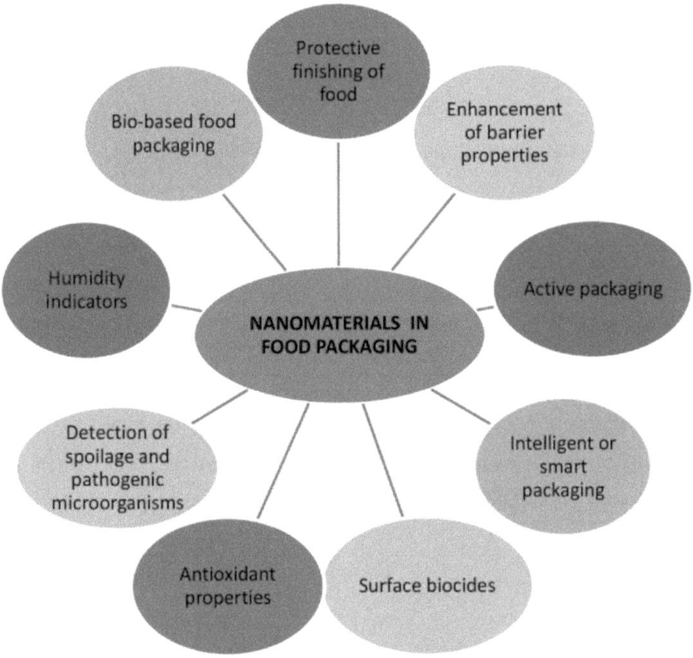

Figure 9.1 Applications of nanomaterials in food packaging.

9.2.1 Protective Finishing of Food

Handling of food packs should be easy, and should appropriate to be used in the food dispensing. Moreover, they also have many other characteristics that are associated to the physical attributes of the packaging stuff [5].

The exploitation of rods and nanofibers (nanosized in two dimensions) can converse advantageous physical characteristics to the food packaging. Specifically, the fibers with the size of >300 nm length to width are useful. Therefore, nanomaterials like carbon nanotubes (CNTs) and cellulose nanowhiskers (CNWs) are integrating innovative materials suitable for packaging purposes [4–6].

9.2.2 Enhancement of Barrier Properties

Food packaging must have properties to preserve freshness of food materials and to shield the food against putrefaction through light as well as oxidation of the products that leads to spoilage. Demands of today are the light-weighting of stuffs and their added shelf life in order to increase suitability and lower the food waste. In this direction, the uses of nano-based coatings are becoming boon, as they can provide enhanced barrier operation [7].

Nowadays, vacuum-deposited aluminum coatings applied on the plastic films are one of the commonly used barrier materials used for the packaging confectionery, snacks, fast food, and coffee. Thickness of this aluminum coating that is permissible cannot be more than 50 nm, as only this limit of thickness makes it formally qualified as a nanomaterial (as one dimension is on the nanoscale). In the similar way, treatments of the glass made surfaces of containers that have been used to contain food and beverages, such as jars and bottles, using organosilanes through plasma or some other high-temperature technologies, are reasonably appropriate for the purpose [8, 9].

9.2.3 Active Packaging

Conventional packaging is projected to be mostly "inactive" that offers a protective approach for foodstuffs as a barrier to and from the outside environment. In contrast, active packaging concept can be understood as the packaging system that provides spoilage protection without alteration in composition of the food items. Nano-enabled materials are being used for substitutes to conventional ones and thus are very promising. Illustrations comprise nanomaterials exploited for rummaging purposes to remove oxygen or taint and/or odor generating chemicals from within the pack. Instead, nanoencapsulation can be applied for releasing the additives for preservatives or to provide instance colors onto the surfaces of food materials, where the additives are required. This could decrease the required quantity of chemical additives [10].

9.2.4 Intelligent or Smart Packaging

Smart or intelligent systems of food packaging enhance the communication properties of the packaged items. These systems exploit numerous advanced communication systems like oxygen sensors, nanosensors, freshness indicators, time–temperature indicators, etc. [10–12]. The insertion of nanosensors into food packaging systems supports the detection of the spoilage-associated changes, chemical contaminants, and pathogens, and hence provides the exact scenario of freshness of the foodstuffs [13]. Nanosensors can be defined as the nanotechnology empowered sensors categorized with a range of variations. In general, nanosensors can be applied as labels as well as coatings in order to enhance an intelligent function to the food packaging to ensure the integrity of the package by the leaks detection (specifically for the food items that are packed in an inert atmosphere or in vacuum), indications of temperature–time variations (i.e., freeze–thaw–refreeze), as well as microbial safety (in order to protect the deterioration of foodstuffs by microbes) [14–16]. Working of optical-O_2 sensors is based on the principle associated with electromagnetic radiation, such as the law of absorbance changes or luminescence quenching, which arises due to the straight contact with the analytes; whereas, opto-chemical sensors are exploited for checking the excellence of the products using sensing gas analytes like CO_2, H_2S, and volatile amines. The exclusive studies have been known for electro-optical and chemical characteristics of nanosized particles [12, 14–18]. The examples of such changes are humidity or temperature in storage rooms, stages of oxygen exposure, and product dilapidation or microbial contagion. Furthermore, smart packaging technology is curiously benefited for common customers, food supervisors, and industry investors, evenly [17–20].

9.2.5 Surface Biocides

The term "surface biocides" is quite different from active packaging. In surface biocides, the chemicals that have requisite biocidal agents have been exploited in order to maintain the standard hygienic condition for the food contact surface of packaging material by preventing and/or reducing the growth of microbial colonies to maintain 'cleanability'. In this process, the main concern that has to be taken, in addition, is that there should not be any preservative impact of such material on the foodstuff [21].

These equipped materials may also have a character to show in recyclable food containers like boxes and crates as well as inside liners of freezers and refrigerators. Their one time use applicability as throwaway packaging is under question. Compounds including nano Ag and ZnO/MgO showed effectiveness, to be a surface biocide, in the food contacted silicones, rubber, plastics, and others [22].

9.2.6 Antioxidant Properties

Some nanomaterials of metal/metal oxides have found less reactive to behave as antioxidant carriers alone [23–25]. Introduction of nanoparticle-polymer composites are recommended to be appropriate for the encapsulation of the bioactive compounds (i.e. vitamins) as well as to discharge them in acidic atmosphere [26]. In the same way, SiO_2-gallic acid-based nanoparticles as novel nanoantioxidants have been established and verified on the basis of scavenging capability of 2,2-diphenyl-1-picrylhydrazyl radicals [27]. In addition to that, employability of antioxidant medications in association with the comestible coating is the most common technique in order to control oxidation and colouration of fresh-cut fruits [28]. The coloration of fresh-cut fruits is an uninvited effect fetched by the transformation of phenolic compounds into the dark-tinted organics during the storage and marketing. This transformation is due to the presence of oxidative environment [29]. Though, there are only a few applications of nanomaterials are known in which they can directly employed as anti-browning agents. Active packaging based on nanosized-ZnO coatings has been described to be a feasible substitute for the betterment of the shelf-life properties of "Fuji" apples as a fresh-cut product [30]. In this study, researchers reported that the activities of pyrogallol peroxidase and polyphenol oxidase were significantly low in fruits stored in nano-ZnO packaging. Furthermore, the initial texture of freshly-cut Fuji apples was reserved, and the browning index was prohibited in fresh-cut fruits stored in nano-ZnO packaging with the browning index value of 23.9, which was found much lower when compared to the control group with refractive index value of 31.7 on day 12. Zambrano and his coworkers [31] also described that action of freshly-cut red colored delicious apples with nanocapsules containing DL-tocopherol with poly-ε-caprolactone biopolymer membrane expressively lowered the browning index.

9.2.7 Detection of Spoilage and Pathogenic Microorganisms

In the USA, according to a report by the Centers for Disease Control and Prevention (CDC), nearly 48,000,000 cases of food-borne infections are reported yearly. It is assessed that little decrease of 1% in occurrence of food-borne infections would help save lives of about 5 lacks individuals [32]. Generally, food-borne sicknesses are occurred due to common microbes like fungi, bacteria, and viruses, and are still a big concern for public health [33, 34]. Currently used detection methods mostly investigate an immunological assay that mainly depends on the selective interaction between the antibodies and the antigens. Antibodies conjugated nanomaterials like quantum dots (QDs) are specifically exploited in order to detect bacteria. The QDs are used specifically owing to their high-distinctive sensitivity, high-fluorescence efficiency, prolonged decay life-time, stability for photobleaching, electronic behavior like continuum, and wide-absorption spectrum and narrow-emission spectrum [34–36].

QDs and organic fluorescent compounds (OFCs) are further categorized into two categories. The first category involves down-conversion phosphors (DCPs) and second category contains up-conversion fluorescent nanoparticles (UCNPs). The DCPs are known to have ability to absorb energy at lower wavelength; whereas they emit radiation at higher wavelength. On the contrary, the excitation of UCNPs takes place as a result of absorption of low-energy radiations in the range near infrared spectrum, and gives emission spectrum at higher energy radiations in the range of visible spectrum. These significant features help them to overcome photodamage and auto-fluorescence-related issues. In addition, UCNPs and QDs have found various applications in labeling, imaging, and medicine. They are the potential candidates to get exploited in the food packaging, predominantly for expected progression in intelligent label for the pathogens and the detection of toxins present in the food matrix [37].

9.2.8 Humidity Indicators

Quality of package and safety and quality of food items in package are indicated by monitoring the humidity inside the package. Zhou [38] developed humidity indicator using iridescent technology. The dry iridescent film emits blue-green color, but when the same film gets exposed to the environment of high humidity, it turns into red-orange colored film.

Tian et al. [39] revealed another type of humidity color indicator. These photonic crystal hydrogels have been designed by combining the photo-polymerized acrylamide with photonic mono-dispersed latex spheres made up of poly (styrene-methyl methacrylate-acrylic acid). Hong and his co-workers [40] revealed that these photonic crystals exhibit confrontation toward degradation and photobleaching that enable them to be one of the best known constituents for photonic crystals. Indicator based on photonic crystals is transparent at 20% humidity and changes color first to violet then blue, then cyan, and then green, and finally changes to red, as relative humidity increases to 100%. Using a similar approach, Zulian et al. [41] developed an emulsion radical method and described core-shell latex particles along with the varied degrees of order, resulting in a range of structural colors and different humidity levels.

9.2.9 Bio-Based Food Packaging

Barrier properties of packaging materials are important in maintaining the quality of food and its shelf life without additional chemical preservatives. Improved barrier behavior of packaging materials thus can help reduce food wastage, and offer more reliable food supply and better nutritive value of the food. Bio-based materials are generally found to be inferior as packaging materials, as compared to artificial polymers made up from oil or gas. Nanotechnology can help understand properties of bio-based materials. The characteristics of the bio-based materials can be upgraded by integrating nanomaterials or by exploiting nanotechnology to improve production methods. This can help in substituting synthetic polymers with locally obtained bio-based materials for the purpose of packaging.

9.3 Nanotechnology: Environmental and Human Safety Prospects

There are potential risks of nanoparticles from packaging being absorbed by the food and thus impact its quality. Also, there are concerns on the environmental impact on how the used package is disposed of and how the presence of nanoparticle impacts recycling processes [42]. Impact of consumption of nanoparticles in the human body at organ level and as well as cellular and subcellular levels is uncertain, as different studies have reported conflicting results [43–49]. This challenge is faced not only by the packaging industry, but by other industries, and thus needs further research to establish safety of nanoparticles in human body [46, 50].

9.4 Conclusion

There are challenges in terms of lack of availability of techniques to analytically measure the movement of nanoparticles from packaging to food materials. There are risks of unwanted chemical reactions during processing and fabrication of nanoparticles, owing to the high surface area and high active chemistry of the nanoparticles. Research needs to focus on the above challenges of adverse responses if any, and characterization of nanoparticles toward human consumption approaches and waste disposal, and thus to develop health-related references for safety of these particles in human body. Though it can be concluded that since nanoparticles are fixed and embedded in the packaging, these are not likely to migrate and thus cause threat to the humans.

References

1. Youssef, A. M., and El-Sayed, S. M. (2018). Bionanocomposites materials for food packaging applications: Concepts and future outlook, *Carbohydr. Polym.*, **193**, pp. 19–27.
2. Majid, I., Nayik, G. A., Dar, S. M., and Nanda, V. (2018). Novel food packaging technologies: Innovations and future prospective, *J. Saudi Soc. Agric. Sci.*, **17(4)**, pp. 454–462.
3. Omanović-Mikličanin, E., and Maksimović, M. (2018). Application of nanotechnology in agriculture and food production: Nanofood and nanoagriculture, *In 5th International Conference on Electrical, Electronic and Computing Engineering-IcETRAN 2018*, pp. 961–966.
4. Forsythe, S. J. (2020). The microbiology of safe food, 3rd Ed., John Wiley & Sons, Hoboken, NJ.
5. Cerqueira, M. A., Vicente, A. A., and Pastrana, L. M. (2018). Nanomaterials for food packaging, Cerqueira, M. A. P. R., Lagaron, J. M., Castro, L. M. P., and de Oliveira Soares Vicente, A. A. M. (eds.), Chapter 1 "Nanotechnology in Food packaging: Opportunities and challenges", Elsevier, Amsterdam, pp. 1–12.
6. Ying, Z., Wu, D., Wang, Z., Xie, W., Qiu, Y., and Wei, X. (2018). Rheological and mechanical properties of polylactide nanocomposites reinforced with the cellulose nanofibers with various surface treatments, *Cellulose*, **25(7)**, pp. 3955–3971.
7. Hernández-Muñoz, P., Cerisuelo, J. P., Domínguez, I., López-Carballo, G., Catalá, R., and Gavara, R. (2019). Nanomaterials for food applications, Rubio, A. L., Rovira, M. J. F., Sanz, M. M., and Gomez-Mascaraque, L. G. (eds.), Chapter 8 "Nanotechnology in food packaging", Elsevier, Amsterdam, pp. 205–232.

8. Zhang, H., Lijun, S., Zhengduo, W., Zhongwei, L., Lizhen, Y., and Qiang, C. (2018). Recent progress on non-thermal plasma technology for high barrier layer fabrication, *Plasma Sci. Technol.*, **20(6)**, pp. 06300.
9. Smolander, M., and Chaudhry, Q. (2010). RSC nanoscience: Nanotechnology series, Vol. 14, Nanotechnologies in food, Chaudhry, Q., Castle, L., and Watkins, R. (eds.), Chapter 6 "Nanotechnologies in food packaging", Royal Society of Chemistry, Cambridge, the U.K., pp. 86–101.
10. Bradley, E. L., Castle, L., and Chaudhry, Q. (2011). Applications of nanomaterials in food packaging with a consideration of opportunities for developing countries, *Trends Food Sci. Technol.*, **22(11)**, pp. 604–610.
11. Kerry, J. P., O'grady, M. N., and Hogan, S. A. (2006). Past, current and potential utilisation of active and intelligent packaging systems for meat and muscle-based products: A review, *Meat Sci.*, **74(1)**, pp. 113–130.
12. Bouwmeester, H., Dekkers, S., Noordam, M., Hagens, W., Bulder, A., De Heer, C., Wijnhoven, S., and Sips, A. (2007). Health impact of nanotechnologies in food production, RIKILT Report 2007.014, RIVM rapport 000200702.
13. Emanuel, N., and Sandhu, H. K. (2020). Food packaging development: Recent perspective, *Journal of Thin Films, Coating Sci. Technol. Application*, **6(3)**, pp.13–29.
14. Liao, F., Chen, C., and Subramanian, V. (2005). Organic TFTs as gas sensors for electronic nose applications, *Sensors Actuators B: Chem.*, **107(2)**, pp. 849–855.
15. Mahalik, N. P., and Nambiar, A. N. (2010). Trends in food packaging and manufacturing systems and technology, *Trends Food Sci. Technol.*, **21(3)**, pp. 117–128.
16. Watson, S. B., Gergely, A., and Janus, E. R. (2011). Where is "Agronanotechnology" heading in the United States and European Union?, 26 Meet on Nat. Resources and Env't 8 (2011–2012).
17. Fuertes, G., Soto, I., Carrasco, R., Vargas, M., Sabattin, J., and Lagos, C. (2016). Intelligent packaging systems: Sensors and nano-sensors to monitor food quality and safety, *J. Sensors*, **2016(4046061)**, pp. 1– 8. http://dx.doi.org/10.1155/2016/4046061
18. Biji, K. B., Ravishankar, C. N., Mohan, C. O., and Gopal, T. S. (2015). Smart packaging systems for food applications: A review, *J. Food Sci. Technol.*, **52(10)**, pp. 6125–6135.

19. Mustafa, F., and Andreescu, S. (2020). Nanotechnology-based approaches for food sensing and packaging applications, *RSC Adv.*, **10(33)**, pp. 19309–19336.

20. Duncan, T. V. (2011). Applications of nanotechnology in food packaging and food safety: Barrier materials, antimicrobials and sensors, *J. Colloid Interface Sci.*, **363(1)**, pp. 1–24.

21. Eleftheriadou, M., Pyrgiotakis, G., and Demokritou, P. (2017). Nanotechnology to the rescue using nano-enabled approaches in microbiological food safety and quality, *Cur. Opinion Biotechnol.*, **44**, pp. 87–93.

22. Braihi, A. J., Ahmed, J. K., and Jabbar, R. S. (2019). Prevent intravenous therapy (IV) contamination by addition of magnesium oxide nanoparticles to silicone rubber, *J. Univ. Babylon Eng. Sci.*, **27(1)**, pp. 409–417.

23. Deepika, S., Kumar, R. H., Selvaraj, C. I., and Roopan, S. M. (2018). Green metal nanoparticles: Synthesis, characterization and their applications, Kanchi, S., and Ahmed, S. (eds.), Chapter 5 "Toxicity of metal/metal oxide nanoparticles and their future prospects", Scrivener Publishing, Beverly, MA, pp. 141–164.

24. Chaudhary, R. G., Bhusari, G. S., Tiple, A. D., Rai, A. R., Somkuvar, S. R., Potbhare, A. K., Lambat, T. L., Ingle, P. P., and Abdala, A. A. (2019). Metal/metal oxide nanoparticles: Toxicity, applications, and future prospects, *Cur. Pharmaceut. Design*, **25(37)**, pp. 4013–4029.

25. Sirelkhatim, A., Mahmud, S., Seeni, A., Kaus, N. H. M., Ann, L. C., Bakhori, S. K. M., and Mohamad, D. (2015). Review on zinc oxide nanoparticles: antibacterial activity and toxicity mechanism, *Nano-Micro Lett.*, **7(3)**, pp. 219–242.

26. Bourbon, A. I., Cerqueira, M. A., and Vicente, A. A. (2016). Encapsulation and controlled release of bioactive compounds in lactoferrin-glycomacropeptide nanohydrogels: Curcumin and caffeine as model compounds, *J. Food Eng.*, **180**, pp. 110–119.

27. Molnar, M., Jerković, I., Suknović, D., BilićRajs, B., Aladić, K., Šubarić, D., and Jokić, S. (2017). Screening of six medicinal plant extracts obtained by two conventional methods and supercritical CO_2 extraction targeted on coumarin content, 2, 2-diphenyl-1-picrylhydrazyl radical scavenging capacity and total phenols content, *Molecules*, **22(3)**, pp. 348.

28. Carvalho, R. L., Cabral, M. F., Germano, T. A., de Carvalho, W. M., Brasil, I. M., Gallão, M. I., Moura, C. F. H., Lopes, M. M. A., and de Miranda, M. R. A. (2016). Chitosan coating with trans-cinnamaldehyde improves structural integrity and antioxidant metabolism of fresh-cut melon, *Postharvest Biol. Technol.*, **113**, pp. 29–39.

29. Kim, A. N., Kim, H. J., Kerr, W. L., and Choi, S. G. (2017). The effect of grinding at various vacuum levels on the color, phenolics, and antioxidant properties of apple, *Food Chem.*, **216**, pp. 234–242.

30. Zhang, M., Meng, X., Bhandari, B., and Fang, Z. (2016). Recent developments in film and gas research in modified atmosphere packaging of fresh foods, *Crit. Rev. Food Sci. Nutr.*, **56(13)**, pp. 2174–2182.

31. Zambrano-Zaragoza, M. L., Quintanar-Guerrero, D., Del Real, A., Piñon-Segundo, E., and Zambrano-Zaragoza, J. F. (2017). The release kinetics of β-carotene nanocapsules/xanthan gum coating and quality changes in fresh-cut melon (cantaloupe), *Carbohydr. Polym.*, **157**, pp. 1874–1882.

32. Scallan, E., Hoekstra, R. M., Angulo, F. J., Tauxe, R. V., Widdowson, M. A., Roy, S. L., Jones, J. L., and Griffin, P. M. (2011). Foodborne illness acquired in the United States: Major pathogens, *Emerg. Infec. Dis.*, **17(1)**, 7–15.

33. Bhuyan, D., Greene, G. W., and Das, R. K. (2019). Prospects and application of nanobiotechnology in food preservation: Molecular perspectives, *Crit. Rev. Biotechnol.*, **39(6)**, pp. 759–778.

34. Valizadeh, A., Mikaeili, H., Samiei, M., Farkhani, S. M., Zarghami, N., Akbarzadeh, A., and Davaran, S. (2012). Quantum dots: Synthesis, bioapplications, and toxicity, *Nanoscale Res. Lett.*, **7(1)**, pp. 480–492.

35. Yang, L., and Li, Y. (2006). Simultaneous detection of Escherichia coli O157: H7 and Salmonella Typhimurium using quantum dots as fluorescence labels, *Analyst*, **131(3)**, pp. 394–401.

36. Mihindukulasuriya, S. D. F., and Lim, L. T. (2014). Nanotechnology development in food packaging: A review, *Trends Food Sci. Technol.*, **40(2)**, pp. 149–167.

37. Zhang, J., Li, C., Zhao, W., Liu, B., Liu, Y., and Aletan, G. (2011). New perspectives in biosensors technology and applications, Serra, P.A. (ed.), Chapter 6 "Biosensing based on luminescent semiconductor quantum dots and rare earth up-conversion nanoparticles", IntechOpen, London, pp. 127–148.

38. Zhou, C. (2013). Theoretical analysis of double-microfluidic-channels photonic crystal fiber sensor based on silver nanowires, *Optics Commun.*, **288**, pp. 42–46.

39. Tian, E., Cui, L., Wang, J., Song, Y., and Jiang, L. (2009). Tough photonic crystals fabricated by photo-crosslinkage of latex spheres, *Macromol. Rapid Commun.*, **30(7)**, pp. 509–514.

40. Hong, W., Huang, L., Wang, H., Qu, J., Guo, Z., Xie, C., and Zheng, Y. (2010). Development of an up-converting phosphor technology-based 10-channel lateral flow assay for profiling antibodies against Yersinia pestis, *J. Microbiol. Methods*, **83(2)**, pp. 133–140.

41. Zulian, L., Emilitri, E., Scavia, G., Botta, C., Colombo, M., and Destri, S. (2012). Structural iridescent tuned colors from self-assembled polymer opal surfaces, *ACS Appl. Mater. Interfaces*, **4(11)**, pp. 6071–6079.

42. Chaudhry, Q., and Castle, L. (2011). Food applications of nanotechnologies: An overview of opportunities and challenges for developing countries, *Trends Food Sci. Technol.*, **22(11)**, pp. 595–603.

43. Maisanaba, S., Pichardo, S., Puerto, M., Gutierrez-Praena, D., Camean, A. M., and Jos, A. (2015). Toxicological evaluation of clay minerals and derived nanocomposites: A review, *Environ. Res.*, **138**, pp. 233–254.

44. Farhoodi, M. (2016). Nanocomposite materials for food packaging applications: Characterization and safety evaluation, *Food Eng. Rev.*, **8(1)**, pp. 35–51.

45. Dimitrijevic, Mirjana, Karabasil, N., Boskovic, M., Teodorovic, V., Vasilev, D., Djordjevic, V., Kilibarda, N., and Cobanovic, N. (2015). Safety aspects of nanotechnology applications in food packaging, *Procedia Food Sci.*, **5**, pp. 57–60.

46. He, X., and Hwang, H. M. (2016). Nanotechnology in food science: Functionality, applicability, and safety assessment, *J. Food Drug Anal.*, **24(4)**, pp. 671–681.

47. Huang, J. Y., Li, X., and Zhou, W. (2015). Safety assessment of nanocomposite for food packaging application, *Trends Food Sci. Technol.*, **45**, pp. 187–199.

48. Sharma, V., Shukla, R. K., Saxena, N., Parmar, D., Das, M., and Dhawan, A. (2009). DNA damaging potential of zinc oxide nanoparticles in human epidermal cells, *Toxicol. Lett.*, **185**, pp. 211–218.

49. Aschberger, K., Micheletti, C., Sokull-Klüttgen, B., and Christensen, F. M. (2011). Analysis of currently available data for characterising the risk of engineered nanomaterials to the environment and human health lessons learned from four case studies, *Environ. Int.*, **37**, pp. 1143–1156.

50. Yusuf, M. (2017). *Food Packaging and Preservation*, Vol. 9, Grumezescu, A. M., and Holban, A. M. (eds.), Chapter 12 "Natural antimicrobial agents for food biopreservation with future prospects", Academic Press, London, pp. 409–438.

Chapter 10

Applications of Nano-Biocomposite Materials as Antimicrobial and Antioxidant Finishing Agents for Packaged Food Products

Md. Aftab Alam,[a] Rizwana Khatoon,[b] Shamsul Huda,[b] Niyaz Ahmad,[c] and Pramod Kumar Sharma[a]

[a] *Department of Pharmacy, School of Medical and Allied Science, Galgotias University, Greater Noida, Uttar Pradesh 203201, India*
[b] *Department of Pharmacy, School of Medical and Allied Science, Galgotias University, Greater Noida, Uttar Pradesh 203201, India*
[c] *Niyaz Ahmad, Department of Pharmacy, Division of Pharmaceutics, College of Pharmacy, Dammam, Kingdom of Saudi Arabia*
aftab.alam@galgotiasuniversity.edu.in

Today, nanotechnology is gaining great attention because of its wide applicability in various sectors, including food packaging. Nanomaterials possess antimicrobial as well as high gas barrier properties with excellent physical and chemical characteristics, and therefore, most market companies have increased investment to develop food packaging alternatives. Innovative packaging systems also facilitate communications at consumer level in the forms of certain transit preservation and effective distribution. The technological advances in food packaging are managed mainly by nanotechnology. It manipulates and produces products of industrial and scientific interest, on a nanometer scale. Specific nano-based materials having active functional abilities increase shelf life of food and its quality and safety. The food packaging industry has embraced

Biomaterials in Food Packaging
Edited by Mohd Yusuf and Shafat Ahmad Khan

ISBN 978-981-4877-98-5 (Hardcover), 978-1-003-25678-6 (eBook)
www.jennystanford.com

the "nano" era to satisfy the growing need for safe and cost-effective goods for customers. The review highlights the use of nanotechnology in the processing and packaging of foods, with a concentrate on food quality and safety.

10.1 Introduction to Nanotechnology in Food Packaging Systems

Nanotechnology means guidance and usage of nano constituents of size between 1 nm and 100 nm, with 1 nm being 1×10^{-9} m [1]. In these nanosizes, the components have unique properties that were not available until they were in their initial size. Nanoscientists across the globe aim to understand these unique characteristics to build new and advanced materials by green processes [2]. Three options are mainly examined by nanotechnological strategies on food packaging. Marketing, successful implementation, and responses to various nanotechnological applications are based on consumer views and the acceptability of newly launched technologies and their submissions [3]. With various uses, nanotechnology has penetrated into the agri-food industry. This is primarily used for production of agri-food products; decreasing the use of consume supplements; reducing the use of agrochemicals; manufacturing of novel food items with enhanced flavors, texture, colors, and sensory properties; nourishment with lowered fat or sugar content; and supplying of supplements (i.e. minerals, vitamins, phenolics, etc.) [4].

Numerous unknown related danger features characterize the latest scientific enforcement in nanotechnology for food applications. The versatility of nanotechnology in foodstuffs defines its application range. Based on their features, food nanotechnology can have an impact on the bioavailability and nutritional benefits of the food. It is understood that nanomaterials have highly dependent on biological characterizations (including toxicological impacts), for their physicochemical parameters [5]. The food industry is always looking for new and inexpensive approaches for food production and conservation. Recent developments in food wrapping relate to nano reinforcement, nanocomposite active packaging, and smart nanocomposite packaging [6].

Polymers are the replacement in food packaging for traditional substances, such as card board, paper glass, metal films, ceramics, etc. Nanocomposites in polymers typically have much stronger interactions with polymer/filler than regular composites. With the assistance of nano-reinforcement strategies, the gaps in packaging substances are packed, which make them more sustainable with improved tensile strength. Food wrapping gets great tensile strength with the system of nano strengthening like nanoclay, cellulose, and other food packaging materials. Improved feasibility and tensile strength are the nano strengthening methods for the final products [7]. Nanotechnology is capable of significantly improving food quality and safety. Nanosensors targeting advanced pathogen identification in food systems that are presently undertaken with considerable efforts. Several e-companies have been using researched polymers that are performing electrically. The same components can also be used to develop sensors capable of detecting extremely low concentrations of spoilage and food-borne microbes within seconds/minutes. This system may identify components per trillion and charges some 50 cents of output, according to scientists involved in the project [8].

To enhance gas transfer into plastic film to remove harmful carbon dioxide, which can reduce the overall life of the product, nanotechnology is being used to build small molecules in the film and offer barrier protection and prevent the deterioration of food by gases such as oxygen and ethylene [9, 10]. Several food and pharmacy investigators are now focusing on advanced nanotechnology to enhance the supply of pharmaceutical drugs, vitamins, or delicate micronutrients in daily foods by generating small, deliverable capsule-based nanoparticles to specific places in the body, on order. The health benefits will, therefore, be substantial for lowering the risks of human diseases like stroke, heart attack, neurodegenerative disorders, and even cancer [11, 12]. Instances of nanotechnology used in the food industry are as follows [13]:

- Enhanced safety in the development, processing, and shipping of food items with pathogens and contaminant identification through sensors.
- Technologies to keep historical environmental data and to track specific transactions in a specific product.
- Systems for monitoring food items (smart/intelligent systems), localization, and remote controlling for improving the efficiency and safety of food manufacturing.

- Encapsulation and distribution networks that transport, safeguard, and supply food products to their specific sites.

Nanostructured foods cover a broad range of materials, from food processing to food packaging. Nanostructures may be utilized in food packaging applications and in particular as antimicrobial agents, anti-caking agents, and fillers to boost mechanical strength and packaging material's durability [14].

10.2 Packaging Technologies Based on Nanotechnology

10.2.1 Nanotechnological Systems for Active Packaging

Effective compounds amalgamation, i.e. anti-microbial products, preservatives, oxygen and water vapor absorbers, and solvent extractors, makes nanomaterials more powerful in improving food products' shelf life and consistency [6, 15, 16]. Nanoparticles with a very large surface region are very likely to trap or release chemicals. They have large outcomes, for example, nano encapsulates may be employed to expel ingredients on the surface of food; whereas, ingredients such as colors and preservatives are required. A new strategy for reducing microbial food contamination on surfaces is the application of active substances in food packaging products. There are certain nanomaterials, such as silver, gold, and zinc, have antimicrobial property [17]. Thus, there are various industrial uses of silver nanoparticles, due to their lower volatility and higher temperature stability, and silver is a strong antimicrobial and anti-fungal agent at a nanoscale.

In comparison with other antimicrobial agents, silver has many benefits. Silver is poisonous to a wide range of fungal, bacterial, algae, and some virus types. Certain antimicrobials, on the other hand, are generally organism-specific (to different degrees). For an extended time, silver is shelf constant. The effect of biofilm penetration, which many molecular antimicrobials fail to attain, is also relatively effective [18]. Nanomaterials are used to directly connect with the food or the environment to ensure higher safety in active packaging advancement. For instance, silver nanoparticles and silver coatings having anti-microbial features are used in other components as oxygen or ultraviolet scavengers. Furthermore, the use of nanomaterials to impart active food packaging properties are commercialized, such as nanosilver, carbon nanotubes, nanocopper oxide, nanomagnesium oxide, and nanotitanium oxide [19–21]. Kodak Company launched and commercialized antimicrobial packaging for food items that consume oxygen [22]. Oxygen scavenging packages are also produced by using enzymes from polyethylene films [23]. Once the product is exposed and re-wrapped with an active film, there is also a need of suitable treatment to stop microbial growth [24].

10.2.1.1 Antimicrobial packaging

The dissolution of silver nanoparticles has documented the antimicrobial activity of Ag^+ ions [25]. The absorption and decomposition of ethylene in silver nanoparticles have also been stated to contribute toward the extension of the fruit's and vegetable's shelf life [26]. Antibacterial characteristics of carbon nanotubes were also reported [27]. Nanosensors designed for this task may detect the complexity of packaged products and the involvement of pathogens. They probably take on the shape of "buttons" on the color-changing packaging [28]. Antimicrobial activity of nanomaterials was researched for use as inhibitors to microbial growth or as antibiotic carriers for use. In this regard, silver nanoparticles have been recognized as popular antimicrobial agents for food packaging applications, on account of their high toxicity for a variety of pathogenic isolates with low volatile and temperature-resistant nature [17, 29–34].

10.2.1.2 Oxygen scavenging packaging

Generally, food and drinks that are prone to oxygen have been packaged so that their sensitivity to oxygen is reduced. Such oxygen may be inserted in the package when the package is closed or entered by permeation or leakage during the storage. Several chemical reactions concerning food degradation, like overcooking and rancid flavoring, involve oxygen. The inclusion of O_2 scavengers will achieve extremely low levels of O_2 in food packaging systems. The literature has documented oxygen scavengers' films with the application of the TiO_2 nanoparticles to various polymers [35].

There is strong clarity about TiO_2 nanoparticles-based nanofilms [10]. Oxygen leads to the degradation of many foods directly or indirectly. Active oxidation reactions, for instance, cause fruit browning and vegetable oils to be rancid. Food spoilage by aerobic microbes is the passive activity of oxygen itself. Consequently, it is possible to keep very low oxygen rate by incorporating oxygen scavenger in the food pack, which is beneficial for many purposes, as it increases food's shelf life [36]. Numerous definitions of active packaging focused on nanocomposite films for O_2 and ethylene scavenging have been examined or the direction of O_2 diffusion has been restricted [37–39].

10.2.1.3 UV scavengers

A film based on nanocrystalline titania (TiO_2) is a generally used content for UV absorption. Metal doping enhances the viewable absorption of TiO_2 and enhances the photocatalytic behavior under ultraviolet rays. Doping TiO_2 with silver was confirmed to have significantly increased photocatalytic inactivation of bacteria. This composition led to excellent antibacterial activities in nanocomposite with PVC of TiO_2/Ag^+ nanoparticles [40–44].

10.2.2 Nanotechnological Systems for Smart/Intelligent Packaging

For the sensation of biochemical or microbial changes in food, nanomaterials in smart/intelligent packaging are used to detect particular pathogens developed in foodstuffs or particular gases caused food spoilage [28] that permit environmental protection [45]. Several companies aimed to develop smart packaging for the optimal protection of products using nanotechnology [46]. The chemical sensors that identify colored changes are presently used by Nestle, British Airways, and the MonoPrix Supermarket [47, 48]. The interactional function of a product is improved by smart or intelligent packaging systems. Such sophisticated packaging identifies all the features of packaged food and uses different methods to document and transmit information of food safety and digestibility as regards to the current quality or condition. Such technology uses various revolutionary methods, for example, oxygen sensors, nanosensors, freshness indicators, time–temperature indicators (TTIs), and others [16, 49, 50].

10.2.2.1 Nanosensors

The existence of pathogens, mycotoxins, and microbes in food can be sensed with nanosensors [51]. Nanoparticle-based sensors, array biosensors, nanocantilevers, nanoparticle in solution, and nanotest strips are all used as various types of nanosensors in the food packaging industry [52]. Advanced nanosensor packages to alert the consumers about the spoilage or pathogens of products use engineered nanosensors like electronic "noses" or "tongues" to "taste" or "smell" aromas and flavors [22, 53–55].

Nanosensor packaging is helpful for tracking of external or internal conditions across the food supply chain of food products, pellets, and containers. In plastic packaging, nanosensors can identify food gases when the food products are spoiled and when packaging itself changes its color, to warn the customers. Also, silicate-packed nanoparticles film can limit the O_2 influx into the container, and it can keep the food fresh while keeping the moisture out of its packaging. Also, the mold development in the refrigerator can be prevented. *Escherichia coli* contamination in packaged food has been detected by sensors. Nanotechnology can minimize packaging disposal in connection with packaged food and can also promote the protection and enhance the shelf life of the fresh food. Modern technology enables food products to identify microbial pathogens within 2–7 days. Consequently, it is also used for the detection of spoilage microbes, pesticides, and toxins [52].

10.2.2.2 TTIs

The major factors that are liable for the decreased shelf life of food are the temperature exploitations that affect food during transit and distribution. TTIs are beneficial for monitoring thermal background during the storage, handling, and distribution of food to counteract different temperature abuses. TTIs enable distributors to assure that food is kept at an appropriate temperature, to benefit consumers to determine the food product quality they are buying, and to help producers monitor the food supplied across the supply chain. TTIs help in improving food products and services [3]. On the enzymatic base and polymeric and biological reactions, various types of TTI activities have been produced. It is important to check changes in temperature and time parameters from the development of food products to till they reach to the customers to assure the safety and quality of the products that require a certain temperature [56].

10.2.2.3 Oxygen sensors

Lee et al. (2005) established an ultraviolet-based colorimetric oxygen indicator, used to decrease methylene blue in a polymer encapsulation medium in the photosensitization of TiO_2 nanoparticles for triethanolamine [57, 58]. Ultraviolet irradiation responds to sensor bleaching that stays colorless up to the oxygen exposure. The level of color change is proportional to exposure of oxygen. Nanocrystalline SnO_2 was used by Mills and Hazafy (2009) [59] as a photosensitizer and colorimetric oxygen indicator. They used glycerol as an electron donor, hydroxyethyl cellulose (HEC) as encapsulating polymer, along with methylene blue dye as an indicator. When the ultraviolet-B light is exposed, the indicator is bleached and photoreduction the coloration by SnO_2 nanoparticles. Another study carried out by Mihindukulasuriya and Lim (2013) [60] developed an electrospinning approach for the activation of the ultraviolet-based oxygen indicator membrane. The encapsulation in the electrospun polyfibers of active ingredients like TiO_2 nanoparticles, glycerole, and methylene blue improves the membrane's oxygen sensitivity [57–60].

10.2.2.4 Freshness indicators

The product quality is demonstrated directly by a freshness indicator. This is generally labeled on the container. The variable is generally focused on detecting the first type of changes, such as pH, composition of gas, etc. These modifications are identified by the indicator and converted into a reaction that can be generally measured by color and correlated to food's freshness value. This reaction may be affected by changes in microbes' metabolism substances, such as carbon dioxide, glucose, ethanol, organic amines and acids, and volatile nitrogen and sulfur compounds in storage, which indicate the growth of microorganisms [61]. Various kinds of gas sensors are constructed, which translate chemical interactions between particles on the surface into a reaction signal, depending on applicable surface properties studies of materials. Many electro-active conjugated polymers were produced that employed either chemical or electrochemical oxidation reactions. Thus, they play an important role in food preservation, due to their electronic, electrical, optical, and magnetic characteristics linked to their conjugated pi-electron backbones [28, 62].

10.2.3 Nanotechnological Systems for Improved Packaging

Many nanoparticles that include polymers of up to 5% w/w nanoparticles with clay nanoparticles with enhanced barrier structures (80–90% lowered) are used in the manufacture of containers and bottles (i.e. edible oils, carbonated drinks, beer drinks, etc.) and films that have a range of nanoparticles (also known as nanocomposites). The United States Food and Drug Administration (USFDA) have authorized the use of nanocomposite in food contact [9, 10, 19, 24, 47, 55]. Table 10.1 describes common nanomaterials for food packaging applications.

Table 10.1 Food packaging applications of some common nanomaterials

Application	Description	Key nanomaterial
Active packaging	Integration of nanomaterials with antimicrobial or other characteristics (e.g., antioxidants, ultraviolet absorption), which are supposed to activate and result in taste, freshness, and shelf-life impact on the packaged food.	Ag, TiO_2

(*Continued*)

Table 10.1 (*Continued*)

Application	Description	Key nanomaterial
Smart packaging	Nanosensors are integrated to track and report the status of food (e.g., oxygen indicators, freshness indicators, and pathogen indicators).	ZnO, CNTs
Improved packaging	Nanomaterials are included in the packaging to boost physical presentation, robustness, obstacle characteristics, biodegradation, biocompatibility, and climate-friendly.	Clays, SiO_2

10.3 Biodegradable Nanocomposites

Nanocomposites, a synthesis of conventional food packaging materials with nanoparticles, are becoming more common in the food packaging industry. It shows a strong mechanical efficiency and difficult resistant, as well as its notable antimicrobial spectrum. In a continuous or discontinuous stage, nanocomposites are typically composed of a polymer matrix [15, 39]. Polymer nanocomposites (PNCs), polymer's mixtures of inorganic or organic additives with specific geometries (fibers, flakes spheres, particulates), are less than 100 nm in size and currently introduced as new packaging materials [63, 64].

A significant function is played by the aspect ratio of packaging filler material (the proportion of the largest filler to the smallest). Further, general surface areas with correlated high reinforcement effects are available in fillers with a higher aspect ratio [65, 66]. Diverse nanomaterials, including silica [67], clay [68], organic clay [69], grapheme [70], nanocrystals of polysaccharides [71], carbon nanotubes [72, 73], chitosan-based [74], other metal nanoparticles, for example, ZnO_2 [75], colloidal Cu [76], or Ti, and other nanoparticles, are thoroughly studied as fillers [77].

It is interesting that nanocomposites based on biopolymers can even have environmental benefits as compared to traditional plastics. Othman (2014) looked at various forms, for the use of biocomposite substances for food packaging activities, of nano-sized fillers and biopolymers [78]. The theoretical production of biodegradable nanocomposites using starch, poly-lactic acid (PLA), or polyhydroxybutyrate (PHB) biopolymers was studied by Goyal and Goyal (2012). Large numbers of nanocomposite biopolymers are being utilized using several organic molecules, including nylons, polyolefins, ethylene vinylacetate (EVA), polyamides (PA), polyimides (PI), polystyrene (PS), copolymers, epoxy-resins, polyurethane (PU), and polyethylene terephthalate (PET) [79]. Ray et al. (2002) investigated that the polylactic acid-based bionanocomposite has a speedier biodegeneration rate than polylactic acid without such additives, when nanofillers are spread within bio-compatible polymer polylactic acid [80].

10.4 Other Polymer Matrix-Based Nanocrystal Systems

Cellulose is a specific and widely available biomaterial possesses a range of beneficial characteristics such as renewability, biodegradability, and non-toxicity. It is a polysaccharide consisting of repeating β-D glucopyranose units and highly functional with three groups of hydroxyl per anhydrous glucose unit. In several industries, like food and pharmaceutical, cellulose is widely utilized in pure form or as a derivative in the manufacture of a broad range of products [81]. In the 1950s, Ranby first obtained a liquid colloidal solution of cellulose micelle by sulfuric acid treatment [82]. Mukherjee et al. reported first in 1953, the images of dried powdered nano cellulose with needle shaped morphology. The term "nano cellulose" was first used to characterize the gel-like substance of wood pulp homogenization at high temperature and high pressure at ITT Rayonier Lab in Whippany, New Jersey, the USA, in the late 1970s by Turbak et al. [83]. In the early 1980s, ITT Rayonier granted several patents to explain how these cellulosic nanomaterials were prepared and applied [84]. Table 10.2 depicts the list of identified nanotechnologies used for food packaging and preservation utilized in the food industry during past few decades.

In recent decades, scientists have identified unique features of cellulose nanocrystals (CNCs) like excellent mechanical properties [97], self-assembling capabilities, surface interface effect [98], photonic properties, thixotropic function, and chemical reactivity [99, 100]. Several types of nano celluloses, including CNF, CNC, and BNC, have been obtained up to now. Acid hydrolysis can produce CNC from native fibers that lead to extremely crystalline and rigid nanoparticles, which are shorter than the nano fibril obtained by homogenization, microfludisation, or by grinding routes [101]. CNC has been studied and a broad range of applications in a wide variety of fields have recently been developed. It may be modified to meet new product enhancement specifications or to produce new applications that have previously been considered impossible. In the food industry, the quality of food and beverage products is constantly increasing with the use of nanocrystals [102, 103].

Various polymers were grafted onto the CNC surface using the grafting-on approach to improve the dispersion and matrix compatibility. Habibi et al. grafted a poly-caprolactone CNC (PCL) with isocyanates coupling agent, and applied high modulus and good ductility were shown in nanoscale films made with PCL-graffed nanoparticles, which represent the strengthening of CNC's polymer matrix. The same approach has been taken by Cao et al. [104] with CNC graft-borne polyurethane chains. In the presence of carbodimide derivatives, Mangalam et al. used TEMPO oxidized CNCs in functional groups using-COOH to graft tranded ADNs [105]. Ljungberg et al. grafted maleate, a low molecular weight functionalized atactic polypropylene on tunicate CNCs extract. Films cast using grafted CNCs have demonstrated strong matrix compatibility and greater transparency and tensile unit. Amina functionally administered polyethylene glycol (PEG) has been grafted by Araki et al. [106] to obtain high-stability CNC suspensions from the TEMPO oxidized CNC by carbodimide chemistry. CNCs modified by the surface have significant properties, including high-colloidal stability, increased dispersion in polar and organic solvents, and thermo-reversible aggregation [107].

There are basic requirements of applicable materials, such as thermal and tensile properties; however, food packaging deserves

Table 10.2 List of identified nanotechnologies used for food packaging and preservation utilized in the food industry

Nanotechnique	Composition	Application	Advantage	Reference
Nanoparticles	Polymeric nanoparticles	Food packaging and preservation	Identified to be well-organized delivery systems and are bactericidal	[85]
			Reduce moisture leakage	**[86–88]**
	Silicon dioxide	Food packaging and preservation	Used as drying and anticaking agents Absorb hygroscopic interaction of water molecules in milk	
	Silver nanoparticles	Food packaging and preservation	Secure food from microbial infestation and act as antibacterial agents Enhance fruit's and vegetable's shelf life through the absorption and degradation of ethylene	**[85,89]**
	Zinc oxide	Food packaging and preservation	Reduce the oxygen flow into the containers	[89]
Nanocomposites	Nanoclay (polymer and nanoparticles)	Food packaging	Used to build gas obstacles to reduce carbon dioxide leakage from carbonated drink bottles	[90]
	Enzyme immobilization	Food packaging	Provide a wider area and quicker transfer	[91]
	Nano core	Food packaging	Used in manufacturing of plastic beer bottles to prevent carbon dioxide from being emitted from beverages	[92]
	NanoCeram PAC	Food packaging	Help quickly absorb unpleasant components that can cause foul smell and repulsive taste	[91]
Nanosensors	Nano-smart dust	Food packaging	Find pollution of any kind	[93]
	Nanobiosensors	Food packaging	Identify viruses and bacteria	[93]
	Abuse indicators	Food packaging	Calculate optimal temperature	[94]
	Carbon black and polyaniline	Food packaging	Detect carcinogens in food materials Identify pathogens transmitted from foodstuffs Detect microorganisms that normally infest the food	[95, 96]

more specific features. Food packaging is aimed at ensuring quality of life and safety of nutrition. Barrier properties are therefore important to slowdown the oxidation of water vapor and gas exchange between the food and the atmosphere [108]. For this reason, the proposed food safety materials are critical to water vapor (WVP) and oxygen permeability (OP). Transparency is also important, as the customer can see by visual inspection about the quality and condition of the material. With, the use of CNCs in the biodegradable polymers, matrix significantly improves the mechanical properties of the neat polymer, which leads to improve the water and oxygen barrier properties. Water resistance is also preferred because contacts with wet surfaces can lead to the disintegration of the film [96].

10.5 Conclusions and Future Prospects

Nanotechnology is becoming ever more important to the food industry. It offers a variety of exciting advantages for food packaging, including improved food quality and sanitation, enhanced food's shelf life, etc. With many innovative applications, its uses for food packaging are promising as nano-based packaging materials that provide excellent protection toward microbial attack, and nanosensors are capable of detecting microorganisms or chemical contaminants at extremely low levels. The future of food security is largely dependent on nanotech progress, the integration of a nanosensor in a food container and the breakdown of IP solutions. This new systems can help detect, monitor, track, record, and communicate through the supply chain.

The application of nanotechnology offers considerable advantages even in the early stages to improve the properties of packaging materials and requires continuing investment to finance research and development to better understand nanotechnology's advantages and disadvantages in packaging system. In the range of advanced functional properties, the use of nanotechnology to make food packaging can offer numerous benefits. It can provide packaging materials with improved processing functionalities, such as health and packaging, transportability, and lower costs.

References

1. European Food Safety Authority (EFSA-Q-2005-041) (2009). Scientific opinion of the scientific committee on a request from the European commission on the potential risks arising from nanoscience and nanotechnologies on food and feed safety, *The European Food Safety Authority (EFSA) J.*, **958**, pp. 1–39.

2. Bajpai, V. K., Kamle, M., Shukla, S., Mahato, D. K., Chandra, P., Hwang, S. K., Kumar, P., Huh, Y. S., Han, Y. K. (2018). Prospects of using nanotechnology for food preservation, safety, and security, *J. Food Drug Anal.*, **26**, pp. 1201–1214.

3. Sharma, C., Dhiman, R., Rokana, N., Panwar, H. (2017). Nanotechnology: An untapped resource for food packaging, *Front. Microbiol.*, **8**, pp. 1–22.

4. Handford, C. E., Dean, M., Henchion, M., Spence, M., Elliott, C. T., Campbell, K. (2014). Implications of nanotechnology for the agri-food industry: Opportunities, benefits and risks, *Trends Food Sci. Technol.*, **40**, pp. 226–241.

5. Srinivas, P. R., Philbert, M., Vu, T. Q., Huang, Q., Kokini, J. L., Saos, E., Chen, H., Peterson, C. M., Friedl, K. E., McDade-Ngutter, C., Hubbard, V., Starke-Reed, P., Miller, N., Betz, J. M., Dwyer, J., Milner, J., Ross, S. A. (2010). Nanotechnology research: Applications in nutritional sciences, *J. Nutr.*, **140**, pp. 119–124.

6. Ranjan, S., Dasgupta, N., Chakraborty, A. R., Samuel, S. M., Ramalingam, C., Shanker, R., Kumar, A. (2014). Nanoscience and nanotechnologies in food industries: Opportunities and research trends, *J. Nanoparticle Res.*, **16**, pp. 2464–2472.

7. Luduena, L. N., Alvarez, V. A., Vazquez, A. (2007). Processing and microstructure of PCL/clay nanocomposites, *Mater. Sci. Eng.: A.*, **460**, pp. 121–129.

8. Chen, H., Weiss, J., Shahidi, F. (2006). Nanotechnology in nutraceuticals and functional foods, *Food Technol.*, **60**, pp. 30–36.

9. Silvestre, C., Duraccio, D., Sossio, C. (2011). Food packaging based on polymer nanomaterials, *Prog. Polym. Sci.*, **36**, pp. 1766–1782.

10. Duncan, T.V. (2011). Applications of nanotechnology in food packaging and food safety: Barrier materials, antimicrobials and sensors, *J. Colloid Interface Sci.*, **363**, pp. 1–24.

11. Koo, O. M., Rubinstein, I., Onyuksel, H. (2005). Role of nanotechnology in targeted drug delivery and imaging: A concise review, *Nanomed. Nanotechnol. Biol. Med.*, **1**, pp. 193–212.

12. Yan, S. S., Gilbert, J. M. (2004). Antimicrobial drug delivery in food animals and microbial food safety concerns: An overview of in vitro and in vivo factors potentially affecting the animal gut microflora, *Adv. Drug Deliv. Rev.*, **56**, pp. 1497–1521.

13. Wesley, S. J., Raja, P., Raj, A. A., Tiroutchelvamae, D. (2014). Review on-nanotechnology applications in food packaging and safety, *Int. J. Eng. Res.*, **3**, pp. 645–651.

14. Ezhilarasi, P. N., Karthik, P., Chhanwal, N., Anandharamakrishnan, C. (2013). Nanoencapsulation techniques for food bioactive components: A review, *Food Bioprocess Technol.*, **6**, pp. 628–647.

15. Arora, A., Padua, G. W. (2010). Review: Nanocomposites in food packaging, *J. Food Sci.*, **75**, pp. R43–R49.

16. Majid, I., Nayik, G. A., Dar, M. S., Nanda, V. (2016). Novel food packaging technologies: Innovations and future prospective, *J. Saudi Soc. Agric. Sc*i., 17(4), pp. 454–462.

17. Kumar, R., Munstedt, H. (**2005**). Silver ion release from antimicrobial polyamide/silver composites, *Biomaterial*, **26**, pp. **2081–2088**.

18. Monteiro, D. R., Gorup, L. F., Takamiya, A. S., Ruvollo-Filho, A. C., de Camargo, E. R., Barbosa, D. B. (2009). *Int. J. Antimicrob. Agents*, **34**, pp. 103–112.

19. Chaudhry, Q., Scotter, M., Blackburn, J., Ross, B., Boxall, A., Castle, L., Aitken, R., Watkins, R. (2008). Applications and implications of nanotechnologies for the food sector, *Food Addit. Contamin.*, **25**, pp. 241–258.

20. Doyle, M. E. (2006). Nanotechnology: A brief literature review. Available: http://www.wisc.xv.edu/ fri/briefs/FRIBrief_Nanotech_Lit_Rev.pdf (Retrieved on 10.02.2020)

21. Miller, G., Senjen, R. (2008). Out of the laboratory and on to our plates: Nanotechnology in food and agriculture. Available: http://www.foeeurope.org/activities/nanotechnology/Documents/Nano_food_report.pdf (Retrieved on 10.02.2020)

22. Asadi, G., Mousavi, M. (2006). Application of nanotechnology in food packaging. Available: http://iufost.edpsciences.org (Retrieved on 10.02.2020)

23. Lopez-Rubio, A., Gavara, R., Lagaron, J. M. (2006). Bioactive packaging: Turning foods into healthier foods through biomaterials, *Trends Food Sci. Technol.*, **17**, pp. 567–575.

24. Brody A. L. (2007). Case studies on nanotechnologies for food packaging, *Food Technol.*, **7**, pp. 102–107.

25. Morones, J. R., Elechiguerra, J. L., Camacho, A., Holt, K., Kouri, J. B., Ramírez, J. T., Yacaman, M. J. (2005). The bactericidal effect of silver nanoparticles, *Nanotechnol.*, **16**, pp. 2346–2365.

26. Li, Q., Wu, P., Shang, J. K. (2014). Nanotechnology applications for clean water, Street, A., Sustich, R., Duncan, J., Savage, N. (eds.), Chapter 19 "Nanostructured visible-light photocatalysts for water purification", William Andrew Publishing, Oxford, pp. 297–317.

27. Kang, S., Pinault, M., Pfefferle, L. D., Elimelech, M. (2007). Single-walled carbon nanotubes exhibit strong antimicrobial activity, *Langmuir*, **23**, pp. 8670–8673.

28. Kuswandi, B., Wicaksono, Y., Abdullah, A., Heng, L. Y., Ahmad, M. (2011). Smart packaging: Sensors for monitoring of food quality and safety, *Sensing Instrumentation Food Quality Safety*, **5**, pp. 137–146.

29. Cioffi, N., Torsi, L., Ditaranto, N., Tantillo, G., Ghibelli, L., Sabbatini, L., Traversa, E. (2005). Copper nanoparticle/polymer composites with antifungal and bacteriostatic properties, *Chem. Mater.*, **17(21)**, pp. 5255–5262.

30. Huang, L., Li, D. Q., Lin, Y. J., Wei, M., Evans, D. G., Duan, X. (**2005**). Controllable preparation of nano-MgO and investigation of its bactericidal properties, *J. Inorg. Biochem.*, **99**, pp. 986–993.

31. Lin, Y. J., Li, D. Q., Wang, G., Huang, L., Duan, X. (2005). Preparation and bactericidal property of MgO nanoparticles on c-Al**2O3**, *J. Mater. Sci.: Mater. Medic.*, **16**, pp. 53–56.

32. Qi, L. F., Xu, Z. R., Jiang, X., Hu, C., Zou, X. (2004). Preparation and antibacterial activity of chitosan nanoparticles, *Carbohydr. Res.*, **339**, pp. 2693–2700.

33. Stoimenov, P., Klinger, R. L., Marchin, G. L., Klabunde, K. J. (2002). Metal oxide nanoparticles as bactericidal agents, *Langmuir*, **18**, pp. 6679–6686.

34. Gu, H. W., Ho, P. L., Tong, E., Wang, L., Xu, B. (2003). Presenting vancomycin on nanoparticles to enhance antimicrobial activities, *Nano Lett.*, **3**, pp. 1261–1263.

35. Xiao-e, L., Green, A. N., Haque, S. A., Mills, A., Durrant, J. R. (2004). Light-driven oxygen scavenging by titania/polymer nanocomposite films, *J. Photochem. Photobiol. A: Chem.*, **162**, pp. 253–259.

36. Kuswandi, B. (2016). *Nanoscience in Food and Agriculture*, vol. 1: *Sustainable Agriculture Reviews*, eds. Ranjan, S., Dasgupta, N., Lichtfouse, E., Chapter 6 "Nanotechnology in food packaging". Springer Switzerland, pp. 151–183.

37. Mihindukulasuriya, S. D. F., Lim, L. T. (2014). Nanotechnology development in food packaging: A review, *Trends Food Sci. Technol.*, **40**, pp. 149–167.

38. Echegoyen, Y. (2015). Nanotechnologies in food and agriculture, Rai, M., Ribeiro, C., Mattoso, L., Duran, N. (eds.), Chapter 7 "Nano-developments for food packaging and labeling applications", Springer, Cham, pp. 141–168.

39. Montazer, M., Harifi, T. (2017). Food packaging: Nanotechnology in the agri-food industry, Vol. 7, Grumezescu, A. M. (ed.), Chapter 16 "New approaches and future aspects of antibacterial food packaging: From nanoparticles coating to nanofibers and nanocomposites, with foresight to address the regulatory uncertainty", Academic Press, London, pp. 533–559.

40. Gelover, S., Gomez, L. A., Reyes, K., Leal, M. T. (2006). A practical demonstration of water disinfection using TiO_2 films and sunlight, *Water Res.*, **40**, pp. 3274–3280.

41. Anpo, M., Kishiguchi, S., Ichihashi, Y., Takeuchi, M., Yamashita, H., Ikeue, K., Che, M. (2001). The design and development of second-generation titanium oxide photocatalysts able to operate under visible light irradiation by applying a metal ion-implantation method, *Res. Chem. Intermed.*, **27(4/5)**, pp. 459–467.

42. Page, K., Palgrave, R. G., Parkin, I. P., Wilson, M., Savin, S. L. P., Chadwick, A. V. (2007). Titania and silver-titania composite films on glass-potent antimicrobial coatings, *J. Mater. Chem.*, **17**, pp. 5–104.

43. Reddy, M. P., Venugopal, A., Subrahmanyam, M. (2007). Hydroxyapatite supported Ag-TiO_2 as Escherichia coli disinfection photocatalyst, *Water Res.*, **41**, pp. 379–386.

44. Cheng, Q., Li, C., Pavlinek, V., Saha, P., Wang, H. (2006). Surface-modified antibacterial TiO_2/Ag^+ nanoparticles: Preparation and properties, *Appl. Surf. Sci.*, **252**, pp. 4154–4160.

45. Bagchi, A. (2012). Intelligent sensing and packaging of foods for enhancement of shelf life: Concepts and applications, Int. J. Scientific Eng. Res., **3(10)**, pp.1–13.

46. Nura, A. (2018). Advances in food packaging technology: A review, J. Postharvest Technol., **6**, pp. 55–64.

47. Pal, M. (2017). Nanotechnology: A new approach in food packaging, J. Food Microbiol. Safety Hygiene, 2, pp. 121–136.

48. Pehanich, M. (2006). Small gains in processing, packaging, Food Proc., **11**, pp. 46–48.

49. Kerry, J. P., Ogrady, M. N., Hogan, S. A. (2006). Past, current and potential utilisation of active and intelligent packaging systems for meat and muscle-based products: A review, Meat Sci., **74**, pp. 113–130.

50. Bouwmeester, H., Dekkers, S., Noordam, M., Hagens, W., Bulder, A., de Heer, C., Voorde, S. T., Wijnhoven, S., Sips, A. (2007). Health impact of nanotechnologies in food production, RIKILT/RIVM Report 2007.014. Available online at: http://edepot.wur.nl/120669

51. Bratovcic, A. L., Odobasic, A. S., Catic, S., Šestan, I. (2015). Application of polymer nanocomposite materials in food packaging, Croat. J. Food Sci. Technol., **7**, pp. 86–94.

52. Adeyeye, S. A. O., Fayemi, O. E. (2019). Nanotechnology and food processing: Between innovations and consumer safety, J. Culinary Sci. Technol., **17(5)**, pp. 435–452.

53. Joseph, T., Morrison, M. (2006). Nanotechnology in agriculture and food. Available online: www.nanoforum.org/nf06~modul~showmore~folder~99999~scid~377~.html?action=longview_publication

54. Scrinis, G., Lyons, K. (2007). The emerging nanocorporate paradigm: Nanotechnology and the transformation of nature, food and agrifood systems, Int. J. Sociol. Food Agric., **15**, pp. 22–44.

55. Sozer, N., Kokini, J. L. (2009). Nanotechnology and its applications in the food sector, Trends Biotechnol., **27**, pp. 82–89.

56. Fuertes, G., Soto, I., Carrasco, R., Vargas, M., Sabattin, J., Lagos, C. (2016). Intelligent packaging systems: Sensors and nanosensors to monitor food quality and safety, J. Sensors, **4046061**, pp. 1–8. doi: 10.1155/2016/4046061.

57. Lee, S. K., Sheridan, M., Mills, A. (2005). Novel UV-activated colorimetric oxygen indicator, Chem. Mater., **17**, pp. 2744–2751.

58. Gutierrez-Tauste, D., Domenech, X., Casañ-Pastor, N., Ayllon, J. A. (2007). Characterization of methylene blue/TiO_2 hybrid thin films prepared by the liquid phase deposition (LPD) method: Application for fabrication of light activated colorimetric oxygen indicators, J. Photochem. Photobiol. A, **187**, pp. 45–52.

59. Mills, A., Hazafy, D. (2009). Nanocrystalline SnO_2-based, UVB-activated, colourimetric oxygen indicator, Sensor Actuat. B Chem., **36**, pp. 344–349.

60. Mihindukulasuriya, S. D. F., Lim, L. T. (2013). Oxygen detection using UV-activated electrospun poly (ethylene oxide) fibers encapsulated with TiO_2 nanoparticles, J. Mater. Sci., **48**, pp. 5489–5498.

61. Arvanitoyannis, I. S., Stratakos, A. C. (2012). Application of modified atmosphere packaging and active/smart technologies to red meat and poultry: A review, Food Bioproc. Technol., **5**, pp. 1423–1446.

62. Ahuja, T., Mir, I. A., Kumar, D., Rajesh. (2007). Biomolecular immobilization on conducting polymers for biosensing applications, Biomaterial, **28**, pp. 791–805.

63. Prateek, Thakur, V. K., Gupta, R. K. (2016). Recent progress on ferroelectric polymer-based nanocomposites for high energy density capacitors: Synthesis, dielectric properties, and future aspects, *Chem. Rev.*, **116**, pp. 4260–4317.

64. Gokularaman, S., Stalin, Cruz, A., Pragalyaashree, M. M., Nishadh, A. (2017). Nanotechnology approach in food packaging: A review, *J. Pharm. Sci. Res.*, **9**, pp. 1743–1749.

65. Dalmas, F., Cavaille, J. Y., Gauthier, C., Chazeau, L., Dendievel, R. (2007). Visco elastic behavior and electrical properties of flexible nanofiber filled polymer nanocomposites: Influence of processing conditions, *Comp. Sci. Technol.*, **67**, pp. 829–839.

66. Rafieian, F., Simonsen, J. (2014). Fabrication and characterization of carboxylated cellulose nanocrystals reinforced glutenin nanocomposite, *Cellulose*, **21**, pp. 4167–4180.

67. Bracho, D., Dougnac, V. N., Palza, H., Quijada, R. (2012). Functionalization of silica nanoparticles for polypropylene nanocomposite applications, *J. Nanomater.*, **263915**, 1–8. doi: 10.1155/2012/263915

68. Schütz, M. R., Kalo, H., Lunkenbein, T., Gröschel, A. H., Müller, A. H., Wilkie, C. A., Breu, J. (2011). Shear stiff, surface modified, mica-like nanoplatelets: A novel filler for polymer nanocomposites, *J. Mater. Chem.*, **21**, pp. 12110–12116.

69. Ham, M., Kim, J. C., Chang, J. H. (2013). Thermal property, morphology, optical transparency, and gas permeability of PVA/SPT nanocomposite films and equi-biaxial stretching films, *Polym. Korea*, **37**, pp. 579–586.

70. Lee, Y., Kim, D., Seo, J., Han, H., Khan, S. B. (2013). Preparation and characterization of poly (propylene carbonate)/exfoliated graphite nanocomposite films with improved thermal stability, mechanical properties and barrier properties, *Polym. Int.*, **62**, pp. 1386–1394.

71. Lin, N., Huang, J., Dufresne, A. (2012). Preparation, properties and applications of polysaccharide nanocrystals in advanced functional nanomaterials: A review, *Nanoscale*, **4**, pp. 3274–3294.

72. Swain, S. K., Pradhan, A. K., Sahu, H. S. (2013). Synthesis of gas barrier starch by dispersion of functionalized multiwalled carbon nanotubes, *Carbohydr. Polym.*, **94**, pp. 663–668.

73. Chang, P. R., Jian, R., Yu, J., Ma, X. (2010). Starch-based composites reinforced with novel chitin nanoparticles, *Carbohydr. Polym.*, **80**, pp. 420–425.

74. Sandquist, D. (2013). New horizons for microfibrillated cellulose, *Appita J.*, **66**, pp. 156–162.

75. Esthappan, S. K., Sinha, M. K., Katiyar, P., Srivastav, A., Joseph, R. (2013). Polypropylene/zinc oxide nanocomposite fibers: morphology and thermal analysis, *J. Polym. Mater.*, **30**, pp. 79–89.

76. Cardenas, G., Díaz, J., Meléndrez, M., Cruzat, C., Cancino, A. G. (2009). Colloidal Cu nanoparticles/chitosan composite film obtained by microwave heating for food package applications, *Polym. Bull.*, **62**, pp. 511–524.

77. Li, R., Liu, C. H., Ma, J., Yang, Y. J., Wu, H. X. (2011). Effect of orgtitanium phosphonate on the properties of chitosan films, *Polym. Bull.*, **67**, pp. 77–89.

78. Othman, S. H. (2014). Bio-nanocomposite materials for food packaging applications: Types of biopolymer and nano-sized filler, *Agric. Sci. Procedia*, **2**, pp. 296–303.

79. Goyal, S., Goyal, G. K. (2012). Nanotechnology in food packaging a critical review, *Russ. J. Agric. Socio-Econ. Sci.*, **10**, pp. 14–24.

80. Ray, S., Maiti, P., Okamoto, M., Yamada, K., Ueda, K. (2002). New polylactide/layered silicate nanocomposites: Preparation, characterization and properties, *Macromolecule*, **35**, pp. 3104–3110.

81. Turbak, A. F., Snyder, F. W., Sandberg, K. R., (1983). Microfibrillated cellulose, a new cellulose product: Properties, uses, and commercial potential, Applied Polymer Symposium United States, ITT Rayonier Inc., Shelton, WA, *J. Appl. Polym. Sci.*, **37**, CONF-8205234-Vol. 2.

82. Ranby, B. G., Banderet, A., Sillén, L. G. (1949). Aqueous colloidal solutions of cellulose micelles, Acta Chemica Scandinavica, **3**, pp. 649–650.

83. Turbak, A. F., Snyder, F. W., Sandberg, K. R. (1984). Nonwovens Symposium, Myrtle Beach, SC, p. 115.

84. Dhar, P., Bhardwaj, U., Kumar, A., Katiyar, V. (2014). Food additives and packaging Vol. 1162: ACS symposium series, Komolprasert, V., Turowski, P. (eds.), Chapter "Cellulose nanocrystals: A potential nanofiller for food packaging applications", American Chemical Society, Washington DC, pp. 197–239.

85. Bouwmeester, H., Dekkers, S., Noordam, M. Y., Hagens, W. I., Bulder, A. S., De Heer, C., Sips, A. J. (2009). Review of health safety aspects of nanotechnologies in food production, *Regulatory Toxicol. Pharmacol.*, **53**, pp. 52–62.

86. Coma, V. (2008). Bioactive packaging technologies for extended shelf life of meat-based products, *Meat Sci.*, **78**, pp. 90–103.

87. Horner, S. R., Mace, C. R., Rothberg, L. J., Miller, B. L. (2006). A proteomic biosensor for enteropathogenic E. coli, *Biosensors Bioelectronics*, **21**, pp.1659–1663.

88. Jones, N., Ray, B., Ranjit, K. T., Manna, A. C. (2008). Antibacterial activity of ZnO nanoparticle suspensions on a broad spectrum of microorganisms, *FEMS Microbiol. Lett.*, **279**, pp.71–76.

89. Arshak, K., Adley, C., Moore, E., Cunniffe, C., Campion, M., Harris, J. (2007). Characterisation of polymer nanocomposite sensors for quantification of bacterial cultures, *Sensors Actuators B: Chem.*, **126**, pp. 226–231.

90. Yotova, L., Yaneva, S., Marinkova, D. (2013). Biomimetic nanosensors for determination of toxic compounds in food and agricultural products (review), J. Chem. Technol. Metallurgy, **48**, pp. 215–227.

91. Burdo, O. G. (2005). Nanoscale effects in food-production technologies, J. Eng. Phys. Thermophys., **78**, pp. 90–96.

92. Flanagan, J., Singh, H. (**2006**). Microemulsions: A potential delivery system for bioactives in food, Crit. Rev. Food Sci. Nutr., **46**, pp. 221–237.

93. Coles D., Frewer, L. J. (2013). Nanotechnology applied to European food production: A review of ethical and regulatory issues, Trends Food Sci. Technol., **34**, pp. 32–43.

94. Su, H. C., Zhang, M., Bosze, W., Lim, J. H., Myung, N. V. (2013). Metal nanoparticles and DNA co-functionalized single-walled carbon nanotube gas sensors, Nanotechnology, **24(50)**, pp. 505502.

95. Vidhyalakshmi, R., Bhakyaraj, R., Subhasree, R. S. (2009). Encapsulation 'the future of probiotics': A review, Adv. Biological Res., 3, pp. 96–103.

96. Yuan, Y., Gao, Y., Zhao, J., Mao, L. (2008). Characterization and stability evaluation of β-carotene nanoemulsions prepared by high pressure homogenization under various emulsifying conditions, Food Res. Int., **41**, pp. 61–68.

97. Kamal, M. R., Khoshkava, V. (2015). Effect of cellulose nanocrystals (CNC) on rheological and mechanical properties and crystallization behavior of PLA/CNC nanocomposites, *Carbohydr. Polym.*, **123**, pp. 105–114.

98. Routara, B. C., Bandyopadhyay, A., Sahoo, P. (2009). Roughness modeling and optimization in CNC end milling using response surface method: Effect of workpiece material variation, *Int. J. Adv. Manuf. Technol.*, **40**, pp. 1166–1180.

99. Guidetti, G., Atifi, S., Vignolini, S., Hamad, W. Y. (2016). Flexible photonic cellulose nanocrystal films, *Adv. Mater.*, **28**, pp. 10042–10047.

100. Buchanan, M., Knaapila, M., Helgesen, G., Hoeyer, H. (2013). Method for assembling conductive particles into conductive pathways and sensors thus formed, US Patent 2, 649,439.

101. Vilarinho, F., Silva, A. S., Vaz, M. F., Farinha, J. P. (2017). Nanocellulose in green food packaging, *Crit. Rev. Food Sci. Nutr.*, **5**, pp. 1526–1537.

102. Salajková, M., Berglund, L. A., and Zhou, Q. (2012). Hydrophobic cellulose nanocrystals modified with quaternary ammonium salts, *J. Mater. Chem.*, **22**, p. 19798.

103. Mu, R., Hong, X., Ni, Y., Li, Y., Pang, J., Wang, Q., Xiao, J., Zheng, Y. (2019). Recent trends and applications of cellulose nanocrystals in food industry, *Trends Food Sci. Technol.*, **93**, pp. 136–144.

104. Habibi, Y., Dufresne, A. (2008). Highly filled bionanocomposites from functionalized polysaccharide nanocrystals, *Biomacromolecule*, **9**, pp. 1974–1980.

105. Mangalam, A. P., Simonsen, J., Benight, A. S. (2009). Cellulose/DNA hybrid nanomaterials, *Biomacromolecule*, **10**, pp. 497–504.

106. Ljungberg, N., Bonini, C., Bortolussi, F., Boisson, C., Heux, L., Cavaillé, J. Y. (**2005**). New nanocomposite materials reinforced with cellulose whiskers in atactic polypropylene: Effect of surface and dispersion characteristics, Biomacromolecule, **6**, pp. 2732–2739.

107. Araki, J., Wada, M., Kuga, S. (2001). Steric stabilization of a cellulose microcrystal suspension by poly (ethylene glycol) grafting, *Langmuir*, **17**, pp. 21–27.

108. Azeredo, H. M., Rosa, M. F., Mattoso, L. H. C. (**2017**). Nanocellulose in bio-based food packaging applications, *Ind. Crops Prod.*, **97**, pp. 664–671.

Chapter 11

Antioxidant Finishing Enabled Packaging for Improved Shelf Life of Food

Meenu Aggarwal,[a] Anjali Gupta,[b] Vanita Sapra,[c] Meenakshi Singhal,[d] and Nisha Saini[e]

[a] *Department of Chemistry, Aggarwal College, Ballabgarh, Faridabad, Haryana, India*
[b] *Department of Chemistry, Galgotias University, Greater Noida, Uttar Pradesh, India*
[c] *Department of Chemistry, GGDSD College, Palwal, Faridabad, Haryana, India*
[d] *Department of Chemistry, Aggarwal College, Ballabgarh, Faridabad, Haryana, India*
[e] *Department of Chemistry, Gargi College, University of Delhi, New Delhi, India*
meenuaggarwal5@gmail.com

Antioxidants are valuable additives capable of preventing the rancidity of food by preserving lipids from oxidation either by annihilating free radicals or stealing oxygen so as to amplify the shelf-life expectancy of the foodstuffs. Synthetic antioxidants regulate the oxidation by sustaining the refined oils and fats within the food products. But, owing to deadliness and cancer-causing properties of certain man-made antioxidants, they could be swapped by natural antioxidants. The increased awareness of consumers toward products has shifted the focus to use natural active agents. Biocomposite constituents, come up with the packaging tool, retain the aspect, assurance, as well as sensual equities of the foodstuffs; all together securing the environment due to their biodegradable and renewable nature. Thus, it is a necessity to include active agents into packingmaterials.

Biomaterials in Food Packaging
Edited by Mohd Yusuf and Shafat Ahmad Khan

ISBN 978-981-4877-98-5 (Hardcover), 978-1-003-25678-6 (eBook)
www.jennystanford.com

Biocomposite reinforced with nanofibrillated cellulose (NFC) has advantageous benefits in food packaging applications by taking the place of petroleum-based typical plastics.

11.1 Introduction

This chapter is focused on the antioxidant compounds that currently are pertinent to the food industry, worth mentioning them as food additives employed in the most auspicious types of packaging. Foodstuff packaging can be defined in many ways. The main job of food packaging is to keep the foodstuff safe, amid repository and shipping over the biological, chemical, and automated responses [1]. Thus, to control, ameliorate, and improve the packaging protective properties, there is a need to first concentrate on food packaging materials. Information about the properties of food products and packaging materials allow to control, utilize, achieve, and experience better packaging alternatives with new technological improvements to enhance the mean life of the food products and protect them from multiple external and internal changes [2].

So as to conserve the food products against deterioration beside proximity to extraneous factors, certain features are connecting to stock like being inactive, non-toxic, and impervious to pathogens, and prerequisite to be explored while packaging. Due to ambiguities in user and natural defense, the handling as well as uptake of food packaging operations gets hindered in spite of the biggest proportion of the running market [3].

11.2 Current Challenges in Food Packaging Industry

Plastic packaging was used, due to its low cost, stable, and oil resistant properties. Consumers should be careful rather than taking a risk and give away the plastic as much as possible as per intensifying evidences collected against the safety of plastics. [4]. Traditional packaging keeps elementary postulate "to stay as passive as possible", i.e. coaction amidst the food and packaging should be minimum, nevertheless simultaneously it needs to offer a defensive blockade to foodstuff from outside effects that guarantee food management, in addition to safeguard the food excellence for a definite time. Food packaging technology [5, 6] is constantly changing in reaction leads to several challenges for the current society, which is becoming more and more heterogeneous because of different type of users like food processors, food producers, logistic operators, retailers, and consumers. The changes include modifications in the usage of materials that are novel and productive in food wrapping to promise food security, value, and traceability.

Conventional packaging can be replaced by edible coverings as well as films that improve the shelf life of foodstuffs. Thus, development of new techniques for food storage act in lessening gas interchange, respiration, moisture movement, etc., in the items like vegetables, in addition to decrease metabolic disorders and depress the proportion of oxidative reaction.

11.3 Smart Packaging Systems

In the 20th century, the development in packaging skills gives the impression as intelligent or smart and active packaging (oxygen scavengers, antimicrobic agents, breathing regulators, and odor absorbers). The smart packing is efficient in sensing, recording, tracing, or collaborating data about the value and/or status of the creation during the whole food chain. The second packing is constructed on the notion of accumulating definite constituents into packaging structures that liberate or consume materials from or into the packed food or the neighborhood area, in order to encompass the shelf-life time along with strengthening the value, security, and sensory features of the food products [7].

The smart packaging is sandwiched amid active and intelligent packaging. Once it draws closer to active packaging, the technology is grounded on the connections between the eatable and packaging materials, along with the interior gaseous atmosphere. Typically, the packing practiced in the food occupation comprises lively enzymes and other constituents that can be the copycat procedures of a live system, such as respiration. Eliminating oxygen from food packages is not an easy-going approach. The oxygen can be detached through machine-driven handling or additional techniques. Recently suggested method for oxygen scavenging is phyllosilicate clay functionalized with iron, which is produced in order to accomplish a tremendously active oxygen scavenger. This process depends on the exclusion of oxygen through passage and responding with the dispersed iron with the purpose of generating iron oxide, which is captured inside the packing and the clay acts as a barricade stopping any migration. The novelty of active packaging is to cut the deterioration of food/beverage inside the package and to influence constructive alterations to shelf life of the foodstuffs inside the package. Active packages with assimilated oxygen blockades or films with selective porous properties also have potential applications in the storing of probiotic food products. This is an open area of research for scientists and industry [8].

With the aim to fulfill the necessities of active packaging, numerous specific additives are to be incorporated like oxygen absorbers, materials producing or absorbing CO_2 or sulfur dioxide, antimicrobial substances, ethylene absorbers, moisture content regulators, packaging releasing antioxidants, packaging releasing or absorbing taste, and aroma compounds [9].

Table 11.1 depicts some commonly used absorber types, compounds, and their applications in improving food's shelf life.

Table 11.1 Some commonly used absorber types, compounds, and their applications in improving food's shelf life [2–4, 10]

Type of absorber	Compound	Application
Oxygen absorbers	Iron complexes, ascorbic acid, metal salts, glucose oxidases	Cheese, bakery products, sweets, nuts, beans, cereals, meat, powdered milk, chocolate, tea
Moisture absorbers	Silica gel, glycerol	Bakery products, meat, fish, poultry, vegetables, fruits
CO_2 absorbers	Calcium, sodium or potassium hydroxide	Roasted coffee beans
Substances exhibiting antimicrobial activities	Encapsulated ethanol, silver containing compounds, sorbic and benzoic acids, components of spices/herbs	Fats, cereal products

(*Continued*)

Table 11.1 (*Continued*)

Type of absorber	Compound	Application
Ethylene absorbers	Aluminum oxide, active carbon, potassium tetraoxomanganate	Fruits (apples, bananas, avocado), vegetables (carrot, potatoes, tomatoes, cucumbers)
Absorbers of aroma compounds	Citric acid, cellulose esters, polyamide	Readily oxidizable products, e.g. adipose tissuein fish products, fruit juices

11.4 Biopolymer Packaging: Edible Films and Coverings

Bio-based products and modern progression are receiving attention in the food industry to reduce the dependency on fossil fuel and transfer to a maintainable constituent's base.

11.4.1 Bioactive Packaging

A novel idea of so-called 'bioactive packaging' has been presented when talking about food packaging. The dissimilarity amid active and bioactive packaging lies in the fact that the active packaging aids to attain as well as recover the state and security of deposited things, whereas the bioactive packaging directly influences the well-being of the user through the creation of healthier food packages [10, 11]. Edible coverings or films are well defined as continual patterns made up of lipids, proteins, and polysaccharides, which are fit for human consumption [12].

To improve security, constancy, sensory, and dietary properties of food [13], functional food substances, such as antimicrobials, antioxidants, flavoring agents, and nutrients, are incorporated. Materials producing palatable films can be classified into four categories: resins, biopolymer hydrocolloids, lipids, and composites. Biopolymer hydrocolloids comprise proteins such as collagen, casein, keratin, wheat gluten, soy protein, and polysaccharides (starch, cellulose derivatives, and plant exudates); resins include wood rosin;

lipids comprise acylglycerols, waxes, and fatty acids; and composites usually comprise lipids and hydrocolloid elements in the form of a bilayer or else an emulsion [14]. Researchers are hitting more efforts in the progress of green products by recycling and selecting eco-friendly products in lieu of the growing moral mindfulness and 'green consumerism' among the consumers [15, 16].

Edible films and coverings are coats of constituents applied to food products, which provide antioxidants, nutraceuticals, antimicrobials, flavoring agents, and numerous flavors that improve the value, management, and reliability of the packaged items. Widespread concern in the world is with the production of protein-based packing by means of durable antioxidants besides antibacterial activities. Suitable gelatin film integrated with carvacrol as appropriate food packing material is made inclusive of strong antioxidant and antibacterial properties [17].

Environment-friendly films can be produced by either wet processing or dry processing. In wet processing method, film forming ability can be produced from materials, which must be soluble in a solvent like water and alcohol, further combined with plasticizers, antimicrobial agents, and colors or flavors in this process. The dry method does not include solvent dispersal as it depends on intrinsic thermoplastic features of some biopolymers [12].

Eco-plastics are both bio-based and decomposable, such as polylactic acid (PLA) or polyhydroxyalkanoate (PHA), poly-4-hydroxybutyrate, and polyhydroxyvalerate, are widely used. The plasticizers, prepared by esterifying tannic acid and gallic acid, can be used for many applications like cosmetics, food preparation, brews, and polymeric matrices [18]. Starch-based polymers are having varied structures forming combinations with other polymer components making the products particularly tough [19].

11.4.2 Antioxidant Properties in Biopolymer Packages

Antioxidant properties in biopolymer packages play a very important role as antioxidant compounds present in the packages to decrease the rate of oxidation of lipids as well as proteins. Thus, the antioxidants [20–23] improve the shelf life of food products by preventing from oxidation, so that the products can be preserved for a longer period.

Thus, antioxidant agent in the polymeric matrix of film is the area of interest in the packaging industry.

The various analytical methods of evaluation of the antioxidant capacity fall into distinct categories like: DPPH, ABTS, FRAP, ORAC, TRAP, etc., which help in determining the potential of an antioxidant agent present in the biopolymer matrix of the film to destroy the free radicals.

The oxidation of unsaturated lipids present in food product results in the decrease of its nutritive value and at the same time makes it tasteless. Further, antioxidants delay oxidation of lipids either by annihilating free radicals or scavenging oxygen. Antioxidants are broadly categorized into two types depending upon their mechanism of action, i.e. primary antioxidants, which break the free radical chain reaction initiated due to oxidation by hydrogen donation; and secondary antioxidants, which control the rate of oxidation by chelation mechanism, scavenging of oxygen, and destruction of hydroperoxides [24].

The most commonly used synthetic antioxidants in the food industry are butylated hydroxytoluene (BHT) and tert-butylhydroquinone (TBHQ). Adverse effects of these synthetic antioxidants on human being have increased the focus on the research of natural antioxidants, which are particularly derived from plant sources. The natural antioxidants can be obtained from varieties of plants by adopting solvent extraction method.

Although plants contain ascorbic acid and tocopherols, but they are also rich in phenolic compounds [25], thus offering a good source of natural antioxidants, which are to be used in various food applications [26]. Antioxidant compounds like anthocyanins, flavanols, hydroxycinnamates, and carotenoids are present in fruits like apple, blueberries, blackcurrant, cherry, cranberries, grapes, orange, peach, plum, raspberries, strawberries, etc. [26]. Grape seed and grape seed extract have the same antioxidant properties as present in propyl gallate, butylated hydroxyanisole, and butylated hydroxytoluene [27]. Apart from grapes, even the rosemary, pomegranate, and cinnamon also possess better antioxidant properties in comparison to synthetic analogs. Thus, plant extracts can be used as natural antioxidants in meat products, particularly for red meat [28].

Nowadays, worldwide, curiosity is increased over the extraction of natural antioxidants from plants to overcome the adverse health effects caused by synthetic antioxidants. Green tea, which is a natural source of antioxidant, is highly useful for food product that contains high percentage of polyunsaturated fatty acid because it is more susceptible to oxidation. The green tea extract is extensively used in dressing of salad, sauces, soups, mayonnaise, fat spreads, meat, poultry, bakery products, frying oils, seafood, pizza toppings, cereals, and snacks [29].

Chitin, a natural polysaccharide, is used for the synthesis of bioactive polymer chitosan via deacetylation process. The biological activities like antibacterial properties and non-toxicity make chitosan and its derivatives a suitable candidate for numerous applications in the food, pharmaceutical, and agricultural sectors [30–32].

Majority of snacks like nuts, cereals, dry fruits, chocolate, etc., that employ ingredients containing high-antioxidant contents fall under the ready-to-eat cereal category. Such snacks are being promoted to target audience [33]. Aerated (black) tea's antioxidant property as well as its flavor is enhanced by the addition of spices. Research finding has shown that some spices are required in higher concentration for optimum taste, but aerated (black) tea give good results at lower concentrations of spices [34]. As the spices are expensive, which can increase the overall cost of the product, therefore in order to reduce the quantity of spices added, focus is on the commercial production of spiced aerated (black) tea by setting threshold levels of spices for product development.

The antioxidative properties of milk and milk products like whey, skim milk, and casein contribute significantly in maintaining high-milk quality by preventing lipid peroxidation. This is all because of high concentration of phenolic compounds in milk, which enhance the oxidative stability of dairy products [35]. The incorporation of plant-based antioxidants further decreases the rate of oxidation of dairy products.

The natural antioxidants are used widely not only to improve the shelf life of the food products, but also to prevent the human beings from the adverse side effects of synthetic antioxidants. The research is going on to find more and more antioxidant molecules from a variety of plants, which can be used very safely in the dairy industry [36].

The deficiency of vitamin C in human beings as well as in some species of birds and fishes, which are not capable to synthesize it, suffer from the disease scurvy. Therefore, the citrus fruits such as orange, kiwi, lemon, guava, grapefruit, papaya, and strawberries, and vegetables such as broccoli, cauliflower, and capsicums, which are rich in ascorbic acid, are highly recommended for daily consumption.

Vitamin C or ascorbic acid is a natural organic compound presents in both plants and animals, enriched with antioxidant properties. Ascorbic acid is a strong reducing agent, which allows it to get oxidized easily to dehydroascorbic acid. This property makes it suitable for scavenging free radicals such as hydroxyl radical, hydrogen peroxide, and singlet oxygen. The free radicals are very harmful for the humans, as they can damage tissues and speed up the destruction of liver in people having hepatitis. This mechanism is responsible for the reduction of major antioxidant, i.e. glutathione, presents in the liver cells. Ascorbic acid promotes the assimilation of the iron in the intestine as well as the detoxification caused by heavy metals such as nickel, cadmium, and lead, to provide cellular defense mechanism. The food industry is looking for some specific additive of ascorbic acid and its salts, which possess strong antioxidant properties to improve the quality of food products. The various additives of ascorbic acid are: E300, ascorbic acid; E301, sodium ascorbate; E302, calcium ascorbate; and ascorbyl stearate. The ascorbic acid also prevents the meat products from getting oxidized and discolored upon storage [37].

Also, fundamental element influencing lipid oxidation in mayonnaise has been discussed in this chapter. In nutshell, oxidative degradation of lipids in mayonnaise can be reduced by controlling all those factors that increase rancidity such as reduction in storage temperature, avoiding direct light, and improving chemical properties of oils.

However, oxidation of mayonnaise can also be prevented by using antioxidants. Keeping in mind the health of people, there is a tendency of using natural antioxidants in food products. After reviewing the literature, it can be interpreted that it could be possible to design a mayonnaise with greater oxidative stability, by replacing synthetic antioxidants with natural ones.

With several researches in the area of using natural antioxidants in mayonnaise, it is decided that now focus is on elucidating the mechanism of oxidation, exploring more natural antioxidants, as well as to study the interfacial area of oil in water emulsion to decrease the rate of oxidation. These all help in finding the best suited natural antioxidant to be used in mayonnaise to decrease the side effects caused by oxidation [38].

Now the consumer preference has been changed, i.e. more inclination is toward organic and green food. So, the food industry is also concentrating on the use of natural compounds, which are the safest alternatives to synthetic products, but at the same time they maintain the quality of food products and prevent them from microbial spoilage, which are utmost importance.

The futuristic approach for the use of antimicrobial and antioxidant properties of natural substances is not only helping in the production of safer and healthier food, but also maintaining the integrity of environment, as harmful chemicals are replaced by natural analogs [39].

Nowadays, there is an increasing interest in nutraceutical compounds based on natural sources being added for food coloring as well as eco-functional properties [40]. However, the limited development of novel nutraceuticals and bio-based food additives is all due to the low production with poor yield upon purification, specifically when these bioactive compounds are synthesized by microbial fermentation. Thus, in the future, more projects will be made compulsory for the research in the area of improvising the production and purification of natural bioactive compounds to maximize the yield for fulfilling the demand of increasing population. It is also essential to establish regulatory authority to define rules and regulations for the use of natural substances in the food, cosmetic, and pharmaceutical industries [41].

According to the present invention, the food packaging materials include polymeric materials and natural antioxidants. Typically, the natural antioxidant is organic in nature and obtained from active ingredient presents in a plant, which is responsible for providing desired antioxidant effect. The active ingredient is obtained from plant products by extraction process using a suitable solvent to dissolve the active ingredient, in order to separate it from residual plant material. The organic solvent extracts can be obtained by single or successive extraction process from the plant. The extracts may be used in aqueous, gel, or oil form in the food packaging material. Aqueous solvents, alone or in alkaline and acidic medium; oils (palm oil, coconut oil); and organic solvents like hexane, ethanol, methanol, acetone, tetrahydrofuran, chloroform, etc., are used in the extraction process [42].

11.5 Nanotechnology in Food Packaging Industry

The elementary opinion in the usage of active packing based on the integration of specific constituents inside the polymer and inherent features of the polymer itself can be used in the food packaging industry. Because of the size of nanoparticles, relatively larger surface area, and cost-effectiveness, nanotechnology is applied to polymer matrices.

Due to the changeable structures and shapes of nanoparticles, these are used in formation of different nanostructured materials, which are accountable as food packing constituents, mainly because of their antimicrobic nature, in addition to oxygen-scavenging capacities. The usage of biopolymers as packing films in the food division has increased, owing to their extraordinary price and mediocre implementations, as compared to synthetic polymeric materials.

Nano-reinforcement that integrates the application of nanoparticles and nanofibers, e.g. nano clays, metal oxides nanoparticles, carbon nanotubes, and other fillers into a matrix of the polymer, thus providing smart or active assets to the wrapping arrangement by diffusing it in the form of nanocomposites [43]. Polymeric elements strengthening through nanoparticles have unlocked a great space of prospects and upgrading by even spreading of nanofillers in the polymer. The contact between nanofiller and polymer, which is vital to accomplish healthier reinforcement, can be augmented through insertion of parameters like amount, functional moiety, quality of filler, polymer performance, and handling method.

However, cellulose is an actual fascinating accepted polymer. An extremely auspicious segment of cellulose is NFC, which is broadly exploited in abundant applications, precisely in the food packing business. To recover the mechanical and barrier properties of the composites, the nanocellulose is chiefly used to reinforce with the polymer, which may supplement the mixture performance like antimicrobic and antioxidant, in addition to mechanical features. The composites with boosted properties may be well-matched for conserving fruits, vegetables, and other foodstuffs in the food packing industry. Consequently, problems mainly met by the food industry will be overcome by the usage of nanocellulose-based polymer composites [46].

Further, there are some shortages restraining their possibility of uses, for instance, agglomeration as well as acquiescence with hydrophobic polymeric matrices. These challenges can be overcome by chemical treatment of NFCs, contributing novel functionalities to these materials. Thus, the interfacial linkage between NFCs and polymer matrices is extremely improved by means of filler functionalization, which allows even nanoscale dispersal of the reinforcing fillers within the matrices, therefore attaining the persistent shelf life [47, 48]. Special emphasis has been given to blends in the frame of packaging application [48–50].

Nanocomposite idea has technically acquainted with uniqueness in production of a novel refinement of innovative polymeric constituents possessing adaptable, exciting, and superior assets, embracing thermo mechanical attributes though sustaining varying polymer matrix transparency. These materials have also established competence with respect to costing and effectiveness in several applications especially in packaging.

11.6 Conclusion and Future Remarks

Among the various types of active packaging, mainly for the food industry, the antioxidants play the promising role to progressively meet the necessities to encompass the shelf life, thereby generating novel potentials for their applications. The use of synthetic antioxidants will be substituted with natural antioxidants, which are good synergists that can enhance the shelf life expectancy of foodstuffs by including herbs and spices. The forthcoming scope of bio-based packaging materials is hopefully very bright, due to the environmental concerns and the global realization that petroleum resources are limited and renewable resources will be the solution. Bio-nanocomposites offer an opening for the use of high performance, lightweight nanocomposite constituents, hence substituting traditional non-biodegradable petroleum-based plastic packaging materials. With regard to the upcoming visions of NFC packaging materials, it is positively expected that simple packaging systems will be substituted with high-tech intelligent packaging systems. Also, biosensing application not only protects food against environmental factors, but also integrates properties to the packaging materials, permitting quite a lot of budding choices in the food packaging industry.

References

1. Bilska A. (2011). Packaging Systems for Animal Origin Food, *J. Log Forum 7*, **1(4)**, pp. 35–44.
2. Ravishankar, R. V., and Bai, J. A. (2018). Nanotechnology Applications in the Food Industry, CRC Press, Boca Raton, the USA.
3. Bradley, E. L., Castle, L., and Chaudhry, Q. (2011). Applications of Nanomaterials in Food Packaging with a consideration of Opportunities for Developing Countries, *J. Trends Food Sci. Technol.*, **22(11)**, pp. 604–610.
4. Saravanan, P., Sathish Kumar, S., Visvanathan, D., Silambarasan, S., and Charles, A. (2012). Safe Food Packaging and Storage for Better Health and Environment, *Int. J. Int Sci. Inn. Tech. Sec. A*, **1(5)**, pp. 1–4.
5. Altaf, U., Kanojia, V., and Rouf, A. (2018). Novel Packaging Technology for Food Industry, *J. Pharmacog. Phytochem.*, **7(1)**, pp. 1618–1625.
6. Mehta, K., Akhilesh, D., and Kumar, S. (2012). Recent Trends in Pharmaceutical Packaging: A Review, *Int. J. Pharmaceut. Chem. Sci.*, **1(3)**, pp. 1282–1292.
7. Arroyo, B. J., Santos, A. P., De Melo, E. A., Campos, A., Lins, L., and Boyano-Orozco, L. C. (2019). Bioactive Compounds: Health Benefits and Potential Applications, Campos M. R. S. (ed.), Chapter 8 "Bioactive Compounds and Their Potential Use as Ingredients of Food and its Application in Food Packaging", Woodhead Publishing, Duxford, the U.K., pp. 143–156.
8. Dobrucka, R. (2013). Application of Active Packaging Systems in Probiotic Foods, *J. Log Forum*, **9(3)**, pp. 67–175.
9. Brunazzi, G., Parisi, S., and Pereno, A. (2014). The Importance of Packaging Design for the Chemistry of Food Products, SpringerBriefs in Molecular Science Book Series, Chapter 4 "Packaging: A Communicative Medium", Springer, Cham, pp. 87–108.
10. Boarca, B., Lungu, I., and Holban, A. M. (2019). Trends in Beverage Packaging, Grumezescu, A. M., and Holban, A. M. (eds.), Science of Beverages Vol 16, Chapter 3 "Bioactive Packaging for Modern Beverage Industry", Woodhead Publishing, Duxford, the U.K., pp. 51–71.

11. Kaur, P., Ghoshal, G., and Banerjee, U. C. (2019). Preservatives and Preservation Approaches in Beverages, Grumezescu, A. M., and Holban, A. M. (eds.), The Science of Beverages Vol 15, Chapter 3 "Traditional Bio-Preservation in Beverages: Fermented Beverages", Woodhead Publishing, Duxford, the U.K., pp. 69–113.

12. Goswami, M., Pathak, V., Bharti, S. K., Tewari A., Chappalwar, A., Tanuja, and Ojha, S. (2018). Shelf Life Enhancement of Muscle Foods with Biodegradable Film Packaging, *J. Animal Feed Sci. Technol.*, **6**, pp. 3339–3346.

13. Pashova, S., Radev, R., Dimitrov, G., and Ivanov, J. (2018). Edible Coatings in Food Industry Related to Circular Economy, *J. Quality Access Success*, **19(166)**, pp. 111–117.

14. Nandane, A. S., and Jain, R. K. (2012). Value Addition of Fruits and Vegetables by Edible Packaging: Scope and Constraints, *J. Res. Rev.: A J. Food Sci. Technol.*, **1(1)**, pp. 1–11.

15. Sam, S. T., Nuradibah, M. A., Chin, K. M., and Hani, N. (2016). Natural Polymers: Industry Techniques and Applications, Olatunji, O. (ed.), Chapter 6 "Current Application and Challenges on Packaging Industry Based on Natural Polymer Blending", Springer, Cham, pp. 163–184.

16. Shaikh, U. A., Deshpande, H. W., and Kulkarni, D. B. (2018). A Review on Probiotic Beverages Prepared Using Vegetables, *Int. J. Chem. Studies*, **6(5)**, pp. 61–65.

17. Kavoosi, G., Dadfar, S. M. M., Mohammadi, P. A., and Mehrabi, R. (2013). Antioxidant and Antibacterial Properties of Gelatin Films Incorporated with Carvacrol, *J. Food Safety*, **33(4)**, pp. 423–432.

18. Adam, G. A., and Needham, A. (2016). Edible Plasticizers for Food and Food Packaging Films, US Patent (US 2016/0145288 A1).

19. Piergiovanni, L., and Limbo, S. (2016). Food Packaging Materials, SpringerBriefs in Molecular Science, Springer, Cham.

20. Fernandes, R., de P. P., Trindade, M. A., and de Melo, M. P. (2018). Alternative and Replacement Foods: Handbook of Food Bioengineering, Holban, A. M., and Grumezescu, A. M. (eds.), Chapter 2 "Natural Antioxidants and Food Applications: Healthy Perspectives", Academic Press, London, the U.K., pp. 31–64.

21. Pabón-Baquero, L. C., Otálvaro-Álvarez, Á. M., Fernández, M. R. R., and Chaparro-González, M. P. (2018). Antioxidants in Foods and Its Applications, Shalaby, E., and Azzam, G. M. (eds.), Chapter 5 "Plant Extracts as Antioxidant Additives for Food Industry", InTech Open, London, the U.K., pp. 87–116.

22. Pandey, R. M., and Upadhyay, S. K. (2012). Food Additive, El-Samragy, Y. (ed.), Chapter "Food Additive", InTech Open, London, the U.K., pp. 1–32. DOI: 10.5772/34455

23. Souza Cruz, Renato, Camilloto, GeanyPeruch, Pires, and Ana, Clarissa dos Santos (2012). Structure and Function of Food Engineering, Eissa, A. A. (ed.), Chapter 2 "Oxygen Scavengers: An Approach on Food Preservation Structure and Function of Food", InTech Open, London, the U.K., pp. 21–39.

24. Jiang, J., and Xiong, Y. L. (2016). Natural Antioxidants as Food and Feed Additives to Promote Health Benefits and Quality of Meat Products: A Review, *J. Meat Sci.*, **120**, pp. 107–117.

25. Shahid, M., Yusuf, M., and Mohammad, F. (2016). Recent Progress in Medicinal Plants, Vol. 41, Analytical and Processing Techniques, Govil, J. N. (ed.), Chapter 16 "Plant phenolics: A Review on Modern Extraction Techniques", Studium Press LLC, Houston, the USA, pp. 2–20.

26. Thorata, I., Dipali, J., Mohapatrac, D., Joshib, D. C., Sutarb, R. F., and Kapdib, S. S. (2013). Antioxidants, Their Properties, Uses in Food Products and Their Legal Implications, *Int. J. Food Studies*, **2**, pp. 81–104.

27. Colindres, P., and Susan, B. M. (2011). Oxidative Stability of Cooked, Frozen, Reheated Beef Patties: Effect of Antioxidants, *J. Sci. Food Agric.*, **91(5)**, pp. 963–968.

28. Shah, M. A., Bosco, S. J. D., and Mir, S. A. (2014). Plant Extracts as Natural Antioxidants in Meat and Meat Products, *J. Meat Sci.*, **98(1)**, pp. 1–54.

29. Senanayake, N. S. P. J. (2013). Green Tea Extract: Chemistry, Antioxidant Properties and Food Applications – A Review, *J. Functional Foods*, **5**, pp. 1529–1541.

30. Muxika, A., Etxabide, A., Uranga, J., Guerrero, P., and de la Caba, K. (2017). Chitosan as a Bioactive Polymer: Processing, Properties and Applications, *Int. J. Biolog. Macromolecules*, **105**, pp. 1358–1368.

31. Yusuf, M. (2018). Food Packaging and Preservation: Handbook of Food Bioengineering, Grumezescu, A. M., and Holban, A. M. (eds.), Chapter 12 "Natural Antimicrobial Agents for Food Biopreservation", Academic Press, London, the U.K., pp. 409–438.

32. Qing, X., Tian, Z., and Huili, L. (2014). Antioxidant and Free Radical Scavenging Activities of Chitosan Materials, *J. Adv. Mater. Res.*, **1002**, pp. 81–90.

33. Carunchia, M., Wang, L., and Han, J. H. (2015). Handbook of Antioxidants for Food Preservation, Woodhead Publishing Series in Food Science, Technology and Nutrition, Shahidi, F. (ed.), Chapter 19 "The Use of Antioxidants in the Preservation of Snack Foods", Woodhead Publishing, Cambridge, the U.K., pp. 447–474.

34. Oduor, S. O., Kanyiri, J. W., and Kipngeno, H. R. (2015). Antioxidant Activity, Economic and Sensory Evaluation of Spiced Aerated Tea, *Int. J. Agric. Food Res.*, **4(4)**, pp. 15–23.

35. Revilla, I., Gonzalez-Martin, M. I., Vivar-Quintana, A. M., Blanco-Lopez, M. A., Lobos-Ortega, I. A., and Hernandez-Hierro, J. M. (2016). Antioxidant Capacity of Different Cheeses: Affecting Factors and Prediction by Near Infrared Spectroscopy, *J. Dairy Sci.*, **99(7)**, pp. 5074–5082.

36. Modi, A. A., Hanan, H. A., Lojayn, S. T., and Amal, B. S. (2017). Antioxidant Properties of Dairy Products Fortified with Natural Additives: A Review, *J. Association Arab Univ. Basic Appl. Sci.*, **24**, pp. 101–106.

37. Varvara, M., Bozzo, G., Disanto, C., Pagliarone, C. N., and Celano, G. V. (2016). The Use of the Ascorbic Acid as Food Additive and Technical-Legal Issues, *Ita. J. Food Safety*, **5**, pp. 1–10.

38. Ghorbani G., S., Smyth, H. E., Sharma, M., and Fitzgerald, M. (2016). Lipid Oxidation in Mayonnaise and the Role of Natural Antioxidants: A Review, *J. Trends Food Sci. Technol.*, **56**, pp. 88–102.

39. Bondi, M., Lauková, A., de Niederhausern, S., Messi, P., and Papadopoulou, C. (2017). Natural Preservatives to Improve Food Quality and Safety, *J. Food Quality*, **1090932**, pp. 1–3. DOI: 10.1155/2017/1090932

40. Yusuf, M., Shabbir, M., Mohammad, F. (2018). Natural Colorants: Historical, Processing and Sustainable Prospects, *Nat. Prod. Bioprospect.*, **7(1)**, pp. 123–45.

41. Moldes, A. B., Vecino, X., and Cruz, J. M. (2017). Current Developments in Biotechnology and Bioengineering, Pandey, A., Sanromán, M. A., Guocheng, D., Soccol, C. R., and Dussap, C. G. (eds.), Chapter 6 "Nutraceuticals and Food Additives", Elsevier, Amsterdam, the Netherlands, pp. 143–164.

42. Withers, P. (2018). Antioxidant Active Food Packaging, US Patent (US 2018 / 0290811 A1).

43. Siddiquee, S., Melvin, G. J. H., and Rahman, M. M. (2019). Nanotechnology: Applications in Energy, Drugand Food, Springer, Switzerland.

44. Rhim, J. W., Park, H. M., and Ha, C. S. (2013). Bio-Nanocomposites for Food Packaging Applications, *Prog. Polym. Sci.*, **38(10** and **11**), pp. 1629–1652.

45. Basu, A., Kundu, S., Sana, S., Halder, A., Abdullah, M. F., Datta, S., and Mukherjee, A. (2017). Edible Nano-Bio-Composite Film Cargo Device for Food Packaging Applications, *J. Food Packaging Shelf Life*, **11**, pp. 98–105.

46. Perumal, A. B., Sellamuthu, P. S., Nambiar, R. B., Sadiku, E. R., and Adeyeye, O. A. (2019). Green Biopolymers and their Nanocomposites, Materials Horizons: From Nature to Nanomaterials, Gnanasekaran D. (ed.), Chapter 4 "Biocomposite Reinforced with Nano Cellulose for Packaging Applications", Springer, Singapore, pp. 83–123.

47. Idumah, C. I., Zurina, M., Ogbu, J., Ndem, J. U., and Igba, E. C. (2019). A Review on Innovations in Polymeric Nanocomposite Packaging materials and Electrical Sensors for Food and Agriculture, *J. Composite Interfaces*, pp. 1–72.

48. Fortunati, E., Mazzaglia, A., and Balestra, G. (2018). Sustainable Control Strategies for Plant Protection and Food Packaging Sectors by Natural Substances and Novel Nano Technological Approaches, *J. The Sci. Food Agric.*, **99(3)**, pp. 986–1000.

49. Luzi, F., Torre, L., Kenny, J., and Puglia, D. (2019). Bio- and Fossil-Based Polymeric Blends and Nanocomposites for Packaging, *J. Structure–Property Relationship Materials*, **12(3)**, pp. 471.

50. Miyagawa, H., Misra, M., and Mohanty, A. K. (2005). Mechanical Properties of Carbon Nanotubes and their Polymer Nanocomposites, *J. Nanosci. Nanotechnol.*, **5(10)**, pp. 1593–1615.

Chapter 12

Active Finishing Agents for Food Bio-Preservation Based on Natural Origin to Prevent Microbial Risks

Smrita Singh,[a] Lalit Prasad,[b] Ashutosh Singh Chauhan,[c] and Shafat Ahmad Khan[c]

[a] *Bioexplore, Dimapur, Nagaland, India*
[b] *Division of Chemistry, School of Basic & Applied Sciences, Galgotias University, Greater Noida, Uttar Pradesh, India*
[c] *Air Pollution & Stone Conservation Laboratory, Archaeological Survey of India, Agra Fort, Agra, Uttar Pradesh, India*
smrita1@gmail.com, lalit.prasad@galgotiasuniversity.edu.in

Bio-preservation technique is used to enhance the shelf life of highly perishable food of plant and animal origin. This process exploits the antimicrobial property of naturally occurring agents and their metabolites to prevent the spoiling of food. Several organisms of biological and natural origin are used for preservation of the food for longer time, without having any adverse effect on human health. This process involves the use of several natural micro-floras of plant and animal origin, as well as of microbial origin, such as bacteria, bacteriophages, etc., under controlled environment. This natural technique helps in reducing the volume of chemical preservatives as well as other processes that may negatively affect the quality and nutritional value of preserved food. Genetically modified micro-

Biomaterials in Food Packaging
Edited by Mohd Yusuf and Shafat Ahmad Khan

ISBN 978-981-4877-98-5 (Hardcover), 978-1-003-25678-6 (eBook)
www.jennystanford.com

floras or their metabolites may also be used as bio-preservatives, but the selection of substances to be used in bio-preservation should be done carefully and judiciously keeping in view of all toxicological and safety issues, is foremost important. New genetically modified micro-floras and microorganisms are being developed for better preservation of food, keeping in view of nutritional value and organoleptic quality of food products. Nowadays, this bio-preservation process is gaining acceptance due to its inert benign approach.

12.1 Introduction

Bio-preservation is a method of extending shelf life of food by exploiting ingredients derived from microorganisms, and metabolites from microorganisms, flora, and fauna. Several ingredients derived from plants, animals, and microorganisms have strong antimicrobial property and inhibit the growth and survival of harmful pathogenic microorganisms in food. Since ancient times, several herbs and spices have been used for the preservation of food and inhibited the growth of bacteria, fungi, mold, and other harmful microorganisms. The choice of natural preservatives is logical one, as they help us to avoid the use of excessive chemical preservatives and physical treatments of the food for preservation. This method helps us to retain the nutritional value of food products and makes them safe for consumption.

Natural antimicrobials and their products/ingredients that can be used as bio-preservatives can be broadly classified into following three categories:

i. Active finishing agents of plant origin

ii. Active finishing agents of animal origin

iii. Active finishing agents from bacteria and other micro world.

This chapter provides a basic overview of natural active finishing agents derived from bacteria and their use as bio-preservatives of food.

12.2 Active Finishing Agents from Bacterial World in Food Preservation

12.2.1 Major Derived Components

12.2.1.1 Bacteriocins

Bacteriocins are minute polypeptide molecules or protein, biosynthesized within ribosomes that exhibit antimicrobial activity against several closely related groups of bacteria and other unrelated groups of microbes. Nowadays, bacteriocins are classified as heat-stable cationic peptides, produced by mainly *Lacto bacillus* (lactic acid bacteria, LAB), which show wide-spectrum antibacterial potential [1, 2]. Bacteriocins are proven toxicologically safe, as they are produced by both Gram-positive and Gram-negative species, which are an important part of the defense system. Bacteriocins produced from Gram-positive bacteria are heat stable and function at the wide range of pH and have broad-spectrum antibacterial activity, while bacteriocins from Gram-negative bacteria are heat labile (exception being Microcin V from *E. coli*) and have narrow antibacterial activity range.

12.2.1.1.1 *Classification of bacteriocins from Gram-positive bacteria*

On the basis of extensive studies performed previously [3, 4], bacteriocins are categorized into four divisions:

Class I (The lantibiotics): These post-translationally modified bacteriocins with molecular weight <5 kDa, having 19–50 (approx.) amino acids, are referred to as variacin, cytolysin, subtilin, and nisin. These lantibiotics are further divided into three types: Class Ia, Class Ib, and Class Ic [5].

Class II (The unmodified peptides): These translationally unmodified bacteriocins have molecular weight <10 kDa, e.g. pediocin PA-1 from *Pediococcus acidilactibci*, enterocin 1071 from *Enterococcus faecalis* BFE1071, and enterocin EJ97 from *E. faecalis* EJ97. These Class II bacteriocins are further grouped into three types: Class Ia, Class Ib, and Class Ic.

Class III: These heat-labile bacteriocins are large proteinaceous molecule with molecular weight >30 kDa. Due to this, they are not widely utilized in food bio-preservation, e.g. enterolysin A from *E. faecalis*, dysgalacticin, streptococcin A-M57, helveticin J, etc. They are sub-grouped into two types: Class IIIa - bacteriolysins and Class IIIb - ornon-lytic bacteriocins.

Class IV (Circular peptides): These bacteriocins are unique in nature. They joined by a circular peptide bond through the ligation of their N-terminus to the C-terminus. Their molecular weight ranges from 5.5 kDa to 7.5 kDa. These circular peptide bacteriocins have lipid or carbohydrate moieties, due to which they are complex proteins, e.g. AS-48 from *E. faecalis* [1, 6].

Furthermore, the classification of bacteriocins based on their molecular mass of Gram-negative bacterial colonies can be separately countered. Bacteriocins produced from Gram-negative are categorized based on their size. For example:

i. Colicins having high molecular mass (HMM) peptides with molecular weight ranging from 20 kDa to 90 kDa.

ii. Microcins having low molecular mass (LMM) peptides with molecular weight ranging from 1 kDa to 20 kDa.

iii. Tailocins having structure like a tail of bacteriophages and have multiple subunits with high molecular weight.

The mode of action of bacteriocins varies depending on their type. Bacteriocins are cationic and hydrophobic in nature and act as membrane permeabilizers [6]; hence, form pores leading to the efflux of ions (disruption of proton motive force) and other molecules out of cells leading to permanent damage and death of cells. The other mode of action of bacteriocins is inhibition of biosynthesis and formation of cell wall.

12.2.1.2 Pediocins

Under the Class II of bacteriocins, pediocins are produced mainly by the genus pediococcus of LAB. The main strains of bacteria producing pediocins are *Pediococcus acidilactici* and *Pediococcus pentosaceus* [7]. Pediocin inhibits the formation of cell wall of target cell by inhibiting the absorption of amino acids required for the formation of cytoplasmic membrane of the target cell. It is thermally stable and active at wide range of pH values and is resistant to several protease enzymes. Pediocin SA-1, produced from *P. acidilactici* NRRL B 5627, shows inhibition activity against bacteria that cause common food spoiling, for example, *Bacillus cereus*, *Micrococcus flavus*, *Salmonella carnosus*, *Clostridium*, *Clostridium thiaminolyticum*, *Leuconostoc mesenteroides*, *E. faecalis*, *Listeria monocytogenes*, *sporogenes*, and *Lactobacillus brevis*. The study carried out by Díez et al. (2012), wherein the effect of PA-1 and combination of PA-1 with sulfur dioxide and ethanol was investigated to assess the growth of a wide collection of pathological strains (i.e. 16 yeast, 18 acetic acid bacteria, and 53 oenological LAB) [8]. It was observed that PA-1 alone can be used as a potential bio-preservative for winemaking. These days, pediocins are being used as bio-preservatives in the preservation of wine, baked products, dry sausage, and meat products, and reducing cheese contamination [9].

12.2.1.3 Enterocins

Primarily *Enterococcus* species produce enterocins, including *E. faecium* (enterocins A and B) and *E. faecalis* (enterocin AS-48). In addition, a few of the entrocins, e.g. enterocin A derived from *Lactococcus* lactis, are broad-spectrum antimicrobial bacteriocins. Major enterocins, such as enterocin AS-48, enterocin 13, enterocin 416K1, enterocin L50A, and enterocin CCM4231, inhibit the growth and survival of several food-spoiling microorganisms, such *as L. monocytogenes*, *Listeria innocua*, *Bacillus cereus*, *Staphylococcus aureus*, etc. [10]. Enterocins such as A, B, and AS-48 were found effective to restrict the growth of *Salmonella* sp., *E. coli* [11, 12], *Stearothermophilus*, several other Bacillus spp., *Lactobacillus sakei*, and other food-spoiling lactic acid bacteria [1, 6]. They are mainly used to preserve canned food, vegetable sauces, soybean sprouts, skimmed milk, yogurt, frankfurters, fresh cottage cheese, raw fruits and fruit juices, alfalfa and bean sprouts, green asparagus, infant rice-based food and drinks, fish spread, etc.

12.2.1.4 Endolysins

Endolysins/lysins are the hydrolytic enzymes and are generally murein hydrolases with strong antimicrobial effectiveness, produced by the bacteriophages at the final stage their replication cycle, kill the host microorganisms by cleavage and hydrolysis of peptidoglycan of host cell wall. Four major types of enzymes that cleave peptidoglycan moiety and degrade cell wall of the peptide glycan molecules by rupturing of targeted areas are:

(i) glycosidases,

(ii) endopeptidases,

(iii) amidohydrolases, and

(iv) lytic transglycosylases.

Endolysin LysH5 has showed a bactericidal nature and used in pasteurized milk to prevent spoilages from *S. aureus* [13]. Similarly, polycendolysin has an eradicating effect on several varieties of food. Endolysin's application controls the spoilage of fruits by inhibiting the growth of *Erwinia amylovora*, the organism responsible for fire blight spoilage in fruits. Several endolysins are found active against zoonotic and food-borne pathogens. In addition, phage-encoded and engineered endolysins are recognized imparting broad-spectrum inhibitory functions toward a number of pathogenic isolates in food [14–16]. Lysing proteins Spp64 and Spp62 from bacteriophage Spp001 show wide spectrum of inhibitory effect on *Shewanella putrefaciens,* which is a common spoilage microorganism for seafood. Endolysin, nowadays, finds wide applicability as antimicrobial, as most of the bacteria are becoming resistant to common antibiotics.

12.2.2 Minor-Derived Components

12.2.2.1 Sakacins

Sakacins, an important class of bacteriocins, frequently come together with other lactic-acid bacteriocins, are derived from *Lactobacillus sake.* The major-isolated sakacins are sakacin A, G, K, P, and Q; out of these, sakacins A and P are commonly used [17, 18]. Sakacins are revealed to inhibit the growth of several food-spoiling microorganisms [19], e.g. *Proteus vulgaris* W7, *Klebsiella aerogenes* N12, *Staphylococcus aureus* N16b, *Bacillus cereus* W18, *B. subtilis* N20, *E. coli* N2, *E. coli* W4, *Pseudomonas aeruginosa* N7, *Listeria monocytogenes* W6, etc. [6, 18, 19].

12.2.2.2 Lacticin

Lacticin, derived from *Lactococcus lactis* sub sp. *Lactis*, is a bacteriocidin with broad-spectrum antimicrobial properties with many significant peptides having generally recognized as safe (GRAS) status. Due to their broad-level antimicrobial prospects, they are generally regarded safe for food bio-preservation and storage [20–22]. Due to the high-inhibitory action against common food-borne microorganisms and wide applicability, they have been utilized in protecting the quality of packaged as well as fresh pork-sausage [23, 24]. Moreover, in combination with nisin, lacticin can be used as an effective bio-preservative.

12.2.2.3 Reuterin

Reuterin is recognized due to its valued safeness and stability over a wide range of pH values, high resistance to heat, proteolytic and lipolytic enzymes, and high solubility in water. Reuterin, composed of non-hydrated, hydrated, and dimeric forms of 3-hydroxypropionaldehyde (3-HPA), is derived from few strains of *Lactobacillus reuteri*. It is found very effective cidal capacities against the molds and yeasts, Gram-negative and Gram-positive bacteria strains, along with some protozoa [25, 26]. Reuterin was reported to have excelled inhibitory activity against many microorganisms, such as *S. aureus*, *S. choleraesuis*, *L. monocytogenes*, *E. coli*, *Campylobacter jejuni*, *Yersinia enterocolitica*, and *Aeromonas hydrophila*, causing food spoilage [1, 6, 27, 28]. Therefore, nowadays, reuterins have been used to impart preservative benefits for meat as well several kinds of dairy products, as a smooth alternate for food bio-preservation [29, 30]. Additionally, reuterin has assessed for its preservative effects and found positive responses in various cooked pork and beef sausage, alone or in combination with other bacteriocidins, and possessed excellent antimicrobial potential and could be an ideal contender to be effective and eco-safe preservative applicable for food and dairy items.

12.2.2.4 Nisin

Nicin, a protein consisting 34 amino acids, is derived from *Lactococcus lactis* and regarded as GRAS category. Some uncommon amino acids are reported in nicin biomolecule, including ethyllanthionine (MeLan), lanthionine (Lan), didehydroaminobutyric acid (Dhb), and didehydroalanine (Dha) [30, 31]. The antimicrobial activity of nisin is attributed to the interaction with phospholipids of host bacterial cell, where disruption of normal cell membrane occurred by nicin. It is found quite effective against the growth of several pathogenic and harmful Gram-positive bacteria such as *Listeria monocytegenes* and *S. aureus*. In addition, it inhibits the spore growth of many species of *Clostridium* and *Bacillus* leading to their death. In combination with heat treatment and chelator EDTA, nisin possesses considerable effectiveness against the growth of spores of several Gram-negative bacteria, e.g. *Pseudomonas aeruginosa, E. coli, Salmonella enteritidis, S. typhimurium,* etc.

Several nisin derivatives have been developed by altering the amino acid sequences, which show enhanced antimicrobial properties for both Gram-positive as well as Gram-negative bacteria [26, 30]. As sole or in combination with other naturally occurring biomolecule, nicin found its potential to be used as food preservative, capable of inhibition of few types of food-borne pathogen isolates [32]. In combination with pediocin, reuterin or organic acids, lysozymes, etc., nicin has shown its improved efficiency against pathogenic and harmful bacteria such as *E. coli, L. monocytogenes, B. thermosphacta,* and others [33].

Consequently, nisin can be used as an effective agent for preserving several food products, such as cheese, juices, beverages, meat, etc., against several pathogenic microorganisms commonly causing food spoilage [34]. On account of brilliant activity for suppressing the growth of food-spoilage generative microbes, nicin is being used in the making of biodegradable polymeric (especially based on polylactic acid (PLA), whey protein isolate (WPI), pea protein isolate (PPI), etc.) films for packaging food materials [35].

12.2.2.5 Natamycin

Natamycin is a naturally occurring compound obtained from *Streptomyces natalensis* comes under GRAS consideration. It is used as a broad-spectrum antifungal in the food industry, and dairy products and beverages. Natamycin shows a high affinity toward ergosterol present in the fungal cell wall. It binds irreversibly to this ergosterol, resulting in the alteration of the permeability of cell wall toward leakage of essential ions and small peptides, thus leading to the cell death [36]. It is not much effective against both Gram-negative and Gram-positive bacteria, as it lacks sterols in its cell wall. A colloidal solution in water is applied either by dipping or spraying on to the food surface. Natamycin is used to control the growth of molds during the fermentation process and it is used as a bio-preservative in beverage products, even it is stable at low temperatures with sufficient effectiveness over a wide range of pH values [37]. Natamycin shows strong inhibitory activity against a number of molds, fungi, and bacteria, such as *Penicillium chrysogenum*, *P. camemberti*, *P. brevicompactum*, *P. glabrum*, *A. niger*, *A. versicolor*, *A. flavus*, *Cunning hamella* spp., *Fusarium* spp., *Pseudalles cheriaboydii*, *Mucorra cemosus*, and *Eurotium herbariorum*. It is also very effective against *Cladosporium cladosporiodes*, *C. tenuissimum*, *Byssochla mysnivea*, *Epidermophyton floccosum*, *Alternaria alternata*, *Candida albicans*, *Acremonium* spp, *Trichophyton* spp., *Aspergillus carbonarius*, *P. commune*, *P. corylophilum*, and *P. verrucosumetc* [38, 39]. Natamycin in combination with nisin [40, 41] is used in polymer packaging films, and it inhibits the growth of *Lactobacillus helveticus*, *Fusarium culmorum*, *Listeria ivanovii*, and *Penicillium expansum*. A mixture containing malic acid, nisin, and natamycin, inhibits the growth [42, 43] of several food-spoiling microbes such as *Penicillium aeruginosa*, *P. commune*, *P. chrysogenum*, *Y. lipolytica*, *Penicillium*, and *L. monocytogenes*. Natamycin in combination with chitosan films shows considerable antimicrobial activity against various food-spoiling pathogens and microbes [44]. As an antimicrobial preservative, the property of chitosan has been proposed at less than 6.0 pH to liquid and solid food products [45]. Chitosan shows poor water solubility, therefore, glucosyloxyethylacrylated chitosan, a water-soluble derivative is synthesized by using the Michael reaction of chitosan with glucosyloxyethylacrylate derivative [46].

12.2.2.6 Bacteriophages

Bacteriophages are viruses that specifically infect and grow in bacterial cell, just like a parasite replicates through the lytic or lysogenic cycle within the cell. They can selectively control the population of food-spoiling bacteria [47]. Their phages are grouped into 13 types, on the basis of their specific sizes, shapes, and types of genetic material (nucleic acid), presence/absence of an envelope (outer capsule), or lipids in their internal structures [47]. Application of bacteriophages to the food is known as phage therapy and this procedure is applied to decontaminate the raw food materials, fresh vegetables & fruits, and other eatables [48]. They may also be used as biosanitizers for disinfecting contact surfaces and equipment. An important application of bacteriophages is to extend the shelf life of perishable food products. Reduction in colonies of *Salmonella, Campylobacter,* and *E. coli* in chicken and meat products has been noticed after phage therapy. It also inhibits the growth of *S. aureus* in dairy products such as milk, curd, etc. Food-spoiling bacteria including *Listeria monocytogenes* may be eradicated completely after surface treatment of phage P100 in ripened cheese. Reconstituted infant milk can be protected against *Enterobacters akazakii* and chicken frankfurters can be protected against *Salmonella typhimurium* after the treatment with bacteriophages. Certain bacteriophages are found to be active against food-borne pathogenic bacteria *B. cereus* and *Clostridium perfringens,* and therefore, can be utilized in making packaging films [47, 48].

12.2.2.7 Organic acids

Since ancient times, several organic acids have been widely used as a preservative in the food of almost all kinds. Organic acids, such as lactic, acetic, propionic, citric, and benzoic acids, exhibit a broad-spectrum enzymatic activity and antimicrobial property against several food-spoiling microbes, and are widely used in food preservation directly or indirectly. These are produced as by-products during fermentation by bacteria [49]. Following organic acids are commonly used in food bio-preservation:

Organic acid	Chemical formula
Acetic acid	$C_2H_4O_2$
Propionic acid	$C_3H_6O_3$
Formic acid	CH_2O_2
Ascorbic acid	$C_6H_8O_6$
Lactic acid	$C_3H_6O_3$
Sorbic acid	$C_6H_8O_2$
Citric acid	$C_6H_8O_7$
Benzoic acid	$C_7H_6O_2$
Phenyllactic acid	$C_9H_{10}O_3$

Ordinarily, the uncharged organic acids exist in un-dissociated form at lower pH, and at certain pH, they exist in equilibrium between un-dissociated and dissociated forms. Organic acids in un-dissociated form can pass freely through the plasma membrane of the bacteria or fungi. Once inside the cell, these organic acids begin to dissociate, leading to the release of negatively charged anions and positively charged protons. These charged ions exhibit toxicological effect in the cell in different ways, such as disruption of essential metabolic reactions, disruption in membrane formation, intracellular stress due to pH change, and inhibition of production of essential enzymes, which ultimately lead to cell death.

Acetic acid and lactic acid are generally cheap, easily available, and environmentally friendly antimicrobial. Vinegar (the main constituent is acetic acid) and lemon juice are used as decontaminants in household for commodities, such as carrot, lettuce, and parsley. Acetic acid is also used as emulsifier, firming agent, stabilizer, flavor enhancer, and preservative. Furthermore, at low concentrations, formic acid is used as a flavoring agent. Dicarboxylic acids, such as adipic, fumaric, and succinic acids, and their calcium and magnesium salts are used in beverage, feed, and food preservation. Alpha hydroxyl acids, like citric, lactic, and malic, acids, have found application in the beverage and food industry [50].

Several sugar acids, such as ascorbic, gluconic, lactobionic, and tartaric acids, are used as preservatives in the milk and dairy industry. Many salts of these organic acids are used as food preservatives and antimicrobials and are used to extend the shelf life of food products, e.g. potassium acetate, calcium and sodium propionate, sodium sorbate, potassium sorbate, sodium benzoate, etc. The use of organic acids in food bio-preservation should be used with the utmost care and should be considered to be acceptable from a toxicological point of view.

12.3 Toxicological and Safety Aspect

Preservatives are used to increase the shelf life of food and maintain the quality for a longer time. In order to determine whether food additives, including preservatives and antimicrobials, are suitable for consumption, their toxicological and allergenic effect should be considered. Naturally occurring and naturally derived compounds are no exception to the harmful effect, as the term "natural" does not necessarily mean "safe". It is very essential to study the toxicological effects and characteristics of these natural antimicrobials along with their biological metabolites. A study should be done for possible toxicity caused by the interaction of natural bio-preservatives with food components, along with their allergic effect to the sensitive individuals. Consumption of a particular type of bio-preservative may not cause any adverse effects up to the certain levels present in nature, but it might become toxic once it reaches at high concentration in order to achieve antimicrobial activity or when consumed continuously over a longer period of time. The adverse reaction of preservatives can vary from very mild to life-threatening. The adverse and long-term effects of preservatives if not suitable to some person include eczema, nausea, vomiting, diarrhea, rhinitis urticaria, angioedema, irritable bowel syndrome, bronchospasmic syndrome, migraine, anaphylaxis, exfoliative dermatitis, hyperactivity, other behavioral disorders, tumors, cancer, DNA damage, etc., and in long run, they may act as mutagens [50]. The process of collecting data and analysis of data about novel antimicrobial compounds should be done keeping in view of their toxicity and acceptable thresholds for food consumption.

In addition, the use of naturally derived materials [51, 52] with antimicrobial characteristics endowed with high potential, not only preserving food and beverages, but also impart color as well flavor, thus extending their shelf life and appealing sense [53, 54]. Thus, various regulating agencies should make laws and ensure that only those food additives that are of GRAS status should be used; and in the case, it becomes essential to use some food antimicrobials that are not recognized as safe (Non-GRAS), the acceptable daily intakes (ADIs) should be clearly mentioned and should not be exceeded the limit and recommended quantity.

12.4 Conclusion and Future Outlook

During these days, there is an increasing demand for natural preservatives, either derived from animal, plant, or microbial sources, to be used as food preservatives, to reduce the consumption of synthetic preservatives. There are a number of natural products being used as bio-preservatives because of their effectiveness and potent antimicrobial, biocompatible, non-toxic, and ecofriendly properties. But still there are many unexplored natural sources and their products that can be used as bio-preservatives. The unexplored resources open up a large possibility of using these natural compounds as novel antimicrobials. While natural antimicrobial opens up a large possibility in the field of food bio-preservation, several factors should be kept in mind before using a natural substance or its derivative as food preservative, such as its compatibility and stability with the target food, effectiveness against the undesirable microbes, safety, etc. Further while exploring and using natural antimicrobials in food bio-preservation, these newly explored natural compounds should not interfere with the original nutritional value of the target food, and in other words, the traditional value of food should be maintained and have GARS status as well as effective against a broad range of harmful microbes. More advance researches should be done to optimize their processing and application in the field of food bio-preservation. These advanced researches should be governed by strict laws and have regulatory approval by concerned authorities, and furthermore they should be cost effective.

References

1. García, P., Rodriguez, L., Rodriguez, A., Martinez, B. (2010). Food biopreservation: Promising strategies using bacteriocins, bacteriophages and endolysins, *Trends Food Sci. Technol.*, **21(8)**, pp. 373–382.
2. Gharsallaoui, A., Oulahal, N., Joly, C., Degraeve, P. (2016). Nisin as a food preservative: Part 1: Physicochemical properties, antimicrobial activity, and main uses, *Crit. Rev. Food Sci. Nutr.*, **56 (8)**, pp. 1262–1274.
3. Xie, L., van der Donk, W.A. (2004). Post-translational modifications during lantibiotic biosynthesis, *Curr. Opin. Chem. Biol.*, **8**, pp. 498–507.
4. Oluk, C.A., Karaca, O.B. (2018). The current approaches and challenges of biopreservation, Food safety and preservation, pp. 565–597.
5. Johnson, E.M., Jung, Y.G., Jin, Y.Y., Jayabalan, R., Yang, S.H., Suh, J.W. (2018). Bacteriocins as food preservatives: Challenges and emerging horizons, *Crit Rev Food SciNutr.*, **58(16)**, pp. 2743–2767.
6. Juneja, V.K., Dwivedi, H.P., Yan, X. (2012). Novel natural food antimicrobials, *Annu. Rev. Food Sci. Technol.*, **3**, pp. 381–403.

7. Niamah, A.K. (2018). Structure, mode of action and application of pediocin natural antimicrobial food preservative: A review, *Basrah J. Agric. Sci.*, **31(1)**, pp. 59–69.

8. Díez, L., Rojo-Bezares, B., Zarazaga, M., Rodríguez, J.M., Torres, C., Ruiz-Larrea, F. (2012). Antimicrobial activity of pediocin PA-1 against *Oenococcusoeni*and other wine bacteria, *Food Microbiol.*, **31 (2)**, pp. 167–172.

9. Grattepanche, F., Miescher-Schwenninger, S., Meile, L., Lacroix, C. (2008). Recent developments in cheese cultures with protective and probiotic functionalities, *Dairy Sci. Technol.*, **88 (4&5)**, pp. 421–444.

10. Khan, H., Flint, S., Yu, P. (2010). Enterocins in food preservation, *Int. J. Food Microbiol.*, **141(1&2)**, pp. 1–10.

11. Park, S.H., Itoh, K., Fujisawa, T. (2003). Characteristics and identification of enterocins produced by *Enterococcus faecium* JCM 5804T, *J. Appl. Microbiol.*, **95**, pp. 294–300.

12. Galvez, A., Abriouel, H., Benomar, N., Lucas, R. (2010). Microbial antagonists to food-borne pathogens and biocontrol, *Curr. Opin. Biotechnol.*, **21**, pp. 142–148.

13. Obeso, J.M., Martınez, B., Rodrıguez, A., Garcıa, P. (2008). Lytic activity of the recombinant staphylococcal bacteriophage phiH5 endolysin active against *Staphylococcus aureus* in milk, *Int. J. Food Microbiol.*, **128**, pp. 212–218.

14. Oliveira, H., Azeredo, J., Lavigne, R., Kluskens, L.D. (2012). Bacteriophage endolysins as a response to emergingfoodborne pathogens, *Trends Food Sci. Technol.*, **28(2)**, pp. 103–115.

15. Schmelcher, M., Donovan, D.M., Loessner, M.J. (2012). Bacteriophage endolysins as novel antimicrobials, *Future Microbiol.*, **7(10)**, pp. 1147–71.

16. Shannon, R., Radford, D.R., Balamurugan, S. (2019). Impacts of food matrix on bacteriophage and endolysin antimicrobial efficacy and performance, *Crit Rev Food SciNutr.*, **60(10)**, pp. 1–10.

17. Galvez, A., Abriouel, H., Lucas, R., Jose, M., Burgos, G. (2011). Bacteriocins for bioprotection of foods, Rai, M., Chikindas, M. (eds.), Natural antimicrobials in food safety and quality, CAB International, Oxford shire, the U.K., pp. 39–61.

18. Mathiesen, G., Huehne, K., Kroeckel, L., Axelsson, L., Eijsink, V.G. (2005). Characterization of a new bacteriocin operon in sakacin P-producing *Lactobacillus sakei*, showing strong translational coupling between the bacteriocin and immunity genes, *Appl. Eviron. Microbiol.*, **71 (7)**, pp. 3565–3574.

19. Mohammed, S.S.D., Damisa, D., Balogu, T.V., Bala, E. (2016). Efficacy of sakacin on selected food pathogenic microorganisms isolated from fermented milk products, J. Appl. Sci. Environ. Manage, **20**(1), pp. 97–101.

20. Fallico, V., McAuliffe, O., Ross, R.P., Fitzgerald, G.F., Hill, C. (2011). The potential of lacticin 3147, enterocin AS-48, lacticin 481, variacin and sakacin P for food biopreservation, Lacroix, C. (ed.), Protective cultures, antimicrobial metabolites and bacteriophages for food and beverage biopreservation, Woodhead Publishing, Cambridge, the U.K., pp. 100–128.

21. Hugo, C.J., Hugo, A. (2015). Current trends in natural preservatives for fresh sausage products, *Trends Food Sci. Technol.*, **45 (1)**, pp. 12–23.

22. Morgan, S.M., O'Connor, P.M., Cotter, P.D., Ross, R.P., Hill, C. (2005). Sequential actions of the two component peptides of the lantibioticlacticin 3147 explain its antimicrobial activity at nanomolar concentrations, *Antimicrob, Agents Chemother.*, **49**, pp. 2606–2611.

23. Scannell, A.G.M., Hill, C., Buckley, D.J., Arendt, E.K. (1997). Determination of the influence of organic acids and nisin on shelf-life and microbiological safety aspects of fresh pork sausage, *J. Appl. Microbiol.*, **83**, pp. 407–412.

24. Scannell, A.G.M., Ross, R.P., Hill, C., Arendt, E.K. (2000). An effective lacticin biopreservative in fresh pork sausage, *J. Food Prot.*, **63**, pp. 370–375.

25. Stevens, M., Vollenweider, S., Lacroix, C. (2011). The potential of reuterin produced by *Lactobacillus reuteri* as a broad spectrum preservative in food, Protective cultures, antimicrobial metabolites and bacteriophages for food and beverage biopreservation, pp. 129–160.

26. Talarico, T.L., Dobrogosz, W.J. (1989). Chemical characterization of an antimicrobial substance produced by *Lactobacillus reuteri*, *Antimicrob. Agents Chemother.*, **33(5)**, pp. 674–679.

27. Arques, J.L., Fernandez, J., Gaya, P., Nunez, M., Rodriguez, E., Medina, M. (2004). Antimicrobial activity of reuterin in combination with nisin against food-borne pathogens, *Int. J. Food Microbiol.*, **95(2)**, pp. 225–229.

28. Arques, J.L., Rodrıguez, E., Nunez, M., Medina, M. (2008). Antimicrobial activity of nisin, reuterin, and the lactoperoxidase system on *Listeria monocytogenes* and *Staphylococcus aureus* in cuajada, a semisolid dairy product manufactured in Spain, *J. Dairy Sci.*, **91(1)**, pp. 70–75.

29. Montiel, R., Martín-Cabrejas, I., Medina, M. (2015). Reuterin, lactoperoxidase, lactoferrin and high hydrostatic pressure on the inactivation of food-borne pathogens in cooked ham, *Food Control*, **51**, pp. 122–128.

30. Montiel, R., Martín-Cabrejas, I., Peirotén, A., Medina, M. (2016). Reuterin, lactoperoxidase, lactoferrin and high hydrostatic pressure treatments on the characteristics of cooked ham, *Innov. Food Sci. Emerg. Technol.*, **35**, pp. 111–118.

31. Hansen, J.N. (1994). Nisin as a model food preservative, *Crit Rev Food SciNutr.*, **34(1)**, pp. 69–93.

32. Allen, H.K., Trachsel, J., Looft, T., Casey, T.A. (2014). Finding alternatives to antibiotics, *Ann. N. Y. Acad. Sci.*, **1323**, pp. 91–100.

33. Govaris, A., Solomakos, N., Pexara, A., Chatzopoulou, P.S. (2010). The antimicrobial effect of oregano essential oil, nisin and their combination against *Salmonella enteritidis*in minced sheep meat during refrigerated storage, *Int. J. Food Microbiol.*, **137(2&3)**, pp. 175–180.

34. Bastarrachea, L., Dhawan, S., Sablani, S.S., Powers, J. (2010). Release kinetics of nisin from biodegradable poly (butylene adipate-*co*-terephthalate) films into water, *J. Food Eng.*, **100(1)**, pp. 93–101.

35. Lule, V.K., Garg, S., Gosewade, S.C. (2016). Encyclopedia of food and health, Chapter: "Natamycin", Benjamin, Caballero, Paul, M. (eds.), Elsevier Ltd., Finglas, Fidel Toldrá, pp. 56–62.

36. Kallinteri, L.D., Kostoula, O.K., Savvaidis, I.N. (2013). Efficacy of nisin and/or natamycin to improve the shelf-life of Galotyri cheese, *Food Microbiol.*, **36(2)**, pp. 176–181.

37. Raab, W.P. (1972). Natamycin (pimaricin): Its properties and possibilities in medicine, George Thieme Publishers, Stuttgart, Germany.

38. Stark, J. (2003). Natamycin: An effective fungicide for food and beverages, Roller, S. (ed.), Natural antimicrobials for the minimal processing of foods, Woodhead Publishing, Cambridge, the U.K., pp. 82–97.

39. Bierhalz, A.C.K., Da Silva, M.A., Kieckbusch, T.G. (2012). Natamycin release from alginate/pectin films for food packaging applications, *J. Food Eng.*, **110(1)**, pp. 18–25.

40. Ollé-Resa, C.P., Gerschenson, L.N., Jagus, R.J. (2014). Natamycin and nisin supported on starch edible films for controlling mixed culture growth on model systems and Port Salut cheese, *Food Control*, **44**, pp. 146–151.

41. Pintado, C.M.B.S., Ferreira, M.A.S.S., Sousa, I. (2010). Control of pathogenic and spoilage microorganisms from cheese surface by whey protein films containing malic acid, nisin and natamycin, *Food Control*, **21(3)**, pp. 240–246.

42. Bierhalz, A.C.K., Da Silva, M.A., Kieckbusch, T.G. (2012). Modelling natamycin release from alginate/chitosan active films, *Int. J. Food Sci. Technol.*, **47(4)**, pp. 740–746.

43. Fajardo, P., et al. (2010). Evaluation of a chitosan-based edible film as carrier of natamycin to improve the storability of Saloio cheese, *J. Food Eng.*, **101(4)**, pp. 349–356.

44. Friedman, M., Juneja, V.K. (2010). Review of antimicrobial and antioxidative activities of chitosans in food, *J. Food Prot.*, **73(9)**, pp. 1737–1761.

45. Wang, S., Marcone, M., Barbut, S., Lim, L. (2012). The impact of anthocyanin-rich red raspberry extract (ARRE) on the properties of edible soy protein isolate (SPI) films, *J. Food Sci.*, **77**, pp. C497–C505.

46. Medina, A., Jiménez, M., Mateo, R., Magan, N. (2007). Efficacy of natamycin for control of growth and ochratoxin A production by *Aspergillus carbonarius* strains under different environmental conditions, *J. Appl. Microbiol.*, **103(6)**, pp. 2234–2239.

47. Sillankorva, S.M., Oliveira, H., Azeredo, J. (2012). Bacteriophages and their role in food safety, *Int. J. Microbiol.*, **863945,** 1–14.

48. Połaska, M., Sokołowska, B. (2019). Bacteriophages: A new hope or a huge problem in the food industry, *AIMS Microbiol.*, **5(4)**, pp. 324–346.

49. Lues, J.F.R. (2005). Organic acids: Promising modern preservatives, Interim: Interdisc. J., **4(2)**, pp. 151–161.

50. Inetianbor, J.E., Yakubu, J.M., Ezeonu, S.C. (2015). Effects of food additives and preservatives on man: A review, *Asian Journal of Science and Technology*, **6(2)**, pp. 1118–1135.

51. Yusuf, M., Ahmad, A., Shahid, M., Khan, M.I., Khan, S.A., Manzoor, N., Mohammad, F. (2012). Assessment of colorimetric, antibacterial and antifungal properties of woollen yarn dyed with the extract of the leaves of henna (*Lawsonia inermis*), *J. Clean. Prod.*, **27**, pp. 42–50.

52. Yusuf, M., Shahid, M., Khan, M.I., Khan, S.A., Khan, M.A., Mohammad, F. (2015). Dyeing studies with henna and madder: A research on effect of tin (II) chloride mordant, *J. Saudi Chem. Soc.*, **19(1)**, pp. 64–72.

53. Yusuf, M. (2018). Natural antimicrobial agents for food biopreservation, Food packaging and preservation, Series-handbook of food bioengineering, Vol. 9, Grumezescu, A.M., Holban, A.M. (eds.), Academic Press, London, pp. 409–438.

54. Yusuf, M., Shabbir, M., Mohammad, F. (2017). Natural colorants: Historical, processing and sustainable prospects, *Nat. Prod. Bioprospect.*, **7(1)**, pp.123–145.

Chapter 13

Upsurgence of Bio-Based Packaging Materials: Emphasis on Biodegradable Plastics as a Greener Alternative

Divya Bajpai Tripathy,[a] Pinki Chakraborty,[a] Munesh Sharma,[b] and Mohd Yusuf[c]

[a] *School of Basic and Applied Sciences, Galgotias University, Uttar Pradesh, India*
[b] *Department of Chemistry, Faculty of Science, J.V. College, Baraut, Baghpat, Uttar Pradesh, India*
[c] *Department of Natural and Applied Sciences, School of Technology, The Glocal University, Mirzapur Pole, Saharanpur, Uttar Pradesh, India*
yusuf1020@gmail.com

In today's world, there has been an enormous demand for biopolymers like poly(butyrates) and polylactides, as they find applications in major industries like pharmaceutical and food packaging. The actual importance and significance of biodegradable polymers have emerged, due to the environment-related issues in the conventional food packing sector, including waste management, and the way of composting these bio-based materials. The biodegradability depends on various internal and external factors, including degradation mechanism, suitability, and level of adoption. Several sectors involve the use of biotechnology, as it plays an important role in the treatment processes of different waste materials. The treatment of hazardous contaminants is a promising application of this technology, as it is cost effective and can result in complete removal of hazard materials and their minerals. One

Biomaterials in Food Packaging
Edited by Mohd Yusuf and Shafat Ahmad Khan

ISBN 978-981-4877-98-5 (Hardcover), 978-1-003-25678-6 (eBook)
www.jennystanford.com

of the basic features of biopolymers includes their recyclability and biodegradable nature. This article is a summary to provide an overview regarding biodegradable polymers that are based on bio-based recourses.

13.1 Introduction

From past few decades, many food packaging materials have been utilized. They can be natural (i.e. wood, pulp, paper, etc.) or synthetic (i.e. glass, plastics, metals, etc.). Over time, semi-synthetic and composite materials have been developed with more environment-friendly characteristics. In addition, these materials not only limited to food packaging, but also are being utilized in other growing life care sectors, including cosmetics, personal care, and domestic products [1]. Many of these polymeric substances possess high stability/non-biodegradability with less or non-compatibility to the environment. Also, the huge amount of plastic litters and wastes (microplastics) are being dumped annually to the environment, which particularly gives rise to severe health-related problems by the action of food-web with detrimental effects [2]. Consequently, the global environmental concerns triggered the minimal usage of synthetic materials for food packaging applications [3], as they enter to the municipal waste streams after the end of their usage.

Recycling of these wastes is found another major problem. According to an estimation, it was observed that during 1950–2015, only 9% of the plastic-based wastes were recycled [4, 5]. To execute this major concern, strict rules and regulations have been developed to use better alternatives to materials produced from non-renewable resources. In this context, several global regulatory bodies are involved in setting standards for biodegradable and compostable plastics, for example, the American Society for Testing and Materials (ASTM), the European Committee for Standardization (CECS), the British Plastics Federation (BPF), the US Composting Council (USCC), the American Certification System of Biodegradable Products Institute (ACS-BPI), the Environment & Plastics Industry Council (EPIC), and the Japan's Green Plaprogram (JGPla) [6].

In the habitat of living beings (land/aquatic), the existence of plastic-fritters has imposed critical issues, which are being vigorously exploited by the environmental lobby to demand solutions from the packaging industry. The solutions for the associated problems have been proposed by scientific community, including incineration, recycling, composting, and environmental degradation. Out of these four, the last refers to both photodegradation (attack by photons of ultraviolet radiation) and biodegradation using microbes (bacteria, algae, or fungi) [2, 7]. In fact, the diversity of biodegradable materials and their varying properties makes it difficult to describe so easily. In this context, a common thought can be drawn as biodegradable products are all 'good' or petrochemical-based products are all 'bad' [8, 9]. The chapter provides a literature on exploration toward biodegradable plastic materials explaining their biochemical characteristics, and economic viability with future trend and applicability (Figure 1).

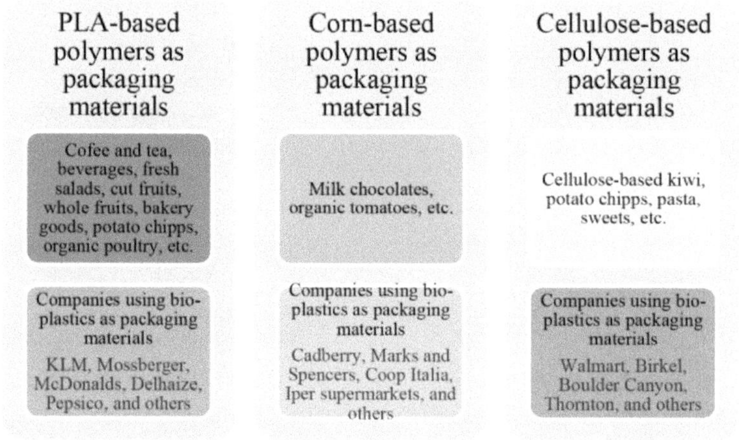

Figure 13.1 Common biopolymers, packaging applications, and companies using them as packaging materials.

13.2 Types of Biodegradable Plastics

Nowadays, plastics are seemingly ubiquitous and their regenerated products are often used globally by consumers [10]. Chemically, plastics are synthetic or semi-synthetic organic compounds in origin that are malleable and can be transformed into a wide range of solid objects. Commonest utilized plastics are depicted in Table 13.1 with their broadly significant applications [11–13].

Table 13.1 Most common plastics and their significant applications

Type of plastic	Abbreviation	Application/use
Polyamide or nylon	PA	Fibrous materials like toothbrush bristles, tubing, fishing line, fibers, ropes
Polyimide	PI	Adhesives tapes
Polycarbonate	PC	Security windows, traffic lights and lenses, compact disks, eyeglasses
Polyester	PES	Fibers, threads, textiles
Polyethylene	PE	A wide range of packaging and carrier thin bags, bottles
High-density polyethylene	HDPE	Strong and thickened bottles, milk jugs
Low-density polyethylene	LDPE	Floor tiles, shower curtains and clamshell packaging, cheap & light weight portable furniture
Polyethylene terephthalate	PET	Transparent drinks bottles, jars, packaging films
Polypropylene	PP	Yogurt containers, appliances, bottle caps, drinking straws
Polystyrene	PS	Disposable cutlery, cups, plates, CDs & DVDs, cassette boxes, plastic tableware, food containers
High impact polystyrene	HIPS	Refrigerator liners, vending cups, food packaging containers
Polyurethane	PU	Leather looking appliances, jackets, belts, shoes, printing rollers, cushion, thermal insulation foams, surface coatings

(*Continued*)

Table 13.1 (*Continued*)

Type of plastic	Abbreviation	Application/use
Polyvinyl chloride	PVC	Electrical wire/cable insulation, plumbing pipes and guttering, home appliances
Polyvinylidene chloride	PVDC	Food packaging containers, strong films
Polyvinyl alcohol	PVA	Adhesives
Acrylonitrile butadiene styrene	ABS	Cases for electronic equipment/appliances (e.g. computer monitors, printers, keyboards), drainage pipes
Polycarbonate+acrylonitrile butadiene styrene	(PC+ABS)	Mobile phone bodies, interior and exterior car parts, etc.
Polyethylene+acrylonitrile butadiene styrene	PE+ABS	Low-duty dry bearings
Polyepoxide	Pepox	Adhesives, potting agents for electrical components, matrix agents
Polymethyl methacrylate	PMMA	Contact lenses, fluorescent/luminescent light diffusers, rear light covers for vehicles
Polytetrafluoroethylene (Teflon)	PTFE	Heat-resistant and low-friction coatings, non-stick surfaces
Phenol formaldehyde	PF	High modulus and fire resistant polymer coatings, thermally insulation foams
Urea-formaldehyde	UF	Multi-colorable alternatives to phenolics, wood adhesives
Melamine formaldehyde	MF	Break-resistance alternatives to ceramic cups, plates and bowls, decorated top surface layer of paper laminates

(*Continued*)

Table 13.1 (*Continued*)

Type of plastic	Abbreviation	Application/use
Polyetheretherketone	PEEK	Strong, chemical & heat-resistant thermoplastic moldings, medical implanting applications, aerospace moldings
Polyetherimide (Ultem)	PEI	Used in applications that require high strength, heat resistance, modulus and broad chemical resistance like making of electric transmission components, medical instruments
Maleimide & bismaleimide	MM & BMM	High temperature-resistant composite materials
Furan resin	FR	Foundry sands, biologically derived composites
Polysulfone	P(S)	High-temperature resistant materials like water heater dip tubes, membranes, filtration media
Polydiketoenamine	PDK	A novel plastic variant that can be dunked in acid and reshaped endlessly (under trial)
Silicone polydiketoenamine	SPDKA	Cooking utensils, sealants, base resin for industrial paints

Nowadays, due to the serious environmental concerns, vigorous researches are going on to explore sustainable, eco-friendly plastics that not only will have the negative impacts on the environment, but also there will be no compromise with the applicability of the products.

Some of the common bioplastic types, originated from natural products are as follows [14]:

13.2.1 Polyhydroxyalkanoates

Polyhydroxyalkanoates (PHAs) are the types of bioplastics, produced by microbes like *C. necator*. Some particular types encompass poly-3-hydroxybutyrate (PHB), polyhydroxyhexanoate (PHH), and polyhydroxyvalerate (PHV). PHA is generally biosynthesized through deprivation of organisms of some nutrients specifically in absence of phosphorus, oxygen, and/or nitrogen like elements. This can be attained through excess supply of carbon sources [15].

PHAs can be further divided into short chain length PHAs (originated from short chain generally 3–5 carbons and produced by various bacteria including *C. necator* and *A. latus*) and medium change length PHAs (originated from short chain generally 6–14 carbons and produced by *P. putida* like microbes) [16, 17].

13.2.2 Polylactic Acid

Polylactic acid (PLA) is a type of thermoplastic aliphatic polyester produced by microbes through fermentation of plant starch like corn, sugarcane cassava, or sugar beet pulp. In 2010, PLA was reported to have the highest utilization volume than any of the bioplastics used, globally. Although as per the American and European standards, PLA is considered as compostable rather than biodegradable, as it is not found degradable outside the artificial composting environment [18].

13.2.3 Starch-Based Plastics

Starch blends belong to the class of thermoplastic polymers originated as starch-plasticizer blends. These are biodegradable, but are not applicable to all plasticizers, so their biodegradability depends on the ratio of plasticizers in blends. Some such blends include starch/polycaprolactone, starch/PLA, and starch/polybutylene-adipate-co-terephthalate [19].

13.2.4 Cellulose-Based Plastics

This class of bioplastics covers cellulose esters and their derivatives along with celluloid. Cellulose needs extensive modification to yield its thermoplastic form. Due to high manufacturing cost of cellulose-based plastics, their exploitation for packaging purpose is not very much promoted [20].

13.2.5 Lignin-Based Polymer Composites

These bioplastics consist of sustainable, biodegradable, and bio-renewable natural aromatic polymers. Lignin can be obtained from the by-product process involves in paper production. Due to its easy availability, low weight, eco-friendly characteristic over conventional counterparts, and comparable chemical properties with current plastic chemicals, lignin-based polymer composites have been produced in great quantities and are available for exploitation as emerging eco-friendly polymers [21–23].

13.3 Bioplastics vs. Compostable Plastics

The basic difference between bioplastic and compostable plastic is that the compostable plastic is a type of plastic that degrades through biological pathways, and during the process, it releases carbon dioxide, water, inorganic compounds, and biomass at a rate coherent with other recognized compostable materials and does not leave any visually distinct or noxious residues. So, this can be concluded that all compostable plastics are bioplastics, but this cannot be claimed in reverse.

Whereas, biodegradable polymers (BDPs) or bioplastics are the polymeric stuff capable to decompose into carbon dioxide, methane, inorganic compounds, water, or biomass, predominantly through enzymatic activity of microorganisms [24].

13.4 Mechanism of Biodegradation

Biodegradation process is classified into three categories, such as bio-deterioration and bio-fragmentation [24]. In the first stage of biodegradation, i.e. bio-deterioration, surface degradation starts that modifies the physical and chemical properties of materials. The surface degradation mainly depends on outdoor and abiotic factors, which involve initial external factors like light, temperature, and chemical environment. The first and the second stages, bio-deterioration and bio-fragmentation, are inter-related that is along with bio-deterioration, the bio-fragmentation process occurs parallelly. The bio-fragmentation mainly involves undesirable action of living organisms on manmade materials, involving corrosion of building materials and metals by microorganisms [23–25]. Aerobic biodegradation is another important stage for biopolymers, which generally means breakdown of organic contaminants into smaller molecules with the release of oxygen. The main function of oxygen in aerobic oxidation is the conversion of carbon containing species into carbon dioxide and the rest of the hydrocarbon gets converted into cell masses; whereas in anaerobic biodegradation, there are many electron acceptor species, which act as substitutes for oxygen, like nitrates, sulfates, and iron [26]. Most common electron acceptors for anaerobic degradation are transition metals and even carbon dioxide in some cases [27]. The anaerobic process is used for treating wastewater with high concentrations of biodegradable organic materials, such as concentrated domestic wastewater, biosolids, animal manure slurry, and food processing wastes [25]. The overall overview of the main stages of the degradation process is represented in Figure 13.2.

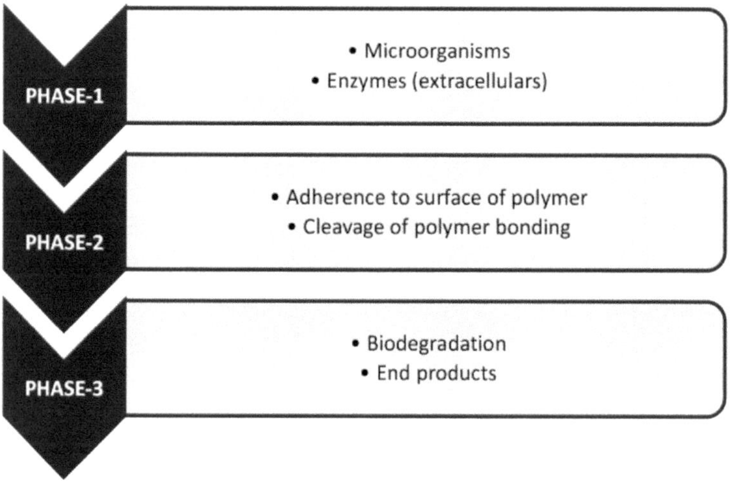

Figure 13.2 Flow chart of the biodegradation process.

13.5 Factors Influencing Biodegradation

13.5.1 External Factors

13.5.1.1 Physical parameters

Physical parameters play noteworthy impact on biodegradation of bio-based food packaging materials. The parameters like polarity of surface area, molecular weight, melting point, and elastic behavior of polymers impart tremendous impact on the biodegradation of the materials. Generally, branched chain polyesters degrade less conveniently than the linear ones [7]. Furthermore, the morphological properties of biopolymers influence greatly to the biodegradation rate. Similarly, the melting temperature of the biopolymers also has great impacts on the enzymatic or microbial polymer degradation. Higher the melting point of the biopolymer, lower will be its degradation [20, 23, 24]. Aliphatic polyesters with esters and/or polycarbonates linkage are the polymers that have great assessments to be exploited as bioplastics. This is because of their vulnerabilities to microbial and lipolytic enzyme degradation. In this series, the aliphatic polyamides and polyurethanes were claimed to be relatively less susceptible to the degradation, as compared to aliphatic polyesters and polycarbonates, due to their higher Tm (temperature of melting) values.

13.5.1.1.1 *Moisture or water content*

Moisture or relative humidity presents in the environment also leads to the significant change in the rate of polymer biodegradation. It is a well-known fact that water is essential for the development and propagation of almost every microbial species [28].

Increase in moisture content presents in the environment raises the activity of microbes, thus increases the rate of biodegradation of polymers. Ho et al. disclosed that there was sharp increase in the degradation rate with the rise in relative humidity of the surroundings [27].

13.5.1.1.2 *Temperature*

Temperature is one of the most vital parameters that influences the polymers degradation remarkably [29]. It is claimed that degradation of most of materials is almost impossible in the absence of suitable temperature, as the extension of optimum temperature for biodegradation is a reaction-specific modality. Generally, the hydrolysis rate and the microbial activity increase with the rise in temperature; but at an extended high temperature, the microbial activity reduces and even can be halted [30, 31].

Usually, the activity rate of microbes increases with temperature, since at a certain temperature range, the microbes start dying. Same trends were obtained in another study, in which the biodegradation of bioplastics has been studied by burying 0.5 g of each bioplastic sample in 30 g of normal soil. Same procedure has been adopted for blank, where same amount of sterile soil has been taken as control [32].

13.5.1.2 Chemical parameters

pH also plays an important role in the biodegradation of biopolymers. Its type specifically depends on the reaction type. pH value can alter the hydrolysis reaction rate as well as propagation of microorganisms [28]. Hayakawa et al. (2014) revealed a progressive relationship between enzyme activity of cellulose and pH that leads to the conclusion that low pH value decreases the cellulose degradation [33].

13.5.1.3 Environmental parameters

Besides the chemical composition and crystallinity, biodegradation of biopolymers is also influenced by various environmental factors. The illustration of "readily degradable" sugars can exist as such in soil organic matter (SOM) for longer period even for centuries; whereas, "intractable" lignin turnover can be somewhat more [34]. Climate can also determine the rate of degradation. In contrast, Prescott (2010) [35] published a review paper, encompassing 70 researches, in which he revealed that carbon-to-nitrogen ratio had greater impact on the decomposition rate. In this study, it has also been claimed that the impact of climatic variations, due to global alterations on the microbial cleavage, is probably to be little; until or unless the change is modified or manipulated related plants species [36, 37]. In this way, the plant species type is the utmost significant parameter to determine the decomposition of biopolymers.

13.5.2 Internal Factors

13.5.2.1 Composition

In addition to the other factors, composition of bioplastics also plays a vital role in their decomposition. For instance, due to the chemical structure and chemical composition, PHA shows tremendous biodegradability [38, 39]. Its capability to degrade in natural environment enables it to get listed in bioplastics. Microorganisms on the degradation of PHA secrete extracellular enzyme, PHA depolymerase. Besides this, chemical additives also influence the biodegradation of PHA in a positive manner. Addition of chemical additives has great impact on the degradation of PHA on both the chemical scale and physical scale by inhibiting enzymatic cleavage. In addition, non-degradable chemicals like co-polyesters subsidized the area available to the interaction of enzymes and biopolymers. For example, chain extenders and anti-fouling agents that are the best-known used chemicals influence the biopolymer degradation by reducing the extracellular enzyme activity or by minimizing the availability of surface area to microorganisms to produce biocatalysts [35]. Moreover, the microbial degradation of polyethylene decreases reciprocally with the molecular size. Linear paraffin molecules with less than or equal to molecular size of nearly 500 Da can be easily degraded through various microbes [40]. In the same way, microbial degradation of starch–polyethylene blend and their chemically modified blends has been examined [41]. Result of this study revealed that the removal of carbon from the blend of starch and polyethylene was relatively low, comparing to the pure form of starch. Same result has been concluded for linear low-density blend of polyethylene with starch in the soil. Outcome of a simultaneous biodegradation study on the cellulose, polycaprolactone (PCL) and starch blend, and aliphatic–aliphatic co-polyester reflected that cellulose was more rapidly degrade than PCL-starch blend, which was consequently more degradable over co-polyester [42]. In the same way, PVA blends-based films were claimed to have very promising biodegradation properties [43].

13.5.2.2 Crystallinity

Polymeric structure is indissolubly associated with crystallinity impact and is also related with the effectiveness of biodegradability [38, 41]. Final sample crystallinity of the polymer is also influenced by the processing techniques of polymers, which also alter the solvent casting and quenching like parameters [44]. For instance, the biodegradation of PHA starts with the rise in mass losses and intermolecular phase erosion on the sites of disordered chains prior to the samples of well-ordered crystalline regions. On contrary, amorphous area allows moisture and enzymes to enter into the active site and results in increasing microbial degradation. This is due to the more availability of surface area [45]. In case of PHA and PHB, due to the greater extendibility to crystalline forms, less degradability is expected.

13.6 Economic Aspects

The biodegradable polymers are widely used in the packaging industry, and there are no mandatory regulations for customers when it comes to the use or post-customer treatment.

The increasing urge for product improvement has opened up new doors for packaging segments, such as cellulose products and PLAs for packaging-related applications [46, 47]. The increasing usage of biodegradable polymers in the packaging and medicinal industries has enormously increased their productivity and these are used widely in the preservation of food items [48]. Moreover, from economic perspective, these biodegradable materials find their applications in the chemical industry, as useful building materials. Among all biodegradable materials, the demand for PLAs is constantly increasing, as these are greener in nature, renewable, and biodegradable products, which are resulting in constant growth over the years of the production and innovation units [40, 44]. However, one of the limitations of biodegradable polymers is that they are expensive, and thus it is important to make them cost-effective. Moreover, from both psychological and economic perspectives, the research achievements in this area have resulted directly or indirectly in effective legislations, manufacturing practices, directives, and sustainable cycles based for packaging procedures [49]. The increased awareness and consciousness among the people regarding the significance of recyclable and biodegradable polymers' usage have also led to the enhancement in the economic aspects of polymers. In most of the recent studies, it has been prioritized that most of the USA and Canada customers generally are involved in buying biodegradable household products rather than other commercial goods and commodities. So, from economic perspective till date, most of the uses of biodegradable polymers have been in the field of fiber and non-fiber packaging; and in recent times, these biodegradable polymers and plastics are combined with non-renewable materials to be used in the production of oil-based plastic materials [41, 48, 49].

13.7 Industrial Aspects

Due to the increasing environmental pollution levels and rising uses of biodegradable polymers, these materials have gained a lot of importance in the automobile, agricultural, and medicinal industries [42]. Biodegradation of automotive waste, such as polyester and polyurethane foam, can take place by using *Pseudomonas chlororaphis* [50]. In these industries, majorly of biodegradable polymers are used for the conversion and production of less toxic and harmful environmental species. Moreover, plastic pollution has become one of major concerns all over the world, which has ultimately resulted in increasing demand for biodegradable polymers [47].

PHA and PLA are the two most promising biodegradable polymers. The PHA is price-competitive and offers a challenge for some commercial petroleum-based plastics. Whereas, the PLA has already replaced a number of non-biodegradable polymers in the food packaging and carrier bag markets, but its biodegradation is still much slower than those of other biodegradable polymers. However, in terms of properties and usability for certain applications, a very few commercial non-biodegradable polymers can be replaced by known biodegradable polymers. It is, therefore, possible to only slightly reduce the current environmental problems by using known biodegradable polymers for some low-level applications. Thus, a lot of researches are still needed to develop biodegradable polymers or polymer blends that will have the necessary properties to replace most of the current non-biodegradable polymers [47–50].

13.8 Conclusion and Future Outlook

This review has covered mainly the biodegradation chemistry and degradability studies. The influence of polymer chemistry on environment can be both positive and negative, and can also result in different environmental issues. This factor has been the driving force for the wide area of researches on renewable resources and their applications. Composting with organic wastes is one of the major ways to enrich soil and also increases the soil longevity and stability. This article reveals that the use of biodegradable polymers is beneficial for both environment and industries. Therefore, in respect to other polymers, the use of biodegradable polymers is of utmost importance from commercial perspective.

References

1. Ciriminna, R. and Pagliaro, M. (2020). Biodegradable and compostable plastics: A critical perspective on the dawn of their global adoption, *Chem. Open*, **9(1)**, pp. 8–13.

2. Wang, W., Gao, H., Jin, S., Li, R., and Na, G. (2019). The ecotoxicological effects of microplastics on aquatic food web, from primary producer to human: A review, *Ecotoxicol. Environ. Safety*, **173**, pp. 110–117.

3. Majeed, K., Jawaid, M., Hassan, A. A. B. A. A., Bakar, A. A., Khalil, H. A., Salema, A. A., and Inuwa, I. (2013). Potential materials for food packaging from nanoclay/natural fibres filled hybrid composites, *Mater. Design*, **46**, pp. 391–410.

4. Geyer, R., Jambeck, J. R., and Law, K. L. (2017). Production, use, and fate of all plastics ever made, *Sci. Adv.*, **3(7)**, e1700782. DOI: 10.1126/sciadv.1700782

5. Eriksen, M., Lebreton, L. C. M., Carson, H. S., Thiel, M., Moore, C. J., Borerro, J. C., Galgani, F. P. G., and Ryan, J. (2013). Plastic pollution in the world's oceans: More than 5 trillion plastic pieces weighing over 250,000 tons afloat at sea, *PLoS One*, **9(12)**, e111913. DOI: doi.org/10.1371/journal.pone.0111913

6. Greener, J. (2007). Biodegradation of compostable plastics in green yard-waste compost environment, *J. Polym. Environ.*, **15**, pp. 1–5.
7. Huang, J. C., Shetty, A. S., and Wang, M. S. (1990). Biodegradable plastics: A review, *Adv. Polym. Technol.*, **10(1)**, pp. 23–30.
8. Song, J. H., Murphy, R. J., Narayan, R., and Davies, G. B. H. (2009). Biodegradable and compostable alternatives to conventional plastics, *Philosophical Trans. Royal Soc. B: Biolog. Sci.*, **364(1526)**, pp. 2127–2139.
9. Kumar, S. and Thakur, K. S. (2017). Bioplastics-classification, production and their potential food applications, *J. Hill Agric.*, **8(2)**, pp. 118–129.
10. Greene, J. (2007). Biodegradation of compostable plastics in green yard-waste compost environment, *J. Polym. Environ.*, **15(4)**, pp. 269–273.
11. Tokiwa, Y., Calabia, B. P., Ugwu, C. U., and Aiba, S. (2009). Biodegradability of plastics, *Int. J. Molecular Sci.*, **10(9)**, pp. 3722–3742.
12. Osswald, T. A., Baur, E., and Rudolph, N. (2019). Plastics handbook: The resource for plastics engineers, Hanser Publications, Ohio, USA.
13. Nazami, H. (2017). Polymers in our daily life, *Bioimpacts*, **7(2)**, pp. 73–74.
14. Young, K., Lenz, B., and Robert W. (2001). Biopolyesters, Wolfgang, B. and Alexander, S. (eds.), Chapter 3 "Polyesters from microorganisms", Book series: Advances in biochemical engineering, Springer-Verlag, Berlin, pp. 51–79.
15. Nicolas, J., Chi-Wei, J., Yu-Hong, W., Ho-Shing, W., and Wang, S. (2008). Isolation and purification of bacterial poly (3-hydroxy alkanoates), *Biochem. Eng. J.*, **39(1)**, pp. 15–27.
16. Philip, S., Keshavarz, T., and Roy, I. (2007). Polyhydroxy alkanoates: Biodegradable polymers with a range of applications, *J. Chem. Technol.*, **82(3)**, pp. 233–247.
17. Mangaraj, S., Yadav, A., Lalit, M. B., Dash, S. K., and Mahanti, N. K. (2019). Application of biodegradable polymers in food packaging industry: A comprehensive review, *J. Packag. Technol. Sci.*, **3**, pp. 77–96.
18. Chaleat, C., Halley, P. J., and Truss, R.W. (2014). Starch polymers: From genetic engineering to green applications, Halley, P.J. (ed.), Chapter "Mechanical properties of starch-based plastics", Elsevier, Burlington, MA, USA, pp. 187–209.

19. Khalid, S., Yu, L., Meng, L., Liu, H., Ali, A., and Chen, L. (2017). Poly(lactic acid)/starch composites: Effect of microstructure and morphology of starch granules on performance, *J. Appl. Polym. Sci.*, **134 (46)**, pp. 45504. DOI: https://doi.org/10.1002/app.45504

20. Avérous, L. and Pollet, E. (2014). Starch polymers: From genetic engineering to green applications, Halley, P.J. (ed.), Chapter "Nanobiocomposites based on plasticized starch", Elsevier, Burlington, MA, USA, pp. 211–239.

21. Thakur, V. K., Thakur, M. K., Raghavan, P., and Kessler, M. R. (2014). Progress in green polymer composites from lignin for multifunctional applications: A review, *ACS Sust. Chem. Eng.*, **2(5)**, pp. 1072–1092.

22. Taniguchi, I., Yoshida, S., Hiraga, K., Miyamoto, K., Kimura, Y., and Oda, K. (2019). Biodegradation of PET: Current status and application aspects, *ACS Catalysis*, **9(5)**, pp. 4089–4105.

23. Calabrò, P. S. and Grosso, M. (2018). Bioplastics and waste management, *J. Waste Manage.*, **78**, pp. 800–801.

24. Das, M.P. and Kumar, S. (2014). Microbial deterioration of low density polyethylene by Aspergillus and Fusarium sp, *Int. J. Chem. Tech. Res.*, **6(1)**, pp. 299–305.

25. Thakur, S., Chaudhary, J., Sharma, B., Verma, A., Tamulevicius, S., and Thakur, V. K. (2018). Sustainability of bioplastics: Opportunities and challenges, *Curr. Opin. Green Sust.*, **13**, pp. 68–75.

26. Zoungranan, Y., Lynda, E., Dobi-Brice, K. K., Tchirioua, E., Bakary, C., and Yannick, D. D. (2020). Influence of natural factors on the biodegradation of simple and composite bioplastics based on cassava starch and corn starch, *J. Environ. Chem. Eng.*, **8(5)**, pp. 104396. DOI: https://doi.org/10.1016/j.jece.2020.104396

27. Verlicchi, P., Al Aukidy, M., and Zambello, E. (2012). Occurrence of pharmaceutical compounds in urban wastewater: Removal, mass load and environmental risk after a secondary treatment — a review, *Sci. Total Environ.*, **429**, pp.123–155.

28. Dahlen, G., Basic, A., and Bylund, J. (2019). Importance of virulence factors for the persistence of oral bacteria in the inflamed gingival crevice and in the pathogenesis of periodontal disease, *J. Clin. Med.*, **8(9)**, pp. 1339. DOI: https://doi.org/10.3390/jcm8091339

29. Karamanlioglu, M., Preziosi, R., and Robson, G. D. (2017). Abiotic and biotic environmental degradation of the bioplastic polymer poly (lactic acid): A review, *Polym. Degrad. Stab.*, **137**, pp. 122–130.

30. Meereboer, K. W., Misra, M., and Mohanty, A. K. (2020). Review of recent advances in the biodegradability of polyhydroxyalkanoate (PHA) bioplastics and their composites, *Green Chem.*, **22(17)**, pp. 5519–5558.

31. Prata, J. C., Reis, V., da Costa, J. P., Mouneyrac, C., Duarte, A. C., and Rocha-Santos, T. (2020). Contamination issues as a challenge in quality control and quality assurance in microplastics analytics, *J. Hazard. Mater.*, **403**, pp. 123660. DOI: https://doi.org/10.1016/j.jhazmat.2020.123660

32. Roy, S., and Rhim, J. W. (2021). Anthocyanin food colorant and its application in pH-responsive color change indicator films, *Crit. Rev. Food Sci. Nutr.*, **61(14)** pp. 2297–2325.

DOI: https://doi.org/10.1080/10408398.2020.1776211

33. Hayakawa, C., Funakawa, S., Fujii, K., Kadono, A., and Kosaki, T. (2014). Effects of climatic and soil properties on cellulose decomposition rates in temperate and tropical forests, *Biol. Fert. Soils*, **50(4)**, pp. 633–643.

34. Müller, G., Hanecker, E., Blasius, K., Seidemann, C., Tempel, L., Sadocco, P., and Bobu, E. (2012). End-of-life solutions for fibre and bio-based packaging materials in Europe, *Packaging Technology and Science*, **27(1)**, 1–15.

35. Prescott, C.E. (2010). Litter decomposition: What controls it and how can we alter it to sequester more carbon in forest soils?, *Biogeochemical*, **101(1–3)**, pp. 133–149.

36. Leja, K. and Lewandowicz, G. (2010). Polymer biodegradation and biodegradable polymers: A review, *Polish J. Environ. Stud.*, **19(2)**, pp. 255–266.

37. Gorassi, G. and Pantani, R. (2013). Effect of PLA grades and morphologies on hydrolytic degradation at composting temperature: Assessment of structural modification and kinetic parameters, *Polym. Degrad. Stab.*, **98(5)**, pp. 1006–1014.

38. Mumtaz, T., Yahaya, N. A., Abd-Aziz, S., Yee, P. L., Shirai, Y., and Hassan, M. A. (2010). Turning waste to wealth-biodegradable plastics polyhydroxyalkanoates from palm oil mill effluent: A Malaysian perspective, *J. Clean. Prod.*, **18(14)**, pp.1393–1402.

39. Vannini, C., Rossi, A., Vallerini, F., Menicagli, V., Seggiani, M., Cinelli, P., Lardicci, C., and Balestri, E. (2021). Microbial communities of polyhydroxyalkanoate (PHA)-based biodegradable composites plastisphere and of surrounding environmental matrix: A comparison between marine (seabed) and coastal sediments (dune sand) over a long-time scale, Sci. Total Environ., **764**, p. 142814. DOI: https://doi.org/10.1016/j.scitotenv.2020.142814.

40. Zimmerling, J., Oelschlägel, M., Großmann, C., Voitel, M., Schlömann, M., and Tischler, D. (2021). Biochemical characterization of phenylacetaldehyde dehydrogenases from styrene-degrading soil bacteria, Appl. Biochem. Biotechnol., **193**, pp. 650–667. https://doi.org/10.1007/s12010-020-03421-8.

41. Peres, A. M. and Oréfice, R. L. (2020). Effect of incorporation of Halloysite nanotubes on the structure and properties of low-density polyethylene/thermoplastic starch blend, *J. Polym. Res.*, **27(8)**, pp.1–10.

42. Spiridon, I., Anghel, N. C., Darie-Nita, R. N., Iwańczuk, A., Ursu, R. G., and Spiridon, I. A. (2020). New composites based on starch/Ecoflex®/biomass wastes: Mechanical, thermal, morphological and antimicrobial properties, *Int. J. Biol. Macromol.*, **156**, pp. 1435–1444.

43. Bajer, D., Janczak, K., and Bajer, K. (2020). Novel starch/chitosan/aloe vera composites as promising biopackaging materials, *Journal of Polymers and the Environment*, **28(3)**, pp.1021–1039.

44. Schawe, J.E. (2014). Influence of processing conditions on polymer crystallization measured by fast scanning DSC, *J. Thermal Anal. Calorimetry*, **116(3)**, pp. 1165–1173.

45. Uthumporn, U., Zaidul, I. S., and Karim, A. A. (2010). Hydrolysis of granular starch at sub-gelatinization temperature using a mixture of amylolytic enzymes, *Food Bioprod. Process.*, **88(1)**, pp. 47–54.

46. Gray, D.G. (2020). Cellulose nanocrystal research: A personal perspective, *Carbohydr. Polym.*, **250**, pp. 116888. DOI: https://doi.org/10.1016/j.carbpol.2020.116888

47. Reichert, C. L., Bugnicourt, E., Coltelli, M. B., Cinelli, P., Lazzeri, A., Canesi, I., Braca, F., Martínez, B. M., Alonso, R., Agostinis, L., and Verstichel, S. (2020). Bio-based packaging: Materials, modifications, industrial applications and sustainability, *Polymer*, **12(7)**, pp. 1558. DOI: https://doi.org/10.3390/polym12071558

48. Yusuf, M. (2018). Food packaging and preservation, Series handbook of food bioengineering, Vol 9, Grumezescu, A. M. and Holban A. M. (eds.), Chapter 12 "Natural antimicrobial agents for food biopreservation", Academic Press, London, U.K., pp. 409–438. DOI: https://doi.org/10.1016/B978-0-12-811516-9.00012-9

49. Hanss, D. and Doran, R. (2020). Responsible consumption and production: Encyclopedia of the UN sustainable development goals, Leal Filho, W., Azul, A. M., Brandli, L., özuyar, P. G., and Wall, T. (eds.), Chapter "Perceived consumer effectiveness", Springer, Cham. DOI: https://doi.org/10.1007/978-3-319-95726-5_33

50. Gautam, R., Bassi, A. S., Yanful, E. K., and Cullen, E. (2007). Biodegradation of automotive waste polyester polyurethane foam using Pseudomonas chlororaphis ATCC55729, *Int. Biodeterior. Biodegrad.*, **60(4)**, pp. 245–249.

Index